Grassroots Ecology

A Call for Help

Ken Greenwood

T0127878

Hamilton Books
A member of
The Rowman & Littlefield Publishing Group
Lanham • Boulder • New York • Toronto • Plymouth, UK

Copyright © 2008 by
Hamilton Books
4501 Forbes Boulevard
Suite 200
Lanham, Maryland 20706
Hamilton Books Acquisitions Department (301) 459-3366

Estover Road
Plymouth PL6 7PY
United Kingdom

Library of Congress Control Number: 2007934788
ISBN-13: 978-0-7618-3855-5 (paperback : alk. paper)
ISBN-10: 0-7618-3855-4 (paperback : alk. paper)
eISBN-13: 978-0-7618-4196-8
eISBN-10: 0-7618-4196-2

Contents

Preface   v

Acknowledgments   vii

The All-American Conservation Team   ix

1   The War of the Maps   1

2   Grandmother's Quilt   13

3   The Ten Ecological Ethics   25

4   The Theoretical Ethics in Practical Use   35

5   The Ecological Model   44

6   Relationships and Connections   55

7   The People of the Dawn   63

8   The Poncas . . Myths and Messages   74

9   The Shoshone   85

10   How Measurement Got Us into Trouble   96

11   Considerations of the Prairie Model   110

12   Evolution of Grassland Ecological Thinking   124

13   Tallgrass Prairie Preserve   135

14   Burning   151

15   Symbols and Symbolism   165

16 The Myths that Hover over the Land   177

17 Social Contracts   186

18 Why Do We Hunt?   198

19 The Model Applied to Wildlife Management   211

20 Can You Put It Back?   220

21 Tom's Dream   233

22 Dry Land   244

23 Beginning with the Beaver   255

24 The Trail of the Wapiti   266

25 Wolves   275

26 Judgment Day for the Cause   289

27 Naturalist Thinking   304

28 The Least of These   314

29 "Where There is No Vision The People Perish"   325

Index   339

About the Author   355

Preface

More than 30 years ago, a love of hunting and fishing led me to begin collecting pieces of information about conservation and the environment. In the back of my mind, I would say something like . . . "someday, I might want to write a book." I ended up with hundreds of articles and interviews.

I wrote several books for the broadcast industry on selling and leadership. These were successfully published. Then a small book about putting value in your life and managing your time a little better sold rather well. Then, I began writing chapters for a book on ecology.

Another fifteen years of writing and revising, talking to many people in the conservation field, gathering more information, verifying facts . . . the result is this book. It is more a labor of love and less an effort to write a heavy book on a subject I have come to feel very deeply about.

As I plodded along, I came to the conclusion that our land labors with at least ten conservation beliefs which I call "ethics". Many of these ideas are in direct conflict with each other. These range from the Pilgrim ethic that nature is hostile and must be subdued to the Native American idea of the "spirit of nature." They explain why our American culture has had difficulty establishing a conservation ethic.

Being an old sports announcer who once did University of Nebraska football games for the State Sports Network, it was easier for me to write like people talk. So, there is the talk of hunters who are lost on a road, to the cowboys of Wyoming and their bar talk, as well as the comments and thoughts of people who work on the grass roots level of conservation and wildlife management.

I have also tried to identify people who made considerable contributions to conservation and environmental thinking. You will find short biographies of these people scattered through the book. This trail begins with George Perkins Marsh and his classic book, *Man and Nature* and ends with people still living like Stuart Udall and Barry Commoner.

My hope is the book will find its way into high school and university class rooms and perhaps even into the mainstream. I hope it causes people of all ages and types to take better care of the earth.

Acknowledgments

To do the research for this book, it was necessary to travel to many locations in the United States and Canada. Consequently, I met many people, talked with them, gathered information from them. I am grateful to them for their time, consideration and interest in what I was trying to accomplish.

In Vermont, I met with Eric Smeltzer, State Limologist in the state of Vermont Agency of Natural Resources. Frederick D. Larsen, Dana Professor at the Department of Geology at Norwich University provided Vermont ecological history and suggested people I should talk with. Mary C. Watzin at the University of Vermont Science Laboratory granted me an entire morning to sketch efforts to clean up Lake Champlin. Frederick M Wiseman, Director of the Wobanakik Heritage Center in Swanton provided deep background on the Abenaki Native American.

Much of the information about Saskatchewan, Canada came from R.J. Bob Santo who was then with Ducks Unlimited as a District Manager. He is now with The Nature Conservancy in Canada.

Scott Sutcliffe, Associate Director of the Cornell Laboratory of Ornithology gave me a great briefing on their work in the world of birds.

In Wyoming, Pete Petera, the former Head of Wyoming Game and Fish, not only provided great information on that area, but we became good friends. Much of the information about wolves and elk came from Pete. Dave Dufek, Regional Fisheries Supervisor, was a valuable resource for information about trout and fisheries management. Richard J. Baldes, Vice Chair, Western Region of the National Wildlife Federation provided names in the Shoshone community as sources. Tyrel Mack, Executive Director of the Wind River Alliance in Ethete, Wyoming was one of those sources. I got to spend an afternoon with David Love, the noted geologist, shortly before he passed away.

Two visits with Dan Smith, Yellowstone Park Service, was the source for wolf information.

Folks from all over the country have been kind to grant me interviews. There was Kathy McKinstry, Wild Horse Specialist with Bureau of Land Management in Elko, Nevada. At the Halsey National Forest I located Herb Gerhardt who is writing the history of the Forest based upon his many years of working there. At the Joyce Kilmer Memorial Park the interpreters were generous with time and showed me around. Dr James Stubbendieck at the University of Nebraska, who has done so much research on grasslands in northwestern Nebraska, and I talked about silica in grass and how that affected animal teeth. Up in Bismarck, North Dakota, John Devney, now a Senior Vice President at Delta Waterfowl, gave me the Delta view on ducks. Closer to home, Harvey Payne, Director of the Tallgrass Prairie Preserve, (and who furnished the beautiful photo for the cover of this book), has been a storehouse of information about the prairie. And Bob Hamilton, who is the Director of Science at the Preserve, has provided me much of the scientific information on the Preserve. Steve Sherrod, Director of the Sutton Avian Research Center was my source of information on eagles and prairie chickens. And in the world of game management, Richard Hatcher, Assistant Director of the Oklahoma Department of Wildlife Conservation has been a great person to compare notes with.

To my wife, Marian, who spent so many hours at the computer and word processor, a very deep thank you.

To all of these people, my kindest thanks for your help.

The All-American Conservation Team

Scattered through the book, at the ends of chapters, are brief profiles of twenty-five people I would list on an All-American team of naturalists, conservationists or environmentalists. They all played, in some way, in the big game of conservation in this country.

What a patchwork quilt they become. They range from a one-eyed walker in the Sierras to a one-armed prophet who found a Grand Canyon. They include rich people and poor people who were actually in debtors' prisons, presidents, artists, cartoonists, and people little thought of as conservationists. Some were old curmudgeons.

They are an old sports announcer's effort to get at the personal side of the myriad of conflicts, contradictions and disarray of our conservation thinking. But all were, in one way or another, the people who produced our conservation culture.

There is a commentary on this group. I had trouble as I reviewed all the conservation games played in the U.S., finding twenty-five players who would rank in an All-American category.

That may speak to our national concern with conservation of our national natural resources. The subject would hardly make a Super Bowl!

Chapter One

The War of the Maps

Two hunters are on a road that isn't well marked. One was a very successful businessman who has made his money in the oil business. The other is in the tree business and runs a reasonably successful small business. He is very active in the Rocky Mountain Elk Foundation on the grassroots level. Their one common link is hunting elk. They are on a hunting trip.

It is dark. They have stopped along a road to look at their maps. They are using a flashlight. Each has brought along a map. It seems their maps are in direct conflict. They have no idea how long the motel in the small town that is their destination will hold their reservations. They are about out of diesel. They don't relish spending the night in the extended bed pickup truck loaded with hunting gear and an assortment of outdoor technology . . . GPS gear, scopes and guns.

One of them had stopped earlier, stepped out of the truck and attended to nature. He drops their only working compass and it breaks. They really have no idea where they are. Except they are in a four-wheel pickup truck.

The driver, Marvin, the business executive, switches on the dome light. "Lemme see the maps together," he says. There is silence. "Whadda ya think?" says the other.

"I don't," says Marvin "How the hell did we get in this mess?"

"Well", says Charley, "you were in a hurry and you wanted to take those short cuts and cut down on driving time."

"Yeah . . . yeah . . . yeah." says Marvin switching into his executive voice. "The bottom line is we are about out of fuel, it's getting late and we haven't the foggiest idea where the hell we are. We can't very well go back and even if we could reach somebody with your phone, we couldn't tell them where we are. You shoudda been watching the maps closer."

"Don't lay it on me," says Charley. "You were in the driver's seat."

"Well," says Marvin, "I'll just shut this damn thing off and we'll wait it out until daylight."

What would it be like if, as the dark slowly lifts in the morning, the hunters awaken and find themselves in a world where there are no living species other than humans? Connections between all living things are severed. All living things, animals, birds, trees, insects, plants . . . all living species woven together in daily existence are gone? As our hunters step into morning, it is more than a "Silent Spring." It is a silence without a soul.

And in this silence, all species in the world are no longer with us to provide insights, connections and even psychic direction. Unlike humans, who tend to rebel against natural systems, repudiate their genes and human tendencies, the natural parts of the great connections that coexist no longer are related. The natural relationships that provide a mirror that refracts our ways of thinking are no longer present. We are hunters, predators surviving in an egocentric ecosystem in which only man exists.

Can that metaphor reflect the courses of American thinking about nature? How often has the corporate executive had one map while the person more attuned to nature has had another? How often has industry taken a short cut that provided business expediency at the expense of nature? How often have two people, riding in the same truck, fallen into conflict and contrasting viewpoints?

And in a deeper sense, as a society, do we have any idea where we are in our relationships with nature? Are we about out of fuel? Are we spending a night in a truck waiting for morning and some light?

CONSIDER THAT OUR CULTURE

. . . has been one of the major societies to separate mankind from animals in our thinking. Our culture has embraced an ethic that makes earth subservient to the economic impulses of the society. Natural resources have been exhausted in the name of progress and growth. We extolled the image of the white pioneer to enable this idea, we nearly eradicated a population of Native Americans who had used the land, and natural resources of this country, without damage to the ecosystems. In short, we used one map to pillage nature and another to daub over our actions with piety.

We have not followed very good maps. Environmental scientist David Wann says. "It may not be flattering to our natural concept, but the present American culture is still a bare field of colonizing weeds, struggling toward something more sophisticated, interwoven and permanent. Until now, we have consistently chosen the resource path of least resistance."

I came at this book as a reporter. I began a career in radio and television as a sports announcer. I learned to find the story, dig it out, report all sides if there was more than one side. In the last thirty years, I believe I have dug out at least ten beliefs about nature that could be qualified as "ethics." I call them "social contracts." Nearly all of them are conflicting, at times almost opposite views. (I will expand on these later.).

You will also find a model in this book. It is a different way to look at ecology. Too often in our natural science thinking, we have isolated natural resources. Only in an absolutely natural state can natural resources be regarded as a single, isolated entity. Ecology is the interaction of four determinants, one of which is natural resources. The other three are socio-economics, government and social contracts.

The model is constantly evolving. Changes in natural resources will impact government as when a hurricane strikes. There will he a socio-economic impact and now all three determinants are in a state of churn. How people react to the event will depend on their social contracts. The reaction can range from passive acceptance to looting and lawlessness.

Quality of the environment depends more upon the stage or passage of any of the determinants than it does simply the stage or passage of only the natural resources. The story of the interconnections of the four determinants produces a model that does much to explain the conflicts, the contradictions and the complexities of ecology. And the irony of the model is the closer to the grassroots the situation is, the more complex it becomes.

In the past several years, environmental seminars and conferences have grown in popularity. They are often high quality quilting bees where assembled authorities, scientists and pundits gather to stitch on a huge quilting frame, often world size. World conferences on ecological issues have replaced the more provincial gatherings. I have tried to avoid the world scene as much as possible; rather to deal with issues in this country and to put those issues in the grassroots perspective.

The big quilt certainly has a place in a global quilting frame. And, well it is there are efforts to reconcile the world ecology. But it seems to me there is a place for the idea you get your own house in order before you preach to the world. In a narrow view, you practice conservation morality first. Then you preach to a larger congregation.

So, I have chosen to report from a grassroots perspective, from down on the field as they say in sports reporting. I have visited or seen every area covered in this book. I have talked or visited with the primary people involved or have read their books or papers, sometimes both. I am guilty of stereotyping both hunters and cowboys. I tweak some hallowed religious idols, sometimes in tears. I have listened to pros and cons, heard people replace common sense with science. I try to report the conflict and contention, not from a distant press box, but down on the grass roots level.

THREE CASE STUDIES

. . . help find similarities between three different parts of the United States, with a few side trips to validate the similarities. The first part of the story is located in Vermont, a state that provides some isolation for the study of regional ecology. Early patterns of Vermont land development can then be extended into other parts of the country.

Vermont was the home of George Perkins Marsh, as nearly as we can determine the first American to study what has become ecology. Marsh was a resident at a farm owned by Frederick Billings. Billings came to put many of the ideas Marsh espoused in *Man and Nature* into practice on his farm. Billings was known for his work in reforestation conservation and progressive farm management. This work is still in evidence.

Later, in 1886, Dr. William Seward Webb and Lila Vanderbilt Webb would establish Shelborne Farm, a 1400 acre unit that became a model for progressive farming and is today a National Historical Landmark.

In Vermont, we use the Abenaki Indians as a pattern for the way Native American people first cooperated with and helped white settlers adapt to a foreign soil. As the white newcomer "settled in," the Indians were pushed out.

And coincidentally, Vermont provides something of a template for a region that has phased from pioneering to timber extraction and exhaustion, to the industrial revolution, to a period of searching for a socio-economic base to some sort of temporary accommodation with the new age of technology. Vermont is still searching for some balance with natural resources.

1 would also add a disclaimer to being a scholar. I respect scholars. I've read the scholarly books, plowed through Thoreau, Marsh and Stegner and studied Harold Ickes and Teddy Roosevelt and been a Rough Rider through a lot of material on conservation and the environment. I have peeked into the philosophy of the Native American, talked with tribal historians and known and visited with some of the chiefs. And been to their funerals.

I've kept notes for 40 some years and have files bulging with news stories and articles from conservation sources. I like to talk conservation with people in the Forest Service or in Wildlife Departments and once I found a fellow in the Department Of Interior who was fairly articulate. And because I think God gave Americans a magnificent gift when he provided us this North American continent . . . or more correctly when we invoked the words of Moses and took it, I have become a pack-rat of notes about the consequences of our stewardship of the land. I have also poked my nose into every Canadian province at one time or another. I include them in this magnificent gift, eh?

Beside several excursions, to Vermont to visit with people on the subject of the environment, I maintain contacts with prairie country. When I report on fishing tournaments in Alma, Nebraska I have seen them. And when I use the Ponca tribe as the model for the plains Indians, it is because I have known these people both in Nebraska and in Oklahoma. My grandmother and mother homesteaded near Egbert, Wyoming so there is a little western wood in my family tree. And the Shoshone people, who are the model for the experience of the Native American in the western United States, live just down the road in north central Wyoming.

HOW WE DECIDED TO TREAT OUR RESOURCES

. . . is shrouded in the history of our culture. By virtue of a bloody civil war, we decided to construct an industrial society based upon the presence of unlimited natural resources. This was a shift from a society that was agricultural, rural and a citizen farmer economy. We set about, with the concurrence of government and social contracts, to construct an economy that demanded the absolute use of natural resources. Agriculture, rather than being a sustaining economy, became a handmaiden of consumption. Our social contracts were based upon "survival of the fittest" and when we didn't exactly like what Darwin said, we modified the context to fit our economic system. We extolled aggressive business models, and tough competition, and used football games to explain business attitudes toward natural resources.

Skip to World War II and another great impact on ecology. Most of my friends were part of what Tom Brokaw has dubbed "the greatest generation" and in his fine book he talks about patriotism and devotion to country and family. He missed the devotion to hunting and fishing and the great contribution this generation made to conservation. He also missed the training they got. They learned to use guns. They hunted other humans and learned to survive when they returned. They did a lot of blocking for the early conservation teams like Ducks Unlimited, The Wild Turkey Federation, Trout Unlimited, The Rocky Mountain Elk Foundation and the Isaac Walton groups.

A LITTLE PREGAME INFORMATION ON HUNTING AND FISHING

. . . might be in order. Relationships between hunting and fishing, state or federal laws or regulations, and the idea of natural resources conservation have gone through three stages or passages. In Stage 1, our model would show abundant natural resources, limited socio-economic consequences (mostly hides), little nor any law or regulation and social contracts based mostly upon the reality of subsistence hunting and fishing.

By 1800, in the second stage, market hunting was impacting natural supplies of animals, birds and fish. As this socio-economic determinant of the model moved up the scale, the supply (value) of natural resources began to fall. Views on conservation were shifting slightly and social contracts were evolving. Government would now be used to counter the exploitation of hunting and fishing.

New Jersey passed a law in 1873 requiring a state hunting license. By 1905, 31 states and nearly all the Canadian provinces had some kind of resident and non-resident license laws. Now, as the idea of conservation grew stronger, there was the socio-economic reality of finding funding for state departments to implement their programs. Hunting and fishing licenses provided early funding. That story would fit most states.

Federal support for natural resources, animals and birds, didn't really come into play until 1937 with the passage of the Pittman-Robertson Federal Aid Act. One of the very important elements of this bill was added by A. Willis Robertson, the House sponsor of the bill. This rider said that money would be available to states that did not divert hunting license money to fund projects outside wildlife conservation. Earmarked money became the life blood of conservation.

HOW HUNTERS/FISHERMEN SUPPORTED CONSERVATION

. . . is important to any study of ecology. This support blossomed after WWII, not just in the sale of licenses, which became the primary support of most state game and fish operations, but also in the increased membership in and support for the non-profit conservation organizations. Ducks Unlimited was actually incubated in 1937, and started its fund-raising in 1938 and by 1940 was raising some $140,000 a year. The war years slowed progress, but by 1947, 25,215 members raised $433,826 and completed 39 new projects.

So the "back from the service" generation not only bought the licenses that supported wildlife conservation in their states, they also gave time and money to conservation organizations that became the backbone for some amazing restoration stories in the world of wildlife. You could cite success with deer, turkey, ducks and geese, fish and even such species as swans, falcons and bald eagles. It was a "user pays" philosophy. It worked for three reasons. First, the money comes from a very committed group of users. Second, license fees are mandatory; if you want to play, you pay. Third, both license fee money and conservation organization money is earmarked for use to restore and conserve wildlife or habitat. It is not likely to get siphoned off for other purposes.

The "greatest generation" of hunters and anglers has harvested "a hellava lot of the crop they helped plant" say the detractors. In the end, their harvest may not have been an unreasonable trade-off. If they hadn't stepped forward there wouldn't have been much to harvest. Their efforts marked the turning point in the American grassroots conservation movement.

Today, there seems to be a growing discontent among these early day "conservationists" who are now pushing the sunset over the rim. They grumble about the contributions of the snowmobilers, the off road bikers, the backpackers and the Kodak moment people, who are enjoying some of the harvest but may not be buying a ticket to the game. The oldsters ask, "Is it fair?" Then some observer will say, "When has it ever been fair when nature is involved? How many people have used it without buying a ticket?"

As our story unfolds, there will be many contradictions. How Ducks Unlimited spent 60 years and $1.5 billion hard raised dollars to restore wetland habitat. Which habitat, in spite of the best efforts of biologists achieves less than a 10% nesting factor. Meantime, ranchers, farmers and developers drain wetlands faster than DU can restore them!

COMPRESSED ECOLOGICAL ISSUES

. . . are not hard to find in Wyoming, an adopted home where I now spend about half of my time. Wyoming is one of only two states, Montana is the other, that gets to cope with rather significant populations of black bears and grizzlies, mountain lions and wolves all at the same time. Wyoming also gets to deal with diseases like brucellosis, chronic wasting disease, and tuberculosis in elk and whirling disease in trout.

Then add this fact. Nearly half the land in Wyoming is either held by or managed by the Federal government. Who has the right to do what, with which and to whom with timber, water, minerals and forage has long been a matter of sharp political debate and no small amount of bar room talk. There has been, as in football, a lot of pushing and shoving. Even when things seem fairly peaceful and the shooting wars have been quieted, the result is a very tentative equilibrium between the maps of rancher and environmentalist, state and federal conservation maps and extractionists and wild lifers.

In a bar, most of which are not all that crowded, except in Jackson, Wyoming, there is space for conversation and difference of opinion. You can order a can of Coors Light and reflect on the big Rainbow Trout hanging in the dim light on the right hand wall. On the left side, is the head of a beautiful mule deer. You sip on the cold can and wonder about people who take great pride in catching the biggest fish and hanging it on the wall. After all, it probably had some pretty good sperm or eggs left. Or, shooting the finest specimen for the horns. What kind of social contract guides the Anglo-Saxon mind to kill the finest and best of the species? What would Freud say?

Would my friend, who has just ambled into the bar in his dusty, black boots and battered, black hat, walk into his herd and pick out his finest bull? Then summarily dispatch the animal with his new rifle, mount the head and perhaps the repositories of the animal's genes above the fireplace in his new log home?

My cowboy friend and I now and then talk about "this new ecology stuff." His concern is forest fires because his herd is on Forest Service land this summer. His views on land use can be strong, even if his ability to count animals is a little suspect. His opinions about environmentalists are even more pungent. I suspect our maps are quite different. In a condensed way we may represent the many issues that exist between consumption and conservation. We are a grass roots version of a much bigger picture, a part of what Stewart Udall has called "the quiet crisis." My cowboy friend applies this phrase to cattle prices. I think in terms of water quality.

You discover, when you dig a little into the story of ecology, that what man did to Vermont, or prairie country or the sagebrush country of Wyoming was pretty much a repetition of what others before him had done to Asia, or the Middle East, from the Stone Age through the passages of time. Man has been a hunter for a long time. He has been a resource predator for a long time. Much of that early record has been lost. In this country, the transition was shorter, faster . . . so much so that when you dig away at the records, many are amazingly complete and in some instances, the folly and fortitude may still be on display.

ECOLOGY IS STILL A NEW CONCEPT

. . . if you consider it against historical clocks. In fact, the first text on ecology, Smith's *Ecology and Field Biology*, was published in 1966. It is already in the 6th edition and now has several thousand references in the bibliography. Here the explorer or student can enter the bewildering world of what is now called "ecology." It can be theoretical or applied, physiological or behavioral, wildlife, biology, population or people, habitats or ecosystems, agricultural, forestry, wildlife, environmental science, anthropology, genetic, atmospheric science, prairies. grasslands, edges, tropical savannahs, tundra, rivers, potholes, lakes, wetlands, watersheds, irrigation, marine environment, intertidal Zones, global environment change, greenhouse gases, carbon dioxide, fossil fuels, forest clearing, changes in earth's climate and not only melting ice caps, but the impact on agricultural production and human health.

WE ARE ALL EMBEDDED IN NATURE

. . . seems to be the logical conclusion to be drawn as more and more we understand the principles of ecology related to everyday life of every living thing, plant, animal, plankton, and human microbe. The simple lessons of the grass roots are that in nature, all things are connected. Ecology then becomes a collection of information about a dynamic, even metabiological network process by which members of every species relate to each other.

And my friend and I sitting at the bar, are not unlike the imagined hunters in the truck with their disparate maps. We are the dichotomy that exists between the various views of ecology. The odds are not favorable we will soon resolve our differences. One of us believes the earth ought to take care of people and the other believes people ought to take care of the earth. If, as individuals, we separate ourselves from the bigger picture and take the position that air quality, or bad grazing practices or the erosion of soil and waste into rivers are not our problem but belong to somebody else then who will solve the problem? Are we really stuck in a truck on a dark night'? And if the individuals involved choose not to see another opinion and we continue to divide into more groups and maintain inflexible positions to attain political leverage or for fund raising purposes, can the problems ever be resolved?

Underneath all the rhetoric and posturing, are not all of us somehow embedded in nature? Are not all things connected by the slender thread that stitches nature into our lives? At the very end argument is not every human being dependent upon nature and natural resources'? And are there not clues in nature that provide insights into how we fit into the comprehensive picture? This book is the search for some of those propositions. We begin with a small story about a small pond.

For many years, I had a ranch with a pond. It was near the world famous town of Hitchita, Oklahoma. When we first bought the place, the evenings were filled with a Mormon Tabernacle Choir of bullfrogs that croaked and grumped as I entertained a vision of Aldo Leopold and Sand County.

The bullfrogs no longer make the evening news on the pond. An old TV set that gets only four channels is a sad replacement. My guttural "saints" are missing. From the TV one rainy evening I learned about amphibians.

It seems this pond is not an exception. For the last two decades, a short space as nature measures frog time, biologists have been measuring the die-off of frogs and toads. The little critters are dying off at an alarming rate. In Colorado, the tube drones on, Boreal Toads in high lakes are down in population 80% in the last ten years. This is true in high lakes in other states where amphibian populations are being measured.

The biological theory is that the ultimate cause of amphibian demise is higher than normal levels of ultra violet radiation. Research indicates this radiation can decay vitamin A which occurs naturally in the upper skin of all animals. (I grab a note pad because I am an animal and have been in the sun most of the day.) This decay becomes retinoids, common chemicals best known for acne treatment. Excess retinoids cause birth defects; or impaired immune systems . . . and dead amphibians.

And all this is related to lupus in human beings and the theory is that exposure to ultra violet light spawns lupus sufferers because it decays molecules in the upper skin which triggers our immune system to attack our own cells. I inspect my arms and the tube drones on. A well modulated voice says, "Could all the deaths of frogs and the problems of lupus patients be similar'?" And I wonder about the missing frogs in my pond. Are the frogs an early warning system related to the thinning of the ozone layer?

Now, the world of nature and the more comprehensive world of ecology is suddenly connected to my present world of reality. The voice drones on. "Researchers don't really understand what is happening." The bottom line to thirty minutes of servitude to the tube is that we really don't know. I am reminded that today our biologists take great care to generate only complicated models. They use terms like "data analysis thresholds, "quantified data summations" and "deciviews." They leave the judgments to other authorities or well modulated voices on TV.

I shut off the tube. Click!

I walk outside and consider the night sky and allow the Labrador to find solace at an oak tree.

It has been a long day, filled with barbed wire, steel posts, cranky come alongs and profanity. Reality stuff! Pragmatic stuff!

It has cleared off after the early evening rain. The air is clean. Rain does that you know. I gaze at the sky. How many stars are up there tonight?

It is very still. A possum scurries by intent on doing his or her part in the food chain.

Where are the bull frogs?

There is a postlude to this opening chapter. I have made the point about the speed with which new information on various aspects of ecology is emerging. In the May 2002 *US News and World Report*, Fred Goterl reported on certain events surrounding Yosemite National Park. An ecologist named Gary Fellers with the U.S. Geology Survey was trying to figure out why half the frogs in Yosemite seemed to have disappeared.

It seems that when he collects tadpoles in the park and releases them in Lassen Volcanic National Park further north, they thrive. When he tries to raise them in Yosemite, they are often born with 1 leg or 3 legs, sometimes as many as 10 legs. He thinks the cause is pesticides drifting over the Sierra Nevada Mountains from fruit and nut farms in the central California valley.

"Ecologists first sounded the alarm about frogs and other amphibians in the early 1990's. Since then, they have stomped around enough swamps and ponds to know for certain the decline is both real and steep: 32 species have gone extinct around the world in the last few decades and 200 more are in decline. The reasons are varied. Climate changes, infectious diseases and new malls and housing developments play a role. But what the scientists have learned recently about pesticides is especially worrisome, not only for the frogs, but what it implies about human health."

The article then explains that frogs live in the water, lay eggs in the water and absorb oxygen through their skin. Consequently they are hypersensitive to water pollutants. This is the question. If frogs are having this much trouble in a protected park like Yosemite, how much worse is their condition where water pollution is more extreme?

THE SEARCH FOR A CONSERVATION ETHIC

. . . or the lack of such an ethic in our American Culture can he better understood if you agree we have improvised our culture as we have gone along. We are a constant work in progress, a song being written as the chords of chance are struck. In the beginning, our founders really had no maps, neither for our government, our social contracts nor our economic systems. Our foundations were laid by people who did not think alike, act alike, worship alike: nor, were they flawed in a like manner. The people who fashioned our foundations left, not so much a history as a catalog.

Americans have never become a race. We became an idea, a culture, even an ideal. Quite often we would profess or preach the ideal in one message and would practice an almost opposite expediency. We were, and are, a people on the move, prodded by a siren of Manifest Destiny, guided by a conviction we were, and are, God's chosen people. When our heroes weren't, or aren't, big enough, we invent myth and symbolism. We became a driven society who used roads, traces, trails and rivers to hurry to someplace we might make better. And when our ideas clashed, we pounded them upon the constitutional rock to determine if they were strong. Some ideas would be deemed as "un-constitutional" and would break on the rock. We even stuck the phrase "it's unconstitutional" in our barroom chatter. Some better ideas would survive the pounding and be incorporated into a larger dream.

And always, beneath the government, the economics or the culture there was the earth. It was to be subdued and possessed, extracted and used, occupied and abandoned. A young, visiting Frenchman was reflecting on the American love of land possession and he said it was beneath "the first causes of prejudices, habits, dominant passions, or all that finally composes the national character." Our founders, for the most part, did not place one particular rock in our foundation. That would have been respect for the earth. We have no conservation rock or earth ethic.

It was as though we would always expect the earth to take care of us: we need not be concerned with taking care of the earth.

The search for the reasons for that absence of a conservation ethic is the story in this book. The grass roots provided no map, only vague patterns. I invented a simple model. It became a work in progress. Taking care of the earth will probably always be a work in progress.

REFERENCES

Belsky, A.J., Matzke, A., Uselman, S. *"Survey Of Livestock Influences on Stream and Riparian Ecosystems in the Western United States,"* Journal Soil and Water Conservation 1999.
Donahue, Debra 1., "The Western Range Revisited," University of Oklahoma
Hawken, Paul. *The Ecology of Commerce.* (Harper Business) 1993
Wann, David, *Biologic: Designing with Nature to Protect the Environment.* Paperback, 1994

Chapter Two

Grandmother's Quilt

"Grandmother Rennau" came to the prairie country as a settler. She claimed she came in a wagon with her sisters. John, her husband, was a railroad engineer. He was considered a "professional man." His run was daily, along the Missouri River from Nebraska City, Nebraska to a point midway between Omaha and his home base. He would turn the engine around at this point and head back home. He died young. Many men of that time did.

Grandma became a widow. She spent her later years in Lincoln, Nebraska. One sister, Sophie, ended up in Deadwood, South Dakota. Her husband was a gold miner. Another sister, Jesse, married a farmer who had land near Tecumseh, Nebraska. When farming didn't go well, they got by when Uncle Nelson allowed timber people to cut down the huge walnut trees that grew along a creek running through the farm. This disbursement of family was common for prairie country in those times, the late 1800s and early 1900s. Railroads, mining and timbering walnut trees all impacted ecosystems.

Grandma spent winters quilting. She would set up a large frame in the living room of her modest, box style home. She had saved rags and scraps of cloth in large flour sacks. From this disarray of disparent resources, she would assemble a quilt, carefully stitching together small patches of cloth. The finished product really had no pattern.

She called it a "crazy quilt."

If some of that cloth had come from Vermont, from Nebraska, some Wyoming, where Grandma homesteaded in 1902, she would have created the panorama of the ecological model I am attempting to stitch together in this book. The quilt is a hodge-podge of different weaves of cloth, different colors, styles, sizes. When sewn together it will cover a bed well enough, with four sides acting as the determinants of the quilt.

Imagine the four sides form a frame of natural resources, social contracts, economics and government. Place the array of patches inside these edges. Try to stitch the patches together. Even the stitches that blend them together may not be exactly the same. Just as Grandma had guests who would join her for an afternoon and patiently sit at the framework and stitch patches together, so have the historians, the chroniclers, the scholars, the immigrants, the tie-hacks, the cowboys, the Mountain Men, the missionaries, the prostitutes, the surveyors, the artists sat at the great frame of the country and stitched together history, ecology, agriculture, industry, guns, technology, commerce, and steam locomotives. It becomes a crazy quilt of data, facts, suppositions, snippets of

THIS MULTI-HUED PATCH represents the myths and symbols of the frontier as it spread west. The myths were often patently false but would justify conquest, intrusion, extermination, extraction, environmental destruction and even genocide. The colors popularized the frontier and created an ideology of expansion, progress, manhood, possibility, virtue, conquest, guns, independence and spurs. Symbols would impact ecology.

THIS RED, WHITE AND BLUE patch is from clothing dumped from a wagon on the St. Vain Trail, under the front of the Laramie Range. It was fall of 1865 and Major General Granville Dodge had failed to impress the northern Cheyenne and Oglalla Sioux with his Powder River campaign, which was ancillary to his main purpose for being in the area. Dodge had been asked by President Lincoln to find a route for the Union Pacific Railroad through Wyoming. Dodge had consulted with mountain man Jim Bridger whose father-in-law was Chief Washakie, great Chief of the Shoshone tribe. Bridger sent Dodge to Lodgepole Creek saying the high ground above was the low point on the crest of the Laramie Range. Bridger's native knowledge was confirmed years later by the theodolites and geologists. Dodge, in another skirmish with Indians, stumbled onto the "Gang Plank" between Crow Creek and Lone Tree Creek and the feasible route west that would move the Union Pacific across southern Wyoming. The route would never have been an emmigrant trail because of lack of food and water. For a railroad route it offered speed and efficiency. This route was the key that allowed the Union Pacific to open the door to the west, and change western ecology.

HERE WE HAVE a parch that glows in the dark. It represents gas flares in oil fields and could have come from Pennsylvania, Illinois, Oklahoma or Wyoming. It represents oil companies burning off gas because at that time gas lacked economic value. This practice represents one of the greatest, gross wastes of natural resources in the history of extraction. Thousands of flares lighting up hundreds of miles of oil field expanse ... sheer arrogance gutted irreplacable natural resources and changed ecology.

THIS GOLDEN PATCH is from a shirt worn in Yellowstone Park, one of about forty so called geologic "hot spots" in the world. It derives its name from rich, golden splashes of chemically altered volcanic rock. It smokes and spits and accommodates millions of "Kodak moments." Some geologists say Yellowstone has somehow been moving east 2-1/2 centimeters a year. Others believe Yellowstone is not moving; rather the earth, huge plates of earth, is moving under it. As nearby Quake Lake in Montana would testify, there is still a whole lot of shaking going on around Yellowstone Park. The park regularly shakes up ecology.

THIS PATCH came from the skirt of Mary Elizabeth Lease, otherwise known as the "populist firebrand of Kansas." She was considered by many the most powerful orator of her time, even better than William Jennings Bryan. She ended her speeches with the stirring line, "Raise less corn and more hell." Raising corn in Kansas raised hell with prairie ecology.

SOMEWHERE IN THE quilt is a patch of two colors. One color stands for the "packers", the people of an area who would pack up and leave, moving on to someplace they perceived as having more opportunity, or going back to where they had come from. Usually, they were people who had no stakehold; they owned no land. The other color is for the "stickers," the people who would stick it out, accumulate the land, become the elite of the emerging community. Evangalistic Christianity and the belief in white supremacy convinced the "stickers" they were God's chosen people. Progress, was their mantra. They cleared, plowed and fenced, turned paths into roads, roads into rails and rails became the threads of the marketplace. Natural systems were changed in many ways.

THIS SCRAP OF GRAY WOOL is a legal patch. It is gray because it is neither black nor white. It is rough wool because it scratches when you rub against it. It poses a question about the use of law as a selfish, manipulative American tool of English origin imposed from the time of the first American Indian treaty through two centuries of abuse of public lands, water rights, environmental issues and development schemes to achieve whatever means a callous and greedy frontier mentality deemed right at the moment. Small wonder, with such a legacy, environmental law would become a tortured venue of litigation. And it impacted the ecology.

THIS DENIM PATCH could have come from an old pair of overalls ordered from Montgomery Ward. Aaron Montgomery Ward, a young Chicago merchant sent his first catalog listing some 163 articles to nearly a million "grangers," many of them in prairie country. Ward used his catalog to battle trusts, including the Barb Wire Trust. He gave farmers an alternative place to buy implements. Less well known than his catalog is his long 20 year court fight to protect Chicago's waterfront on Lake Michigan. His stubborn fight to keep the lake front free of buildings was one of the early citizen efforts on behalf of conservation. Grant Park is his memorial for the environment; his catalog a memorial selected by the Grolier Club of New York as one of the books having the greatest influence on American life. Ward directly and indirectly impacted the ecology.

THIS COPPER COLORED PATCH represents mining and extractionist interests of the prairie country and the west For example. Anaconda Copper Mining Company controlled more than half the newspapers in Montana in the 1910 to 1930 time frame. Their activity drew the wrath of such authors as Bernard DeVoto and Joseph Howard whose written hostility was directed at the corporate exploiters who came west to get rich and then got out. Their impact on ecology was vast.

THIS BLACK AND WHITE patch came from the coat of George Catlin, an early chronicleer of prairie and western wildlife. He may have been the first observer and writer when, about 1835, he suggested a national park or refuge of some sort to protect wildlife. Sixty some odd years later, another observer and writer, Emerson Hough probably saved the last 100 bison in Yellowstone Park by writing about the plight of the animals under the guns of the poachers in the park. Writers have impacted our ecology.

THIS GREEN PATCH came from Magnum, Oklahoma where, in 1935, the first tree was planted in the USDA soil conservation service windbreak project. It became known as the Shelterbelt Program and stretched from northern North Dakota to northwestern Texas. In the spring of 1938, approximately 52,000 cottonwood trees were planted in one severely, sand blown area south of Neligh, Nebraska. Government programs have impacted ecology.

THIS SMALL PATCH of flannel is about the size of cloth that would have been used to clean a Colt revolver. Between 1836 and 1842, Samuel Colt had manufactured about 6,000 of these guns. He closed his plant due to bad financial planning and began to experiment with electrical submarine mines. But his gun had been tested in the Seminole War in Florida and had proven to be a highly effective weapon for mounted cavalry. The cavalry prompted a small revolution in frontier warfare. The Colt and provided a gun that gave the frontier soldier, the cowboy and the mountain men a means to combat the plains Indians. When Colt closed his factory in 1842, most of the work on the guns was done by hand. When he reopened in 1846, most of the manufacturing was done by machine. Colt, Elias Howe and Cyrus McCormick furnished the technology for the National Fair of 1846. But more than that, these names changed the social and economic fabric of this country. What they provided in terms of new manufacturing technology would spark the emergence of the United States as an industrial power. Moreover, what they manufactured in 1846, as the era of the mountain man was winding down, would change the ecology of dry plains country and the arid west forever.

THE MOTTLED PATCH sewn in here comes from the Jacket of a man who homesteaded in 1877 on a piece of land he never really wanted but his boss who was a cattleman talked him into filing for it under the Desert Land Act of March 3,1877. This act offered 640 acre tracts in arid parts of the west for the price of $1.25 per acre provided the buyer agreed to irrigate. This man later "sold" his homestead to his employer who added the 640 acres to his holdings. The resultant large holdings changed the ranching ecosystems forever.

THIS PATCH might have come from the blanket that belonged to the grandmother of my friend Lou Ketchem. He was the chief of what was left of the Delaware Tribe. Once the Delawares were a part of the Algonkian tribes and part of that famous peace treaty with William Penn, which treaty was observed only until Penn died. Then the Delawares were forced out and took refuge on Iroquois soil. Five times thereafter the white man made treaties with them which provided the Delawares sell their land to them and move further west. Finally, the Delawares bought land in Kansas with their own money. They fenced their farms, built houses and a church. But once again they were told they would move. In a seventh treaty, they were moved to Oklahoma to begin again, this time on much less fertile land. Indian treaties impacted ecology.

OVERALLS, or as some called them, overhauls, made good patches. This one could have come from a pair worn by a 14 year old Ohio farm boy, Press Clay Southworth. He lived in Pike County. One day in the spring of 1900, (you can locate different dates in research) he talked his mother into using the family shotgun. She gave him the gun and one shell. He went to the farm feedlot and shot a strange bird. His mother recognized it, when he brought it to the house, as a passenger pigeon. She sent him to a Mrs. Barnes who did taxidermy work. Mrs Barnes mounted it using shoe buttons for eyes. They dubbed the bird "Buttons." Years later, this bird was enshrined in a dusty case in the Ohio Historical Center in Columbus, Ohio. It was the last known wild passenger pigeon. The setting in Columbus for Buttons is not as auspicious as the edifice in nearby Canton, Ohio that enshrines our pigskin heroes. We seldom enshrine ecology.

The Southworth clan had once homesteaded in Clay County, Nebraska but drought convinced them to return to Ohio. In their Nebraska farm home they had hosted a fellow named William F. Cody who was at that time a hunter for the Union Pacific Railroad and was engaged in reducing the bison herds on the Nebraska landscape and reshaping ecology.

THIS PATCH might have come from a worn shirt , perhaps from John Colter. In October of 1803, Colter enlisted in the U.S. Army as a private and joined the Lewis and Clark expedition. He gained a reputation as a hunter. His discovery of the Flathead Tribe enabled the expedition to find Lolo Pass which led to Oregon. Colter left the expedition with the blessing of Lewis on the way back to St. Louis. He helped Manuel Lisa build his trading post at the mouth of the Bighorn River in Wyoming. In 1807, he took some trade goods and went looking for customers for Lisa. He walked over some of the highest mountains in Wyoming, saw the Stinkingwater area of what is now Yellowstone National Park, and saw country no white man had ever seen and recorded. He figured he had used up all his chances with the Indians and the grizzlies and in 1810, after another narrow escape from the Blackfeet, he paddled down the Missouri River in a canoe to St. Louis. Shortly after, he settled on land fifty miles north of St. Louis, married and became a farmer. Three years later, he died of jaundice, not quite 40 years old, and was buried on a bluff now called Tunnel Hill. Later the Missouri Pacific Railroad would dig up the graveyard to widen their track. His remains are now somewhere in the bed of the railway. At least they overlook the Missouri River. Coulter impacted the ecology.

THIS CANVAS PATCH might have come from a surveyor in Wyoming, something like David Love, a later day geologist might have worn. Love liked to look at the rocks in Wyoming better than he likes to talk. But he said some very wise things. Like... "We could see that the natural phenomena were not random. That they were controlled, that there was a system. The processes of erosion and deposition were things we grew up with. An insulated society does not see how important terrain is to someone who has to understand it in order to live with it. Much of it means life or death for the animals, and therefore survival for us. If there was one thing we learned, it was you don't fight nature. You live with it. And you make accommodations ... because nature does not accommodate." This from one of the premier geologists of the world. He just preferred the rocks in his home state of Wyoming to the fast track.

THIS PATCH IS from the jacket of Alice Fletcher who one day sat on the great plains with Chief Sitting Bull. She later related that he said, "You are a woman, take pity on my women because they have no future. The young men can be like the white men, till the soil, supply the food and clothing. They will take the work out of the hands of the women. And the women, to whom we have owed everything in the past, will be stripped of all which gave them power. Give my women a future." Six years later, Fletcher was administering the Dawes Act, allotting land only to men.

THIS GOLD PATCH was in a lot of quilting that went on west of the Mississippi River because ecology was shaped by technology as minerals, water and wood were the lures that would dictate the use of natural resources. Mineral extraction and mining, from Sutler's Mill to the Black Hills and back to Virginia City; water, dams and hydraulics from Fort Peck to Bonner and Hoover Dams; great forests of spruce, pine and redwood blended into a social order that was both regional and national based upon the social contracts that paid scant heed to environmental impact. Gold may have provided the glory, but water and wood did not escape the age of plunder.

THIS PATCH is for the mystery buffs. It is red, standing for a long, sinuous, wavering red canyon. It has the look of a John Wayne movie. To the west is the Wind River range. But in the red, triassic rock is a white line, very consistently five feet thick. It is limestone. It covers fifty thousand square miles and is one of the most unusual rock formations anywhere in the world. With the absence of any diagnostic fossils, no one can tell if the water it was formed in was fresh or salt. Some day an anthropologist may find a similar mystery when exploring the formations caused by the erosion of silt into the lake once behind the Fort Peck dam.

NOW THIS PATCH could also have been from a pair of overalls. They would have been worn in the year 1867 when a law was passed in prairie country in eastern Kansas. The grass covered a land where limestone was close to the surface. The law had abolished the open range on what was tallgrass prairie country. Settlers were paid 40 cents a rod, that's 16-1/2 feet, to build and maintain 4-1/2 foot high native limestone fences. Some of these fences still stand near Highway 99 and Highway 4 in eastern Kansas.

When you locate a nearby graveyard, the stones attest to the short life of stone fence builders, much as gravestones at Sutler's Mill attest to the short life of the gold miner. Both fencers and miners impacted natural systems.

THIS PATCH has a whole lot of holes punched in it but still manages to hold together. In a way, it's like Wyoming, punched full of holes for energy, oil, shale, coal and minerals or natural holes in Yellowstone that bubble their sulfurous gases. So often, land is at the mercy of the appetites of the people in a quiet battle between extraction and conservation that makes holes in the earth and makes a few headlines. In southwestern Wyoming stands modern day Jim Bridger, a coal fired electric plant built in the 1970's as a tribute to our quest to conquer nature. The plant can generate two million kilowatts, the sign says, four times the need of Wyoming. It is 24 stories high. According to a local sage, "it is smoking hell out of the country. Nuthin' grows where the smoke goes."

THIS PATCH comes from a shirt worn on April 29, 1869. It was the "glory day" for bare hands and coordinated backs. James Storebridge and Bull Crocker supervised building ten miles of track in one day across Rosell Flats to Promontory Summit. They started at dawn, quit at 7 pm, took one hour for lunch, 12 hours and 45 minutes. They laid 10 miles and 200 feet of track. For those who like numbers ... that's 25,800 ties, 3,520 lengths of track in 120 foot lengths, 55,000 driven spikes, 7,040 screwed on fishplates. Eight men comprised the "ironmen crew." At the end of the day, they had carried more than 2,000,000 pounds of rail from a flatcar to the road bed. By this time, the crew working west on the Union Pacific toward the junction with the crew working east from California seemed to have gotten the hang of how to lay rail. The railroad would change the natural systems.

HERE'S A PATCH that has some numbers on it. It was cut from a bigger piece of cloth that had a lot of numbers about growth and what has propelled growth in this country and how growth has effected the ecology. From 1900 to 1910, the state of Washington grew 120% ... the reason? The Northern Pacific Railroad made it easy to get there. Alexander Graham Bell placed the first transcontinental phone call in 1915. Better communication encouraged growth. More phone lines encouraged more mining for copper wire .... Montana and Arizona would grow. Los Angeles would complete a 233 mile aqueduct from Owens Valley. Los Angeles boomed and bloomed. In 2000, Nevada had 10 times as many people as it did in 1950, more than Maine or West Virginia. The reasons ... cheap land and air conditioning.

THIS BLACK PATCH is interesting. It is for all the bad guys who fell victim to vigilante groups. Part of the frontier mentality was if things got too intolerable, you took the law into your own hands. Owen Wister mentions that when he writes about the Virginian who becomes a prototype for the Marlboro man. These dusty tales can be found in places like Helena where a gambler named Keene shot another gambler named Slater. Keene tried to take refuge with the sheriff, but was handed over to the crowd who formed a jury who found him guilty. They put him on a wagon, adjusted the rope, gave him a final glass of whiskey. He consumed this and then said, "Let 'er rip." They did. His body hung from the tree for several hours. Today, most of the legends out of Montana revolve around big trout.

This white patch may have come from the petticoat of Helen Hunt Jackson, an Eastern born, well educated lady. She married Edward Bissell Hunt, an Army captain and they travelled widely. In 1879, she happened to attend a lecture in Boston. The speaker was a Ponca Indian Chief. He spoke of the removal of his people to Indian Territory. Jackson was so moved she began to research and then write about Indian mistreatment. Her book, *"A Century of Dishonor,"* drew national attention to the plight of Standing Bear and the Ponca tribe.

THIS BLEAK PATCH came from a shawl of a woman who was a member of the Willie Handcart Company that got caught in the great Wyoming Blizzard of 1856. She froze beside her tent, arms cradling wood for a fire never tended. She had pulled a handcart from Independence, Missouri. She had been secured in England by the Mormons and was destined for Salt Lake as a part of the last vanguard of women who would become wives of the Mormon men already in Salt Lake. She was wearing her third pair of shoes which were nearly worn out. She was nearly starved as were most of the party of 404 people. Of these, 67 are said to have died in the worst calamity of the Oregon or Mormon trail. The Mormans changed several ecosystems.

HERE IS A STEEL COLORED PATCH that could have been taken from pants worn by J.F. Glidden who, in 1874, invented barbed wire. The sharp tips of his invention would become the nails in the coffin of open, western range land. With the wire, land could be fenced and divided up and could be under control. It would encourage cattle wars, sheep wars and be used for water wars. It would impact the ecology of the entire landscape west of the Mississippi.

THIS PATCH is a sandstone color and came from the shirt of an early day camper on Cottonwood Creek in Wyoming. Long ago, Indians camped there because the stream was so full of trout you could reach in under the ledges and catch them with your hands. This water was clear and potable, the result of no shale upstream with no fines to contaminate it. It flowed in a watershed of little or no erosion. Geologists say if you look in a stream, you can see in the sediments, the whole history of the watershed. Today, in some of the streams, if you want a sandstone color patch, you just wash your shirt in the stream. And it is hard to find trout under the ledges.

THIS SHINY BLACK PATCH is from a coat of an early time railroad conductor. While much of the history of the west has been told in terms of military quest and racial violence, it was the railroad that tipped the balance of power from outpost to community. It became the transcontinental link Lewis and Clark were hoping to find as a water road. It would link city to city, coast to coast, rural land to the marketplace. More than that, it allowed national interest to replace local autonomy. It shifted whole patterns of demography and opportunity. It allowed poor immigrants to reach remote places and it eroded two refuges that had protected the early prairie country and the west. Gone was the safety of time and distance.

THIS PATCH could have come from the coat of Josiah Gregg who wrote Commerce of the Prairies, a two volume set published in 1844. He is often referred to as the historian of Santa Fe trade. He drew maps. His writing was acclaimed for its support of the expansionist policies of the United States and the doctrine of Manifest Destiny. His writing also reflected the cultural biases of the mid-nineteenth century and his intolerance of Native Americans, Catholics, Hispanics and Mormons. He also took Protestant America to task for what he felt was a fallen nature. His maps were used for years by travellers who were less interested in religion and more concerned with direction. His writing would mark the beginning of the use of books, pamphlets and even artwork to extol the possibilities of the west.

THIS IS ANOTHER INTERESTING PATCH because it illustrates the many connections between history, economics and happenstance. This patch might be from the trousers of John L. Mason. In 1858, he patented a unique thread necked jar with zinc caps and rubber rings. After his patent lapsed, other glass companies, including the Ball brothers, followed his lead. His jar changed ecology for all time because now fruits, vegetables and meat could be preserved for long periods of time, reducing the immediate need for food consumption. The life of food supply had been extended. This simple fruit jar would encourage production and consumption which in turn would put more pressure on natural resources.

THIS PATCH HAS some ink stains on it because it was thrown atop an old copy of the Cherokee Phoenix. It would be dated 1827. A Massachusetts foundry had forged the syllabary based on an alphabet designed by George Guess, a young Cherokee better known as Sequoyah. The newspaper would be printed in both English and Sequoyan. It was so effective in spreading news that Georgia authorities arrested its editors and confiscated the press. It would be 1844 before another press rose from the ashes. The Cherokee Advocate's first edition was September 26,1844. The new editor was William Potter Ross. The Cherokee people would become the better known of the many tribes moved into Indian Territory. Ecology was changed.

THEN HERE IS A TAN COLORED PATCH that could have come from the sleeve of a shirt, the right sleeve never worn by John Wesley Powell. He would have tucked it in when he went down the canyon he named The Grand Canyon. Powell didn't need a right sleeve. He had lost an arm during the Civil War. Later the Major would head a survey expedition of western lands that would compete with the work of Dr. Ferdinand Hayden under the Department of Interior. In 1878, the DOI got a new secretary who tried to introduce some reforms in the scandal tainted department. It was to him that Powell presented a report, "Report on the Arid Regions of the West." Some historians rank the report as the most revolutionary book ever written by an American on land use. Like many things that espouse new ideas, the report met with little public favor, especially among the western "boosters." Powell's work would eventually lead to the formation of the Bureau of Reclaimation and would find its way into the Taylor Grazing Act of 1934. Powell understood the marginal value of arid land.

THERE WOULD BE PATCHES from Connecticut too. This white patch could have come from the vest of Asa Whitney, a Connecticut merchant and trader who, in 1845, asked congress to deed him a swath of land 60 miles wide. The swath would start at the Mississippi River in northeast Iowa and extend to the Columbia River valley in Oregon. It would be a grant of 75,000,000 acres. Whitney proposed to sell or give land to homesteaders, offer it to new industries, realize cash from minerals, wood, land sales and irrigation projects. This would finance the Lake Michigan to Oregon Railroad. The idea was considered a bit daft. In 1865, most of the daft ideas would be used by the Union Pacific railroad and Congress to build the nation's first western railroad.

AND THEN THIS GRAY PATCH, rather well washed wool, that came from the jacket of a man who had been nearly blinded by an industrial accident. Despite having a number of inventions to his credit, he decided "botanizing" was the road for him so he left Wisconsin and took a thousand mile walk to the Gulf of Mexico. By that time he had become a naturalist/philosopher who concluded that "Nature (God) had made humans, animals and plants for the happiness of each one of them and not for the happiness and use of humans alone." He sailed to the west coast. When his boat docked in San Francisco he asked directions to "any place wild." John Muir started hiking toward the Sierra Nevada wilderness. That was in 1868.

history, journal entries, personal narrative; and if you dig too deep or get into too much detail, you would never get the quilt stitched together. So, you start sewing someplace.

ALL THE PATCHES

. . . in Grandma's crazy quilt had only one commonality. They were cloth. They would be stitched together, with care and patience, to produce a cover. In this case, I hope they cover a subject; how little known events have impacted ecological thinking in this country.

In a way for many years, I have been collecting patches of information, from books, articles, periodicals; from personal experience and first person observation. Some sources could disfigure the facts and some were liars who could figure. In time, I had sacks of patches of information. Each patch, however disparate, had some connection to natural resources or ecology. I had a sack each of government patches, social contract patches, and economic patches all with connections to natural resources. Some got stitched into this report.

The model I have proposed in this book became the thread that stitched all these patches together.

ONCE, WHILE ON A FISHING EXPEDITION

Searching for trout with an especially learned friend, I was off on the subject of connections and contradictions. He rather laconically commented to me, "Ken, you are a throwback, a Ectopistes Migratorius." Needless to say, that brought a pause to my monologue. "Let me explain that," he said in a slightly superior tone, "that means you are a wandering wanderer. You are always looking for the offbeat stuff and trying to connect that to some other offbeat stuff."

He was probably right. Most conservationists are travelers on an endless journey. Most of the people I know of that ilk began their trek as rather passionate hunters or fishermen. Audubon was a helluva shot and he hunted most of his specimens. The pursuit of wild things began as a sporting satisfaction and ended as a lifelong affair of heart and soul. Of course, the people I knew were usually travelers on a road that went through a great transformation sometime between World War II and the turn of the century. The world of natural resources upon which the United States had depended for economic improvement and sporting satisfaction became a world in transition. There was new talk about caring for the earth.

Try to transpose feelings into words on paper, to stitch the patches together, and it is easy to get enmeshed in history and dissonance. Dwell too much on history and the past and you can be disillusioned about the future. Memory looks backward and hope looks forward. Looking back, it is easy to get caught up in the odd, the hidden, the offbeat. The hunter, the fisherperson carries the burden of memory. The hunter is often a wandering predator who collects mental snapshots of grassy draws at sunset, or edges of forests basking in high noon silence or the rush of snapping wind in a marsh blind. That person, in fairness, is more likely to dwell on memories and history. Their reprise of man's treatment of natural resources can pick up smudges of cynicism or sadness. You need to understand they began their search using a slightly different needle to stitch their patches together.

NOT ALL OF THE PATCHES PRODUCE

. . . a final pretty picture. Nature is not all pretty. There are some inevitables. Last summer, an osprey became a companion of mine. I would walk down a winding, rocky road where the bird had a nest. It would have been more likely there were two ospreys, but at a respectable distance it is hard to tell whether you see two ospreys once or one osprey twice. The osprey would fish on a road side lake. I could see the bird dive, the talons lowered, the overdrive of power as wings shifted to ascent. I could watch the awesome moment of predator and prey on a wide screen but I could not know the smaller screen of the predator that surely must be focused on an inner drive of hunger.

I think, as the osprey climbs to a naked limb dining table in the barren fir tree that all living things must know hunger. The raptor, the deer, the duck, or gopher . . . all present on my big screen view, must know hunger. The evening before the osprey show, I had watched Victoria Secret's TV show on the 25" inch screen and marveled as a model, her discreet parts covered by feathers, waddled down the runway. There must be times when she knows hunger. She can land at a deli. My osprey faces a more primitive challenge. Marketing fashion is not an issue. Among the contradictions you collect with conservation patches are the conflicting patches of the pretty and the practical. The painter or the sculpturer can invoke powerful images of grace, beauty and the esthetic. He or she can make the osprey, the eagle or the wolf appreciated for the pretty. In the reality of daylight, all are predators and how they survive is not always pretty. There are times things get messy. There is a grim moment when they strike to kill.

There is a raw underside of ecology. The inevitable side. The word extinct is such a final word. There is no encore. Still, quality in nature is taken to a higher level when it is appreciated for hope and birth rather than apathy and death. And this works so long as hope and birth find the balance with the understanding and reconciliation that nature is a long term process where the tree of life will always be pruned.

Stitch together a crazy quilt of ecology today and it is every bit the jumble of Grandma' s quilt; colors, facts, data, opinions, lawsuits, paintings, books, research conclusions, public hearings, bureaucratic pronouncements, E-mails, observations and TV shows . . . it has no shortage of sacks from which information can be selected and stitched together. Sometimes, the patches clash and seem to contradict each other, so it is with ecology.

Nature becomes a version of Grandma's crazy quilt.

Chapter Three

The Ten Ecological Ethics

TRYING TO PUT ECOLOGICAL VIEWPOINTS

. . . into a single sack can get confusing. It might remind a person of the Mark Twain story about the bird cage. He is supposed to have a put a canary into the cage and it got along fine. He added other birds and there was still reasonable harmony. He wondered about adding a Baptist.

He joked that a hard shell Baptist couldn't get along with anybody, in or out of a cage. He put the hard shell Baptist in the cage. All hell broke loose.

It is possible Twain's allegory fits the way we have accumulated viewpoints and hard positions on conservation issues. A small congregation adopts a certain viewpoint and then expands this position or viewpoint, searching for a broad, denominational consensus. Failing to find consensus, the small group leaves the bird cage.

Now stretch this allegory to a very big bird cage and you put birds in the cage from various environmental viewpoints. There is a U.S. Fish and Wildlife bird, a Bureau of Land Management bird, a Park Service bird, a parrot from the Interior Department and maybe a mallard duck to represent Ducks Unlimited.

Now represent each conservation viewpoint with a Blue Jay to be the attorney. You would need a bird with strong opinions and a strident voice. What do you suppose would happen in the bird cage? What would be the odds of reaching any consensus?

And it has happened. In fact, is happening now. We have filled the cage with strong opinions, feathers, loud words, bird do-do and all kinds of expert testimony on the subject of natural resources. Any condition of consensus, accommodation or compromise would come as a surprise because we have been squabbling about how we would think about conservation for four hundred years.

Many years ago I took the brave step of opening the Twain cage to see if there were any conservation birds that could be identified. I was searching for ideas or viewpoints or even moral principles that would say to me, "This is what the majority of American people think conservation is about . . . this is how they would define it. When they use the word, this is what they mean."

What I located were viewpoints that ranged from dogmatic scripture to mystical uses of broad ideas of what was right and what was wrong. I had a sack full of beliefs, some of them opposite beliefs. But, these viewpoints were often strongly held. And there was really no consensus. About the only commonality was the subject of the viewpoints, natural resources.

THE SEARCH BEGAN WITH THE TIME OF SETTLEMENT

. . . when the early arrivals experimented or flirted with what became at least ten different variations of some sort of "social contracts," ethical or cultural beliefs that could be applied to the use of, or consumption of, natural resources. To go back to the opening analogy, we have put at least ten different birds in the cage, including an amazing number of Blue Jays. These ideas, for lack of a better definition, we will call "Ecological Ethics."

THE FIRST VIEWPOINT is that nature is hostile and an enemy. This is the message from the Old Testament. Nature must be subdued and conquered if man is to survive and prosper. Only mastery of natural forces will release man from the restraints of nature and allow for human use. Nature, then, is the wild adversary. This was the Puritan ethic.

Our language is laced with the residue of this idea. We have "wild game dinners." We have "wild and scenic rivers." We build dams to "tame" a river and even the term "wildlife" has the connotation of untamed. In early times, it was not uncommon for a person who "lived in the wild" to be referred to as a "savage."

THE SECOND VIEWPOINT OR ETHIC is the concept that nature is a resource that should be used for man's needs and if extraction is necessary, then it is justified for the social and economic gain of man. This is the extractionist view. Nature is a commodity. This view was prevalent from 1850 to 1900 and would be disguised as "progress or Manifest Destiny."

THE THIRD ETHIC is that happiness depends upon material satisfaction and nature can be used or consumed in any way that accomplishes this end. This is the entitlement view. Most forms of nature/sports recreation would fall into this "social contract." This ethic is involved in the controversy surrounding snowmobiling in national parks. The "snowmobilers" believe it is "their right" to use their machines in the national parks.

THE FOURTH ETHIC is that nature is property. Land, and what is on that land is property. The individual can own it, buy it, sell it, acquire it and inherit it. It can be bartered like any material property. So, it follows you can own nature and do with it whatever ownership implies. This would be the property right viewpoint. This view has been extended to water and has created a vast reservoir of "water rights" issues.

THE FIFTH ETHIC is that nature belongs to a common good and should be used to the highest and best purpose for all persons who are parties to the given social contract. This would be the utilitarian viewpoint. Frederick Law Olmsted would be a good example of a conservationist who espoused this view. Another later example would have been Gifford Pinchot who initiated the Bureau of Forestry which became the Forest Service.

Some of the ten ethics can be associated with a person who first espoused that viewpoint. There can also be a distant association with some sort of conservation organization.

THE SIXTH ETHIC is that nature is a property, or condition, that can be quantified and qualified. Therefore, it can be managed or manipulated as part of a scientific process. This viewpoint has grown in acceptance as the role of science has been expanded. It is a view that is very prevalent in state game and fish departments.

This ethic, in fact my casual use of the word "ethic," creates a problem for the person who walks the path of ethical philosophy and the issue of the relationship of the words ethics and morality. I would argue that what is missing in conservation "ethics" is not so much the knowledge of how to control nature as it is of a human being to control oneself. BUT . . . and it is a big BUT . . . when one applies modern technology to address the lack of self control, we begin to devise schemes for manipulating and controlling human lives much as we employ devices to control insects, bugs and even smoke.

The embracing of science has allowed us to lose the responsibility of controlling ourselves. We devise systems for controlling innate objects. The individual now becomes a collection of vital functions . . . eating, sleeping, smoking . . . or, of social functions such as producer, consumer, citizen or parent. We reduce value judgments to factual judgments. A shrouded result of this type of thinking is that we leave no room for the idea that in nature, many facts are obscure or even hidden.

To illustrate with a simple analogy that really isn't so simple. A man is walking through the woods. He picks up an acorn, examines it, scuffs a hole in the soil, drops the acorn in it and covers it with earth. The goal of this directed behavior is an oak tree. The acorn also has a directed behavior; the goal is an oak tree. Many things can happen from the germination of that acorn and a large oak tree. Many are variable, not always visible as facts and subject to many whims of nature. If a large oak tree is the end result, we can shrug it off and say, "Nature took its course." In the end, we have very little "factual" information. Our conclusion is based more on faith than on science, but is it any less valid? How would John Muir or Thoreau argue this point?

THE SEVENTH ETHIC is that nature is an ethic and is subject to a moral code. "A thing is right when it tends to preserve the integrity, stability and beauty of the biotic community. It is wrong when it tends otherwise." It becomes a matter of personal or individual stewardship. This is the Aldo Leopold "Land Ethic." Interesting enough, you can locate this idea in the thinking of John Quincy Adams as he was wrestling with political morality.

THE EIGHTH ETHIC is that nature is subject to the laws of any harvest and humans are entitled to a share of that harvest so long as they replace or restore the means to the harvest. You have the undertone of sustainability. The interpretation of this ethic is often found in the grass roots philosophy of the "hook and bullet" guys lurking in groups like Ducks Unlimited, Rocky Mountain Elk Foundation and the Wild Turkey Federation. You might call it the Sports/Harvest Ethic.

For several years, I thought in terms of eight ethics. Then one fall day I was sitting in the office of Eric Palmer, Director of Fisheries for the State of Vermont. "You know," he said, as he was reviewing the eight ethics, something I have asked any number of wildlife people to do, "I think you have a ninth ethic. I would call it 'The Bambi Ethic.' There was a pause in our conservation. I asked him to go on with his idea.

"I agree with your eight points," he said. "But in the last few years, we seem to be having more trouble with people who believe you shouldn't kill anything or harvest anything. Maybe they have seen too many Walt Disney movies where animals become more like people. Anyhow, those people, as a group, are taking some very strong positions about many of our fish management policies. For instance, they don't think you should have catch and release. They just believe there shouldn't be any catching."

So we added a ninth ethic after many years of devoted introspection of what we thought were eight. The Bambi ethic would argue that any taking of wildlife, waterfowl or birds, even for the purpose of harvest or sustaining human life, is not a desirable social contract.

PAUSE NOW BEFORE WE COME TO THE TENTH ETHIC

. . . for consideration of stages or passages in a life span where we put aside one particular conservation ethic and adopt another. Many ardent conservationists began as pure sportsmen. They evolve into another belief and these often involve their view of conservation ethics. Suppose we consider five stages.

The beginning Stage I is when the novice or beginner or newcomer is introduced to hunting, fishing, trapping, any of the harvest sports. They are just thrilled to be there. Whether it is the fly rod, spinning rod, shotgun or rifle, the main idea if they are being introduced to something that is a new world of exciting experiences. If they are successful in bringing some quarry to hand, that is just frosting on the cake.

In Stage II, it begins to get more important to bring the quarry to hand. Now, getting a limit, be it one deer or six trout, becomes an important part of the experience.

In Stage III, getting a limit becomes the mark of achievement and the Freudian complex may be a part of the mix. This is bringing in the biggest deer or landing the biggest fish. Being excellent at what you do, especially in the opinion of the peer group, is also part of the mix.

In Stage IV, two character changes are likely to occur. There is more thought given to the possibility of "catch and release" in the case of fishing, or shooting only the male of the pheasant population or even some thought given to the amount of work involved in getting a large carcass out of the hills, dressing it and lugging it home. Additionally, and more practically, there isn't any more space for racks of horns or fish with gaping mouths. And the freezer is full. There is a growing mindset that just enjoying the companionship and a good day is really what the sport is all about. Value is the quality of the experience which replaces the quantity of meat brought back.

Then there is Stage V. Now, judging the quality of accommodations becomes very important. Good food and a soft bed in a nice motel have more value than a sky full of pheasants on a rainy morning at the end of a corn stubble field. There may be those days, when the weather is just plain nasty, that the value of the accommodations outweighs the value of the outdoor experience.

If this is said rather tongue in cheek, there still is a kernel of truth to it. It serves as another confirmation there is nothing constant in nature and human beings are part of nature. Change or evolution is the only constant connection. Many passionate harvester/sportsmen have become equally passionate conservationists. They have come to embrace a modified or even a different viewpoint. The frame of reference may continue to be that of the harvester/sportsman, but the application of new information has modified the original belief.

Neuroscience represents to us that each of the concepts we have, the longer term ideas that structure how we think, are instantiated into the synapses of our brain. Those concepts cannot be changed by someone telling us a fact. Or, as the old bromide goes. . . ." don't confuse me with the facts." But tell the average human being a story that fits with a story already in our heads and we can, if we choose to, make the connection.

THE REASON TO INCLUDE THE NATIVE AMERICAN

. . . began as a vague feeling many years ago. Slowly, as I learned more about their social contracts and conservation ethics, I decided they had a rather distinct conservation ethic. For lack of a better title, I now believe there is a spiritual conservation ethic. Aldo Leopold comes close to this when he advances the idea of a "land ethic." He is skirting the edge of morality. Native American spirituality is ecological morality.

Originally, most Native American culture was oral. Tribal custom was oral. History was oral. Tribal social contracts were oral. Respect for nature was woven into all this. Most Native American cultures had calendars and generally these revolved around twelve moons. The moons would relate to the periods of tribal life cycles. Most concern dealt with survival because the time of the moon related to food, harvest or conversion of the harvest into food. The spirit of the tribe, or family, was woven into the calendar.

In this spiritual chronology animals often had a special place. Before a hunt there might be ceremonies. Due homage was paid to the quarry. There was a social respect for any harvest. The harvest paid homage to the spirit of the animal for the food it produced for survival. The Native American in most tribes had a spiritual, ecological ethic. But none of this morality was written down. It was only handed down orally, often in animal stories. We can only assume it changed a bit as it was handed down from mouth to mouth and ear to ear.

Only in rare instances has the American culture considered this aspect of the Native American culture. Perhaps our Abrahamic approach to any negotiation with the Indians that involved land couldn't reconcile any position but that of the Euro-American. Maybe we only suspected that possibility. This may explain the many spires that were erected after we had finished fencing off Indian property.

OUR TENTH ECOLOGICAL ETHIC IS SPIRITUAL

. . . both because of the connection to the Native American culture and roots in the Euro-American bible culture. Among the ten ethics we have listed are six with the "strong father" morality. Four ethics contain "Common Good" as the core theme. This is more evidence that our conservation ethics do have a philosophical bent toward morality. We need to be mindful that for many years, extraction as it was practiced in the U.S. was not considered immoral nor ungodly. Again, it was not until the Civil War and George Perkins Marsh that gross extraction was even called into question and then the issue was what we were leaving for the coming generation.

Out of this basic framing comes a moral order of sorts. God is above man, man is above nature, adults are above children. Father teaches the children right and wrong. Western Culture is above non-Western culture, so in this thinking, Euro-American was always the parent and the Native American was the child. When all else failed, the "Great White Father in Washington" would tell the Indian what was right.

So if a person were to ask, "Where did our conservation ethics come from?" a reasonable answer would be, "Well, you certainly had some religion mixed in, you had some morality and regard for the coming generation, and you had a subject that was pretty much formulated by the male mind. In fact, Rachael Carson is the first woman's name that appears on our list of well known conservationists. Her book, *Silent Spring* did not appear until 1962. And other than her presence, the subject of conservation is pretty much a man's world."

THERE IS ANOTHER WILDCARD

. . . in the big game of conservation. There is the murky issue of economics and greed. For Americans, conservation or environmental thinking is not so much a statement of morality as it is an attempt to locate some consensus thinking on the limit of appropriation. The idea we would "use" nature was a given. We have directed our conservation efforts to finding acceptable limits on the use of natural resources. It is not so much a question of using water from a river as it is how much water will be used. It isn't a question of shooting ducks each fall, it is a question of how many do we get to keep. Consumption is our mandate, sustainable ecology is our mantra.

Examine the mission of many conservation organizations. I would argue that consumption is the goal-directed behavior of most groups. The end goal is restoration or protection to enable future or continued harvest. Hence, our argument that conservation is really a consensus opinion on the limits of consumption. Even John Muir wanted to ultimately "save" his valley so "it could be especially adapted for pleasure seeking." He wanted other people to sit on some rocky crag and consume, or enjoy, the beauty and the solitude of what he saw. His ethic of preservation still hasn't solved the problem of locating the limit of the size of the congregation so as not to defeat the purpose of the cathedral.

And if that idea is too pristine for the grass roots, switch to the DU member who wants to see the marsh full of ducks or the Wild Turkey Federation member who wants to see the woods full of gobblers. Harvest of natural resources is the wild card, and the driving energy behind the screen of the conservation organization.

In fact, so often it seems, human kind decides there is not time for evolution and they want to hurry the process. There is a rush to channel a river to prevent flooding or enhance navigation or a hurry to drain a wetland to grow more milo. And what man has done to natural resources to serve a social cause, or government has promoted by issuing a degree for some imagined economic benefit, or even there is psychic satisfaction because "we have conquered nature" is not a neat, simplistic story line.

But our muddled conservation ethics, which I am attempting to extend into ecological ethics, plagued as they are with differences of opinion, saddled with divergent views on the use of natural resources, with all their trial and error are not an obverse issue. Because if we inventory what has happened to our natural resources in the industrial age and consider we started with a relatively new and untapped closet of resources, we need to ask what will happen to us in the new, technological age? In this age we are playing with a closet of resources that have already been badly worn and some of our land, water and air really should go to the consignment shops. The ugly truth is we have fewer natural resources to use in a faster paced world.

ONE OF THE CENTRAL THEMES

. . .of the book by George Perkins Marsh, *Man and Nature,* is that Americans seem to be a little unique in that we accept the idea there can be reform and things can be made better. He also feels man has the power to accomplish the reforms. The whole force of the book would seem to be that the welfare of future generations is more important than any immediate considerations. Marsh was the first ecologist to question American patterns of extraction. In his words, Americans who disdain "a better husbandry for themselves, feel morally obliged to do so for their offspring."

This idea provides another frame for the question of why we think about conservation as we do. Man may be the only living creature to believe in a force, or forces, that provide continuity and perpetuation. This force may drive man from the primal stage of sheer survival to the need for continuity and community. The beginning of this passage is cohabitation with a mate or the results of that mating. Families may bond together to accomplish a broader sense of community. In animals, this results in flocks, herds or packs. In humans, this bonding requires a code, or rules or social contracts, even laws, whether they are written or oral.

When mankind phased from what was primarily a hunting society into more of an agricultural society, the spirit of the community would evolve from individual faith into religion. Man also would decide how society would use natural resources. Out of these organizations Gods would usually emerge, acceptance for the individual and shared social responsibilities.

These ideas would spread into larger economic and agricultural communities and there was increased spiritual bonding. If religion developed, it might range from benign spiritualism where the social contract was the most good for the most people . . . or it would become military in nature. The result could be tribal wars, holy wars, civil wars or national wars. Wars would depend upon weapons.

Whether weapons were developed for the purpose of harvest . . . and the plow or saw can be as deadly as the rifle . . . or, for human eradication, any study of the evolution of ecological ethics must include the issue of weapons. Weapons impact ecology. Wars impact ecology. Every war fought, in or by this country, has produced an after burn in ecology. That afterburn would produce new settlement patterns, new technological advances and new social contracts. If technology produced better methods for killing people, it would later be transferred to use of natural resources. Much of the frontier mentality that produced new social codes for prairie country or the west came from Civil War veterans. I am convinced the surge in hunting after World War II came as a result of veterans returning home. In a broad generality, several great struggles, starting with Lexington, trained young men to be predators. Inevitably, the gun is in the genes of ecological ethics.

So to establish ecological ethics, it is necessary not only to inspect the viewpoints themselves, it is also important to consider the many evolutions of society and culture that framed the ethic. In the last fifty years of this evolution of framing, we have proceeded to replace the often baffling aspects of nature with the omni competent rationale of science. Now, we are discovering empirical knowledge is not the only truth. The physical universe also holds many truths. The issues can range from where small frogs have gone to the extent of global warming.

One of the great tenets of our American culture is that America is a huge melting pot of nationalities and races. We mix all this diversity in a giant pot and proceed to cook liberty and freedom soup. But this very diversity has inhibited the search for any national conservation ethic. Mix in the sheer size of the country and the great variety of ecosystems and the result has been a soup with many flavors.

CONSIDER AGAIN THE TEN ETHICS

we have identified. You will find direct conflict, even before we consider
the impact of framing produced by government, socio-economics and social
contracts. Simplified, this is what they look like.

 1. Nature Is Hostile and Must Be Subdued.
 2. Nature Is A Commodity And Can Be Used As Such.
 3. Nature Can Be Used For Human Satisfaction.
 4. Nature Is A Property Subject To Property Rights.
 5. Natural Resources Belong To The Common Good.
 6. Nature Can Be Managed And Manipulated.
 7. Nature Is Subject To A Personal Moral Code.
 8. Nature Can Be Harvested but Should Be Restored
 9. "The Bambi Ethic" That Precludes Any Killing.
10. Nature Is A Spirit To Be Revered.

The search for a national conservation ethic is not an easy search. You can
locate small, local examples . . . social contracts in small communities.

Now and then you come upon a regional ethic, one of our ethics that is on
display along a river or in a forest. Usually, the issues have become more
complex and will be articulated as rules or regulations, even local laws. Legislation, such as laws governing snowmobiling, can evolve or change or the
law is put on hold depending upon the legal process. And of course, various
causes may produce reasons for action or review or repeal. These written
statements look so neat when they are all organized on paper. They look like
fast, single answer solutions.

It is entirely another dance to put wolves back into Yellowstone Park.

Chapter Four

The Theoretical Ethics in Practical Use

One of the nagging problems presented by ecology is the tendency of things to look rather good on paper but when they are extended to actual implementation, the theory unravels. The idea of using the Missouri River for barge traffic and also accomplishing flood control looked good in the planning stage and reams of supporting data made economic sense.

In the last 12 years, the U.S. Fish and Wildlife Service had repeatedly advised the Corps of Engineers that dam operations had eliminated seasonal flows in the river and seriously degraded the river fish and wildlife habitat. The National Academy of Sciences, in their three year study of science along the river, concluded that "the Missouri River ecosystem will continue to degrade unless the river's natural water flow is restored."

In this case, the arguments go far beyond "social contracts" and get into federal law such as the Endangered Species Act. Meantime, the barge industry, according to the South Dakota Wildlife Federation, has dwindled to an economic impact of less than $7 million year while the tourism industry along the river has swelled to an $85 million industry each year.

There are three of our suggested ethics on display. (1.)The view that nature is a resource and should be used for the social and economic gain of man, (2.) the view that nature belongs to the common good and should be used for the highest and best use for all persons who are parties to the agreement and (3.) the scientific viewpoint that nature is a property, or condition that can be quantified and defined and therefore can be understood, managed and manipulated as part of a scientific process. And because the ethical viewpoints are in discord, if not conflict, you have a mess along the Missouri River.

One of the strange contradictions we have hung in our ecology closet is the part of our professed American creed that we "leave it better for the next generation." Evidence of this creed as part of our culture can be found in the letters of Thomas Jefferson, John Quincy Adams, the writing of George Perkins Marsh, Henry David Thoreau, Theodore Roosevelt, Aldo Leopold and countless other exponents of conservation and ecology. That is a passive creed.

Teaching our youngsters to say the pledge of allegiance to the flag and to sing America the Beautiful, with spacious skies and purple mountains majesty is an important part of our culture. But teaching them any allegiance to harmony with nature does not reach that level of importance. Where do young people learn about relationships between man and nature? There is no text on the subject. Sadly, very little about the relationship is taught in the schools. It takes a great leap of imagination to transfer castles in the sandbox to the erosion of sand dunes on an overdeveloped coastal area.

Some few grade schools are now involved with "outdoor classrooms," generally an area outside the school in a natural setting that provides the space for some discussion of nature. The use of this facility is not mandatory for teachers. The result is some use it and some don't.

Some conservation organizations are now including programs for young people. The Wild Turkey Federation has their Jakes program and Ducks Unlimited has their Greenwing program. And in the past few years more state wildlife departments have begun to include educational and informational sources which are aimed at the younger generation.

For the most part, these programs are preaching to the kids who are already in church. Education on a larger scale just hasn't been developed. As one observer told me, "If we do no more than we are doing now in education, we will never get it done." This person then went on to point out how in a space of ten years we had developed a computer literate generation. "What would have happened," he said, "if that sort of effort had been directed to caring for our natural resources?" Ecological education simply isn't very pro-active.

Through the years I have asked a wide variety of people, from the birdy people who wear binoculars for glasses to the people whose walls are covered with horns and stuffed animals, where they learned about our conservation ethics. Usually, there is a moment where we sort out what we are talking about. There will be expressions like, "Oh, you mean the rules we follow?" Or, "how we do it because we were taught to do it that way?" They all say pretty much the same thing. They learned from a family member, older brothers, relatives who were more like mentors or teachers.

In my case, I learned at least four of the ethics from an uncle and I learned some adaptations of those ethics from a pool room operator. Between the two, they instilled something of a sense of stewardship for natural resources. This education involved the fourth ethic which I call the property right ethic, the second ethic which is the extractionist ethic, the fifth ethic which is the utilitarian viewpoint and the eighth ethic or "social contract" which is the sportsman/conservation viewpoint.

My education began when the various parties in the family decided I was old enough to own a gun. I could have a gun, provided it was single shot, provided I bought it from my own savings, that I took good care of it, that it was never loaded in the house and I hunted only with an older person. This litany of stipulations was not uncommon for teen age boys growing up in Nebraska prairie country in depression times.

THE SOCIAL CONTRACTS OUTSIDE THE FAMILY

Not all the social contracts were within the family. Some were with the country neighbors. One involved having adequate relations with neighbors, especially the one to the east who was considered to be fairly well off because he owned a tractor. That "social contract" involved stacking his barley and oats when they were bailed, helping him thresh the grain and then bail the straw and stack it near his barn. No money changed hands.

He would then come over to my uncle's place and the pattern would be repeated on our twenty acres of oats. When threshed, the oats went to the barn, were stored and would be ground up for turkey feed. It was a simple enough economic system. Again, no money changed hands.

When the neighbor climbed down from his tractor, my uncle and he would discuss the other "social contracts" of the time. . . . the church roof that needed repair, who was related to whom in the area, any recent marriages or deaths, the Farm Bureau, an up-coming pot luck dinner to benefit the volunteer fire department, what President Roosevelt said in a recent "Fireside Chat." I sat on the shady side of the tractor and swatted flies.

All this information had a value to me only in that it would provide contacts we might use after the first snowfall. The night after a good snow, we would carefully check our guns. My single shot 410 didn't take nearly the preparation my uncle's Remington pump gun took. But both got wiped down with a kerosene rag. Then the gun oil was located and carefully dabbed on the critical parts. My uncle would give his gun a few trial pumps and would pronounce it "ready to go."

My uncle was a patient teacher. "You don't," he would say, "ever hunt on another man's property without his permission. It's his property. You always ask. I'll show you how to do that tomorrow."

I had no inkling he was a diplomat. We would pull into a farm yard after he had checked the name on the mail box if he didn't know the resident. The always present farm dog, or dogs, of questionable parentage, would announce

our presence. After a snow, the farm owner was usually in the barn and usually working on machinery.

There was an exchange of names of mutual acquaintances, a couple of "git
down there" to the dogs, then a discussion of weather, some talk about farm
prices, further talk about the lineage of the dogs and maybe some references
to the Governor or government in general. By this time, we were almost
friends of the family.

This led to questions about finding rabbits in a good looking draw, which
led to some waving and finger pointing about the boundaries of the land under discussion. My uncle would never ask where game was. . . . he would
sometimes joke rabbits were where you found them . . . and that would further cement our relationship with the farmer. By this time my feet were beginning to hurt in boots that were too small and I wondered if we were ever
going to go hunting.

But several of the ecological ethics had been fairly well set in place.

THE GRASSROOTS ETHICS

My uncle was a purist. He never poached, never shot anything from the road.
He always walked into the wind, while hunting or in life; he could track a cot-
tontail rabbit with no wasted motion. He dispatched any cripple quickly. You
"always ate what you shot."

He had another diplomatic skill. Sometimes, when we had finished, he
would return to the farm house. He usually did this to say thank you but
sometimes he would offer to share our good fortune. Sometimes this gesture
was accepted. More often, it was not. "We've got plenty," was the usual response. I asked him about this.

"Well," he explained, "these folks are around the game every day, they see
a lot of it and pretty soon they begin thinking there is plenty of it."

I wasn't so sure because I had just walked five miles into a brisk wind to
bring three cottontails to hand.

"Funny thing," he would say, "some folks won't let you hunt on their land.
It's their property. Some have family that hunts or they just don't like hunters.
But it is their land and they have a right to their opinions. Those that let you
hunt I figure have had a hand in raising the rabbits or pheasants and ought to
be entitled to some of the crop. Usually, when you offer to share, you are welcome back."

These attitudes were common wisdom in that stage of the evolution of my
growing up. Today, I believe I could put those ideas and experiences on the

model I have proposed. In those times, the government didn't have many policies that related to hunting rabbits. Most folks thought of the land as something of a sustainable resource. It was tough but with any kind of care at all, it would always be there. You had to work it, control it. This was prairie country in the early '30's. The dust bowl would change that idea.

My uncle's beliefs, however simplistic they were, involved the realities of my theoretical ethics. They were grassroots ethics.

By the time I got back from the service, he no longer hunted. The dust bowl broke him. He had sold the farm and bought a bottle. Hunting no longer interested him.

MY OTHER MENTOR

. . . was a bowlegged fellow who needed a "hey boy." Joe ran the local pool hall when he wasn't hunting. He believed the purpose of ducks was to provide food for dinner, that limits were something you ought to be aware of, they were worthy of discussion, but could be regarded with considerable flexibility when the occasion demanded it.

Joe had a collection of tin silhouettes, cut with snips, mounted on short sticks and painted black. He had a dozen or so Karo syrup cans, lids tightly sealed, painted black. And a dozen kapok decoys which he had kept in a gunny sack for many years. I would set the decoys out under his explicit direction each time we hunted. The rest of his collection remained in front of his blind during the season. Nobody ever bothered them. Now and then a can would sink and I would be told how to replace it.

Joe ate one piece of dry whole wheat toast and drank one cup of black coffee for breakfast. He would fill a thermos of black coffee for late morning. If he stayed all day, he never ate lunch. God put ducks on this earth to be harvested and he intended to shoot his share and mine too. My job was to pick up the dead ones and chase the cripples. We always looked carefully for cripples. We wasted no meat!

Joe also had a conservation ethic of sorts. He had his license. That and waiting for red lights while driving to the blind comprised his recognition of governmental authority. His license gave him the right to shoot ducks. . . . and if a rabbit was sitting in a hedgerow as we headed to the blind, rabbits were good eating too.

What conclusions come from these folk stories and early day experiences which so many of my generation had. We muddled our way around with a lot of conflicting viewpoints. In some cases, laws replaced or formalized what had used as "social contracts. Game managers would learn it is one thing to pass laws regarding conduct. Enforcement of those laws is another thing. The "social contracts" may have been very strong.

Certainly the idea of land as property would have been a strong viewpoint. Most folks agreed it belonged to the owner. A few poachers would not accept that view. Another view would have been the abundance of game. That wasn't always true. In Nebraska, deer had been nearly eliminated and quail were in scarce supply. But the perception was one of abundance, at least in what were called "the good years." Both land and what was found on it were inexhaustible resources.

The game and the land were there to be extracted. That is still a strong ethic in many parts of prairie country and the arid west. When economic conditions get tough, the small operator doesn't leave a lot of cover for wildlife. The corporate operator is miles away and just wants the money.

SOME OF THE OTHER ETHICS

Game was meant to be harvested so long as it was not wasted. Just in my lifetime that ethic has softened considerably especially in fishing where "catch and release" has become a stronger "social contract." Today, even in most conservation organizations, the code seems to have shifted so it is only the "sports" who hunt for the ego satisfaction of just "killing for the sake of killing." This shift would have pleased George Catlin, that old chronicler and prairie artist who wrote of killing too many prairie chickens in 1835 and suggested there should be a national prairie park established.

In the time between 1930 and World War II, most hunting ethics revolved around the harvest. You didn't let game suffer, you dispatched it as quickly as possible, you dressed it efficiently and put it to a worthwhile purpose. This was closer to the Native American idea of respect for the resource than we might have imagined at the time.

And then you had the pure hunters like my small, bowlegged friend. They didn't know much about ecology; they did know how to bring home a limit. Their idea of conservation was a full freezer of meat. They could find justification for bending the laws a little if that served their purposes. They bought a license. That was the extent of their contribution to conservation.

My education was, as was the education of many others of my generation, a blend somewhere between two mentors. It wasn't all wrong on one side or the other and it wasn't all right. The truth might have been someplace in between. On the other hand, some of the ideas in the mentoring may have been in direct conflict. It's just that it seemed like it was a little simpler in those times.

The applications of theory to actual experience that I have just sketched are very parallel to rural traditions in prairie country and the western frontier. But today there are, based on what I have seen, heard and inhaled, some monumental changes taking place in our prairie lands and the arid west.

THE SHIFT IN ECONOMICS

As agriculture, mining and even forestry have declined in economic intensity west of the Mississippi river, many areas are trying to replace those economics with tourism and recreational growth. This effort may be only a temporary timeout, as we grapple to locate a sustainable economy. If tourism and recreation offer some hope for conservation of the natural resources, then how will we modify the economics of this shift to account for the emergence of mostly minimum wage jobs?

You hear an argument, more in Wyoming and the arid west than in the prairie country, that goes this way. There is no question that increased tourism puts different stresses on the environment. And the movement of people from crowded metropolitan areas to some of the more pristine areas puts new demands on land and water. But, once the outsider is exposed to the benefits of the unspoiled ecosystem, they are more likely to oppose such issues as clear cutting, dam building and destruction of riparian areas. They become new converts in the cause of conservation. This movement may be the lesser evil as compared to unrestrained economic development.

There is still something of a spiritual pull in the west. You often hear people say "they love the land." But this esthetic statement may be in direct conflict with the reality of the economic interpretations. What were once grassroots ethics have evolved into complex laws about property rights, water rights, animal management, who can use public land for what purpose and whether drilling for natural gas is good or bad for the ecology.

This complexity can extend the length of a river watershed. It will be a tangle of social contracts, regional interpretations and a jungle of federal and state legislation and regulation. Economic values shift, opinions shift, whether it is the Missouri, the Snake or the Colorado rivers.

Through the years, after trying to put the different viewpoints or ethics under the flashlight of study, reading, research and listening to the views of some well qualified scientists and humanists, I'm not sure any of these sources has an absolute compass that always points to true north. One of the penalties of playing in the game of ecology is that you are a little like the football referee who has just thrown a yellow flag. The crowd doesn't like it and the players just want the game to go on. You pace off 15 yards, blow a whistle, wave an arm and the game resumes. Much of the damage to natural resources or the infractions to the rules of nature are hardly visible to the crowd. Both the referee and the ecologist need thick skin.

THE SEARCH FOR HARMONY WITH NATURE

Not far from my small cabin on a Wyoming River is a place called "the Headwaters." It was so named because smaller creeks or streams originated there. They then flowed into the Gulf of California, the Pacific Ocean and the Gulf of New Mexico. It is hard to gaze upon this diorama of spectacle, this place of origin, and comprehend the scope of the land controlled by these waters. The ecological conflict that flows along these rivers is also hard to grasp. The small river, the Gros Ventre, wanders into the Snake River which makes a contorted journey through dams, hydroelectric systems and irrigation districts until it reaches the Pacific Northwest.

It finally reaches the land of the sock-eye salmon. In 1991, only four sock-eye salmon could be located that had completed the trip up the Snake River. On the Columbia River, over 200 salmon runs are extinct.

Proponents for power have argued for dams, the Native American people have argued for "fishing rights," the timber industry has taken the position it is their property and they have the right to do with it what they want. And the sixth viewpoint that nature is a property, or condition that can be quantified and defined and therefore can be understood, managed and manipulated as part of a scientific process can't really be seen from "The Headwaters." You can only suspect these conditions exist downstream.

And the economics? There are numbers available that would seem to show that in the last ten years, over $1 billion dollars has been spent by the Federal Government on salmon recovery. It is an effort that has not stemmed the decline of the salmon population. "Biologists now conclude that this attempt at renewal has failed because it did not focus on the cause of the salmon's decline: habitat and ecosystem degradation."

Strangely enough, whether it is rabbits or rivers, the ecological ethics are involved. There were probably boys in that salmon watershed who were taught to fish by an uncle. They were exposed to some ethics. Some may have even fished with a pool room operator who had a license he believed gave him the right to take about what he wanted.

Rabbits have never created a great deal of wealth. But certainly, development along the rivers which originate as "The Headwaters" has done so. Looking at it that way, the rivers probably deserve more attention than do the rabbits.

But to this time we have not found a way to reinvest or sustain that wealth. Quite simply, we are losing those resources. The governments nor agencies involved, the utilities involved nor the corporations have been able to establish any long-term strategies that would provide either renewal or a sustainable resource.

THE MYSTERY OF RESTORATION

Can you restore a Snake River? Or, a Missouri River? What will be the cost of undoing what has been done? The restoration of both land and water resources would seem to carry a ponderous cost. One small example. This news item from *USA Today*. "Federal officials said an agreement was reached spelling out who's responsible for the $100 million cleanup and removal of the Milltown Dam east of Missoula. (Montana) The EPA decided last year that the dam should come down. It said about one third of the 6.6 million cubic yards of sediment laced with arsenic should be dredged for disposal." It doesn't take long to get into big numbers.

I understand not many boys hunt rabbits in Nebraska anymore. So there is no "Rabbits Unlimited" calling for restoration of habitat. In fact, on my last visit to central Nebraska I was told by a service station operator there just weren't many rabbits anymore. "Not the places there used to be for rabbits. Too much clean farming. Economics, you know. Even them big government programs didn't help much. Tough to make it farming these days. Times change. You be cash or card?"

I drove away from the pump. Heading west, I reflected on those words. If we folks aren't very concerned about the rabbits, the frogs, the salmon or Dusky Seaside Sparrows, why be concerned with what happens to our rivers? I could hear a service station operator in Oregon. "Not the places there used to be for salmon. Too much development. Economics you know. Even them big government programs haven't helped much. Times change."

Chapter Five

The Ecological Model

THE ORIGINAL IDEA

. . . for using a model that visualized relationships that were interconnected came from leadership workshops I did for many years for companies in the broadcast industry. It was my premise that an organization was a living entity. It was composed of four determinants or frames. Those were: *Value Decisions, Structure, Standards and Systems.*

The difference between management and leadership was basic. Management tried to control and manipulate the four determinants. Leadership provided alignment of the four frames so the organization was in sync. If one of the frames was changed or modified, that would likely cause some reaction in one or more of the other frames.

Further, this premise then suggested the organization would be in one of five stages or passages. These ranged from FORM to STORM to NORM to PER-FORM and then a possible fifth frame of REFORM. This last stage resulted in the demise of the organization or the reinvention of the organization. In essence, this stage V resulted in the organization going back to the FORM stage, if only briefly. Visually, this model looked like the one on the following page. My premise was that every organization is in one of the stages in the picture.

THAT ORGANIZATIONAL MODEL

. . . might have been in the background the day I followed range experts from the Noble Foundation in Ardmore, Oklahoma around the prairie that was to become The Tallgrass Prairie Preserve north of Pawhuska, Oklahoma. The Noble Foundation is a private, non-profit organization that has done ex-

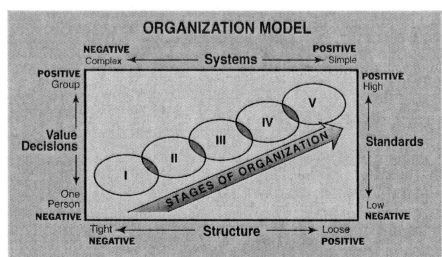

ORGANIZATION MODEL

VALUE DECISIONS. This determinant becomes the mission statement for an organization. Value decisions should state the purpose of the organization, the process by which the purpose is achieved and the desired end result or payoff. As an organization evolves or matures, the Value Decisions may be modified. They serve as the guideposts for management or leadership decisions. Group Value Decisions are more powerful or positive than those made by a single individual.

STRUCTURE. This determinant describes how the organization is put together or assembled. This could be teams, regions, divisions, departments ... all titles commonly used to describe the structure of an organization. The structure will have characteristics. It can be very tight. People do not cross lines. Usually, this is negative. Or, it can be very loose where people can work in more than one area if that becomes necessary to get a job done or accomplished.

STANDARDS. These are nearly the social contracts of the unit in the organization. The continuum can run from very negative and virtually nonexistent to very high, even nearly non-attainable. Sports teams have standards ... or some of them do. Pride is often involved. Standards can be Codes of Conduct. For salespeople, they can be goals. Standards often become the measurements for performance.

SYSTEMS. This determinant would include the economics of the organization and how performance, production, cost and pricing is measured. They could include such areas as benefits, vacation or leave time, compensation plans. They range from very simple systems to very complex systems. Usually, the more complex the systems, the more rigid the organization. Each of these components or determinants can be in different stages. A low stage in one determinant can pull the whole organization down or backward. None of them can be put in a free frame. A high performance organization would have all determinants in Stages 3 or 4 and would be guided by superior value decisions, have a loose structure, high standards and rather simple systems.

Organization Model

tensive research and study of grasslands, grazing patterns and compatible wildlife use of prairie country. They had been invited to inspect The Tallgrass Prairie Preserve and assess the present condition of that land.

These experts would walk an area of grassland and study the condition. Then they would agree upon the quality of the prairie. They would use four stages, I thru IV, with I being poor and IV being high quality. Said another way, Stage I was a negative condition and IV was a very positive condition. It was a simple system. Their model had been a combination of semantic differential and numbers. They were using standards they had agreed upon. They explained these were based upon experience, comparisons to other grasslands they had evaluated and what had become a consensus of opinion.

I asked what they were looking for specifically. They explained it was the general condition of the soil . . . was it relatively intact or had it been disturbed by plowing, oil production or erosion? Another determinant was the use of the land. Had it been grazed and by what? How intense had the grazing been? Had it ever been burned? And if so, when and how often? How many of the original species of prairie grass remained?

I asked if a similar approach could be used with an ecosystem? But one that would be based on the frames of natural resources, economics, government and social contracts? They asked me to define "social contracts" and I explained this was an individual or peer group deciding how things would be done. The conclusion of this visit was the birth of the idea of an ecological model.

I tried the idea of the model on a number of people in conservation work. The general conclusion was the model did have a value but it might work better with five stages. The middle stage would be normal or NORM. The first two stages would be FORM and STORM. The upper level would be HIGH PERFORMANCE and the final stage would be REFORM. The ecosystem would be at the stage of HIGH PERFORMANCE in only limited instances. The idea of REFORM was harder to grasp until you consider that ecology does have such a passage.

You can locate this stage in cattle breeding where one strain of animal is bred to another strain. Both strains may already have value. But cross breeding creates a hybrid which theoretically will produce an improved strain. You find the same general process in grass or corn. Strains are crossed to improve the resultant hybrid. The result is actually a new FORM. It then goes into STORM where the results are evaluated. There might even be further efforts to improve or modify the resultant strain. Eventually, the hybrid goes into the NORM STAGE.

In a way, you find this evolution in wildlife. The common phrase is "the strong will survive." A herd of elk may suffer from disease. The weaker animals die off. The stronger survive. Generally, in the herd, the strongest bull will be the primary breeder. Consequently, if a herd is subjected to unusual hunting pressure of the specimen animals, the general condition of the herd declines. Take this to an extreme, and the result can be a herd of smaller animals.

Expand this line of thinking to an ecosystem. One of the basic assumptions of ecology is that a culture will interact with the ecosystem. This interaction creates a new ecosystem which is at the FORM stage. Now the culture will adjust to the new ecosystem. If "social contracts" are involved, they too will need to adjust. This process happens every time the "rules" or laws of hunting are changed. If the limit on ducks had been five drakes, and is changed to two drakes, both the hunters and the world of the ducks will adjust to a new system. In essence, the hunters go back to FORM and will shoot more carefully, or stop hunting or will do a lot of bitching about the new limits (STORM) before things settle down. On the following page you will find the model of the ecosystem.

IN THE ORIGINAL TALLGRASS PRAIRIE

. . . the essential determinants were FIRE . . . PLANTS. . . . BISON (wild animals) . . . SOIL. Each was a primary part of what was then a "normal" prairie ecosystem. Actually, the prairie was never original. More accurately, it would have been "virgin" or pristine. Even in the original "normal" state it was slowly evolving even if the condition was high performance where sustainability was at a maximum. Enter the human into this pristine condition and the ecosystem would adjust. However, the presence of the human usually quickened that pace of change.

The settler would stop the burning. Notice, the Native American did burn and thus was in closer tune with actual events. The settler stopped the nutrients supplied by burning. This increased herbal growth over larger areas. The Native American was more interested in creating "edges." In prairie country, the euro American removed the bison. This took away the bison manure. The settler replaced this with cattle manure. Where bison grazed the better grass and moved on, cattle were fenced in and there was more intense grazing.

The final stage in what had been a natural ecosystem was plowing of the land by the settler. When the settler did this, they brought a new determinant or frame into the equation, economics. In fact, our entire ecosystem model was now in play. The settler thought he could make more money growing crops than grass. No government decree told him to do this. His decree was "to prove up the land." But, the settler either subjected the land to grazing or plowing. It was the social contract of the time. "Everybody did it that way."

Of the four frames in the prairie model, fire and soil would produce the fast modifications or changes. In the original condition, soil quality depended upon moisture. Where there was more rain or snow, up to 30 inches, there will be tall grass. Essentially, Interstate 35 which runs north and south through

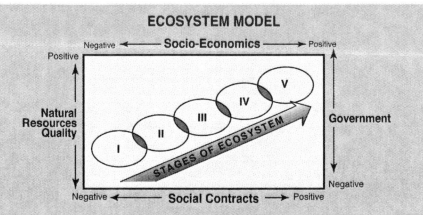

ECOSYSTEM MODEL

NATURAL RESOURCES (quality). This could be habitat quality, moisture, sun or level of intrusion. If it were intrusion, it might be pollution, silt from water, road construction, roads grazing or plowing, predators or even disease. It would be the balance between wildlife, waterfowl or birds. It could be the age of the species involved. Quality is defined and criteria established. It is the sum value of the determinants selected that produces natural resources quality.

GOVERNMENT. This could be any level of government. This determinant has a continuum that ranges from oppressive or negative to enlightened or positive and in the case of our subject, pro-conservation. There is a government model that looks like the organizational model. A Stage I government would be rigid, bureaucratic, have low standards and would not be proactive to conservation causes. The model can also fit government organizations whether local, county, state or federal. In the case of a tribe, it would be oral or written law.

SOCIAL CONTRACTS. These can be written or unwritten rules, traditions, customs or types of behavior that reflect a community consensus. They do not carry the weight of law, but can often involve peer group approval or disapproval. They are the commonly accepted way of doing things in a given community.

SOCIO-ECONOMICS. This is a hybrid of pure economics for which the determinants are supply, demand, cost and price. Socio-economics are more complex because they involve the long term costs of extraction, construction, manufacturing or development. Socio-economics also involve the society costs of product, project or goal. In a sense, the element of social concern has been added to the total equation. This, the impact of this determinant would also have a continuum that would range from immediate cost ... (extraction or production) ... to the long term, sustained condition.

Ecosystem Model

central Oklahoma, would be a line that marked the shift from tall grass to mixed grass. This would have been true in a line extended from northern Texas to the southern border of Canada. Short grass country is more in the shadow of the Rocky Mountains. This is an area of 18 inches or less of average annual rainfall.

FIRE ON THE PRAIRIE

. . . produced the most rapid change. Early settlers dreaded prairie fires. There was no protection. Consequently, fire was an enemy! Subdue it! Our first ethic was on display. If the prairie did not burn for longer periods of time, when it finally did burn, various species of grass or herbage that had been dormant would emerge and fight for survival. In a sense, fire took the prairie

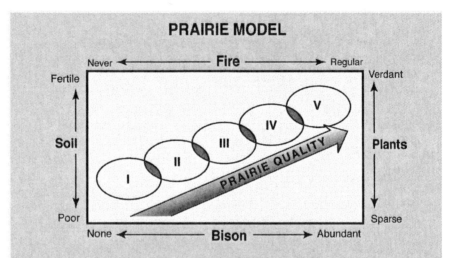

SOIL. The quality of the soil is one of the determinants of the prairie model. "Soil" would need to be defined, because on the prairie, it can often be three dimensional. It can have a flat, top soil meaning, an understory meaning ... rock, limestone, whatever is underneath the soil. It may be qualified as to porous quality. In the case of some prairie soil, the limestone understory would give grass a richer quality but made the soil more vulnerable to erosion if it was broken.

Moisture content, the ability of the soil to gather and store moisture, the presence of prairie potholes would all provide standards for soil quality. In essence, the soil determinant becomes a micro-ecosystem of the soil macrosystem.

FIRE. It is now known that burning has a beneficial impact on the prairie or grassy areas in a forest. What's more, spring burning produces one type of regrowth and fall burning another. Burning also removes trees and brush. Burning is a part of a complex process of prairie rejuvenation that is still being studied. The original prairie was sometimes burned by lightening and Native Americans were known to have set fires.

PLANTS. In an original model of the Tall grass Prairie, I showed this determinant as grass and intended it more for prairie grass. When regular patch burning was introduced by The Nature Conservancy and became a pattern, that activity restored so many flowers and herb plants that had been long dormant, the determinant title was changed to plants and included grasses.

BISON. These animals were the primary grazers on the prairie, but elk, deer and antelope also grazed. The grazing patterns of the bison followed the good quality grass produced by the burns. Bison grazed and moved on. Their manure was part of the enrichment pattern.

Prairie Model

back to form. That happened on the Tallgrass Prairie when some flowers, long thought to have been dead, emerged after a burn.

When settlement spread, new cultures emerged. Folks used prairie grass for building sod houses. They dug, by hand, wells for water. They built fences if material was available for horses, cattle and pigs. They helped build the church. The roots of social contracts in our culture dig into morality, ethics and trust. Character is one of our social contracts. There is a thread of religion in the prairie patches. Churches were often the first permanent structures. Our point is this. Spirituality plays a role in ecology even when it is one thing for the Native American and another for the euro American settler. Even when black settlement began after the Civil War, churches were an important part of the community.

Social contracts also go through stages or passages. Often when a law is passed that conflicts with social contracts, there is a period of storm. Many of our conservation conflicts have been with us since the time of the Pilgrims. In fact, after landing and a short period of accommodation with the Indians, the Pilgrims started pushing and shoving. They pushed the Indians off the land and shoved the pigs onto it. That may have been a portent of football.

WHEN THE PILGRIMS

. . . began to stake our their land, the issue of honey was a social contract. Honey belonged to them that found it. If a "honey miner" found a tree with honey in it, they could claim the honey regardless of who owned the tree, if the tree was on anybody's land. This social contract was later transferred to the gold fields of California, Nevada and Montana.

However, in New England, a property owner decided that because they owned the land, they owned the tree on it, and therefore owned the honey. This idea would be used in establishing "water rights." Eventually, if there was a court case, the property owner would prevail. In time, property right became law. In those early times, the question of who owned the bees did not come up. If it had, that case might still be pending in some court. So our conservation ethics come from interesting backgrounds. Small wonder they are often in conflict with each other.

PURE ECONOMICS BECOMES SOCIO-ECONOMICS

In the ecological model you will notice we use the term socio-economics. In a simple definition for business it is: 1. what do we buy? 2. what do we sell?

3. what is our profit?" It actually can be as simple as barter or trade. Since the early 1990's, there has been some thought given to the business relationship to ecology. The Paul Hawken book, *The Ecology of Commerce* is a good example of this. His book plows some new ground because he adds the consequences of the extraction of natural resources, or what might be called the afterburn to the standard equation for business. His thought is there is a social cost to anything manufactured or made, especially when natural resources are involved. "Business," he says, "has three basic issues to face: what it takes, what it makes, what it wastes, and the three are intimately connected."

Until recently, this aftercost has been almost a hidden issue. Now, there seems to be a growing or emerging view that our natural resources are actually a form of national currency. There is more discussion about the extent people are willing to go to protect a resource or at least conserve it. Perhaps as we are presented with the bills for cleanup work, whether it is lead mining, pollutants in river water or restoring habitat along trout streams, we have more appreciation for the cost of afterburn in ecology.

THEN ALONG CAME ECO-TOURISM

Hardly off the screen, but on the side we have created another ecosystem. It also fits our model. This is eco-tourism, another use of natural resources in areas that have fallen on the blade of extraction. Whether that blade was thin soil in Vermont, the poor use of land in prairie country or exhaustion of minerals in the arid west. Economics is really the dominant frame in our model in this case. The problem is that tourism creates mostly minimum wage jobs in motels and restaurants. Still, the economics are impressive. The latest figures released in Wyoming from 2005 show an industry that generates $30 million a year. It is the second largest industry in the state.

But social contracts also play a role, especially if unbridled development is part of the economic mix. In Vermont there is citizen resistance to big development. Certain businesses are denied zoning or there are very strong restrictions. There is the argument that tourism puts too much pressure on the natural resources. The value of uncrowded conditions, less intrusion of forests by hikers or bikers, less crowded fishing conditions along the rivers takes on an increased value.

There is another scenario. Give government or some federal agency more power, scope or authority. In the model, government moves up the scale and whether that authority is more positive or negative becomes a local issue. Forests and spotted owls assume a higher value as natural resources. Jobs of loggers and community growth assume a lower value. And the frame of social contracts becomes more polarized. While the determinants of natural resources and government move up the scale, the frames of economics and social contracts go down. Very little is normal and there is plenty of STORM.

The issue then becomes one of management or alignment. To this point, in handling natural resources, we have tried management more times than we have tried alignment. The result is that all of the frames have seldom been in more than some semblance of NORM or normal. Only rarely has an issue achieved a stage of high performance. Because there may be no perfect solution in ecological issues, no simple, single solution answer. In fact, we may not be able to manage nature. The best we can hope for is enlightened alignment.

This point is difficult for the scientific mind to accept. As our conservation efforts in the United States have come to be dominated more by the scientific mind, the scientific and engineering mind has had great difficulty accepting the fact it is terribly difficult to manage nature. It is one thing to design and build a dam. It is entirely a different can of worms to figure out what to do with contaminated silt that gathers behind that dam. The pursuit of the "perfect solution" has resulted in many course corrections in our American conservation efforts. It has produced a bevy of studies. One study begets another study. As in the Old Testament there has been a lot of begetting.

If you accept evolution as a core tenet of ecology, then eventually you must accept the morality or ethic that death is part of the evolution. If there is such a thing as "the tree of life" then it is always in some state of pruning. If it is not the natural cycle of death, then it is predator/prey pruning. If it is not a natural occurrence, then it can be man made. Death can ultimately take any of the four determinants in our model back to FORM. In severe instances, a hurricane and the resultant flooding can take a major city back to FORM.

Passing through the stages can also happen to conservation organizations. The Sierra Club has evolved though the stages of our model. So have The Izaak Walton League and The National Wildlife Federation. The organization is formed, it struggles through the stage of storm, things get close to normal, it gets bureaucratic or loses sight of the original mission, membership may decline, and it either reinvents itself or disappears. The conservation organization is always evolving. Cases in point. Ducks Unlimited evolved from "saving ducks" to "saving habitat." The Nature Conservancy went from a species oriented, small group to saving the "last great places" and more of a world wide focus.

We are now coming to understand that wildlife populations will also pass through the model stages. Many years ago, Denali National Park was first enabled by Congress. The National Park System commissioned three separate studies to determine what to do with it. The three studies came back and all three said essentially the same thing. "Leave it alone." Leave it as close to a natural system as possible. The cycles will be slower, they predicted, but eventually the caribou population will cycle up and down as the wolf population goes up or down. Keep the social contract determinant as unobtrusive as possible. Limit the roads. Let the natural forces take their slow, sometimes painful, course of action. Of course this action has made it tough on tourists who come seeking instant gratification on a four lane highway that takes you directly to Mount McKinley. More on this story later.

HOW WELL HAS THIS SYSTEM WORKED?

Well, Denali has been the least controversial of any of our National Parks. People who go there and don't expect to see much except nature are not disappointed. A contrary example would be Yellowstone Park. *Playing God in Yellowstone Park* was a pretty good read on the travails of the many trails in Yellowstone. Without going into full detail at this point, the park has probably operated more in the full glare of public opinion than any other national park. Certainly you have the natural resources, how to use them, manage them, display them, protect them and even restore them.

The animal dimension has been left alone, managed, mismanaged, fed in the winter, eradicated and then restored or reintroduced. Only the skunk has managed to maintain a low profile. You have the social contracts of crowd control, trash control, who gets seated in the dining room at Old Faithful Inn. The economic dimension has been argued by proponents of more roads since the time of horse drawn carriages, by those who want fewer roads, by the ATV folks and the snowmobilers. Their have been proponents of government management of lodges and concessions. Board rooms of free enterprise groups have plotted the cause of privatation.

And the opinions of the political forces have pitted the Park Service against the Forest Service and state against federal and one head of the Department of Interior against another and probably Alston Chase had it about right when he said, "there has been a lot of playing God in Yellowstone." If that is too heavy for your liking, then something lighter.

If Andy Rooney were writing this book, he would probably pause here and say something like . . . "I have often wondered why we stopped call them Commons and started calling them parks. We started out calling them Commons in New England. That word had a certain appeal because it said they were owned in common, used in common and even common people could use them. Then we started calling them parks. I don't know why we did that. Parks sound like someplace you go to play, or watch people throw Frisbees, a place you pay for entertainment or a water slide. Commons is just a place you go. It's a place. I think I like Commons better. It doesn't create as much squabbling."

IN THIS CHAPTER,

. . . we have attempted to validate a model by showing how it fits the business world, the world of marketing, and the many connections between the world of human resources and the world of natural resources. Neither grouping is a static condition. Both are constantly evolving. We have tracked this idea back to the world of the Native American, how they also had a system of economics, different social contracts for different tribes, and tribal organization that produced a government of sorts. Different tribes made slightly different impacts on natural resources, but essentially it was a life style based on sustainability and renewal. Most of them hunted, used fire to renew the land and most grew crops of some kind. They showed the newcomers to this land how to grow crops. There is a symbolic story of placing a fish in each hill of corn for fertilizer. In reality, they showed the newcomer how to grow corn, squash, tomatoes and beans. A less well known story is that in Virginia, in 1613, the Indians showed one John Rolfe how to grow tobacco. He began experimenting. He found that he could grow 500 pounds of tobacco on good bottom land. It was easy to grow and sold for five shillings a pound in England.

One source that I located said that for a time, even the streets of Jamestown were plowed up to plant tobacco. The colony could not keep up with the demands from England. The ecological implications? Virginia, for a time, became a one crop economy, the first American settlement to export an American product other than beaver hides. The economic frame became dominant. Care of the land was of little concern. And all because an Indian showed an Englishman how to grow a funny, long brown leaf! It was another patch for Grandmother's quilt.

The relationships created by our model may be viewed as too simplistic. They do, however, show interrelated and often conflicting conservation ethics and how these have resulted in strong, even polarized, opinions about the use of our natural resources. How the advocate of one frame or interest could argue their case with little nor any regard for the other determinants. How often one position would focus a flashlight only on one map, and then disregard another point of view. Small wonder our conservation efforts in this country have been more than a little discombobulated.

Chapter Six

Relationships and Connections

A Federal study of Jackson elk and bison will never get off the ground unless Wyoming participates, a mediator told the Wyoming Game and Fish Commission.

"There's no sense in proceeding unless the state and the feds work out your differences," said Kirk Emerson of the Institute for Environmental Conflict Resolution. "You've got to figure out a way to work together."

"Commissioners have criticized the U.S. Fish and Wildlife Service as it prepares to study elk and bison in Jackson Hole.

"A top concern is the study would end an artificial feeding program for elk and bison, which could send the animals onto private land, spreading disease to cattle and eating up haystacks.

"The study was ordered by a judge after an animal rights group, the Fund for Animals, sued to stop the federal government from allowing bison hunting on the National Elk Refuge in Jackson.

"Federal Wildlife managers intended to allow hunters to shoot bison to lower the chances of brucellosis infection of cattle, a disease that can cause cows to abort their calves. The Fund For Animals has suggested that bison be allowed to die off naturally and say a supplemental feeding program on the refuge creates an artificially high bison population.

"Commissioners have not committed to cooperating but they recently indicated they may participate in the study's initial planning.

"Emerson, who is working for the Fish and Wildlife Department as a third party mediator, said both sides 'got off on the wrong foot.'"

The wire story is about natural resources. How to best use a refuge? How best to feed animals on that refuge? What are the implications if elk were removed from the refuge? What impact would that have on the Jackson Hole economy? What do the people want? Elk viewing? A nature experience? And finally, what government entity should be involved in managing and manipulating this scenario? Every determinant of our model is involved in this conundrum.

Is the model germaine to times past? You start digging back. You trace the ethics of land use back to Puritan times. That isn't far enough back because the model was already in evidence when the Puritans got here. More spade work. You discover the stories of the French trappers in Northeast New England and Canada. This would be the first "rush" on this continent, the "beaver rush."

You discover another connection. People already on the land were living in what might be called "a sustainable, economic condition." You journey to Vermont and locate the Abenaki history. You find considerable ecological implications. You find parallels to plains Indians and then more parallels with tribes in the more western areas of the country. You conclude that no study of ecology would be complete without including the Native American. That's a big story so you select three tribes as case studies.

THE NATIVE AMERICAN BACKGROUND

The political issue of the Native American was goes back to the days when George Washington was forming his cabinet. Previously, as Commander of the Virginia Regiment during the French and Indian War, he had come to regard the natives, not only as formidable adversaries, but as people who acted very much as any people would act when their independence was threatened. Early letters from several of the Indian chiefs to the new President indicated they now viewed Washington as their personal protector.

"Brother," wrote one Cherokee Chief, "we give up to our white brothers all the land we could anyhow spare, and have but little left . . . we hope you wont let any people take any more without our consent."

Washington would work closely with Secretary of War, Henry Knox. The idea was the Native American tribes would be treated as "foreign nations." He felt the tribes should have multiple sanctuaries under their control. These would not be open to white settlement. Over time, the Native American would be assimilated as a full U.S. citizen. But Washington's best efforts were lost in an escalating cycle of violence as first one state and then another engaged in military action against the tribes in that state.

If a just solution, or an accommodation, with Native American tribes was a singular preoccupation of his first term, Washington's efforts to bring about some semblance of Indian policy were a major political failure.

Washington seems to have believed the behavior of nations was driven more by interests than by ideals. Thomas Jefferson, on the other hand, would argue that ideals were indeed American interests. Washington was convinced the interests of America did not rest on any alliance across the Atlantic. Rather, they rested on a national concern for interests across the Alleghanies. He seems to have had a vision of the consolidation of the American continent. Washington believed the future lay to the west. The Native American stood in the way of that westward expansion.

When the early arrivals to this land wrote or spoke about the North American continent, they often used the phrase "New World." When their reports reached Europe, they were received with little credibility. The European could not cope with the abundance described in those reports. Some commodities like wood, so abundant in the "New World" were nearly exhausted in places like England. Coal arrived in the nick of time to provide heating and energy for the English. The tall trees of America went to England to build wooden ships for the English navy.

Europe had no compunction about exporting troublemakers to America, both criminal and religious, as well as the unemployed. So instead of a country composed of members of any establishment, we began with residents who not only heard a different drummer, but they also had their own ideas of who could beat the cadences of freedom and happiness. The result has been a society that is both devoutly divided and devoutly commercial. On a grass roots level, our economic system is based on getting and spending. These interests are as basic as breathing. We may breathe our ideals, and even fight for the ideal of freedom, but we have achieved world power by exporting Coca Cola, music, movies, blue jeans and automobiles and exploiting natural and human resources.

Couple the "Old World" reality of diminished resources with the fact that most of the European immigrants were the poorest of the poor, and you have the background of the social and cultural philosophy of the early immigrants when they were suddenly thrust into the "abundance of the New World." Conservation, with all its contradictions and complexities, would not be among the priorities of the new arrivals. Small wonder that an extractionist mentality would quickly develop. Natural resources quickly became commodities. Always, it seemed, our model fit the events.

The pattern of settling religious and political differences, as well as settling boundaries and the right to trade, would ultimately be accompanied by a gun in the "New World." When Americans stopped dueling because of social pressure after the infamous Burr-Hamilton face-off at the ledge beneath the plains of Weehawken, the resolution of difference of opinion was shifted to the street of a frontier town. Protagonists usually met at high noon and if the issue of right and wrong was clouded, the good guy got the white hat and the bad guy got the black hat and somebody got shot.

Essentially, newcomers encountered two great relationships in the "New World" with which they could find no satisfactory accommodation. The first of these would be the relationship with the Native American. The second would be the relationship to natural resources and the opportunity provided by the enormous supply of those resources.

WE MAKE TWO FUNDAMENTAL CHOICES.

The newcomers made two choices. They chose consumption over conservation. This choice almost dictated they had to go through the Native American to get to where they wanted to be. The Indians were expendable.

Chronology is the study of events, people or actions that follows a timeline. It is supposed to be organized and logical. Too many times, ecology defies chronology. There are too many relationships and delayed interactions. The best you can find are connections and relationships.

Eventually, if you dig deep enough, you find a system in ecology. All things are connected. The connections may not always be evident. The cause of certain results is not always evident. Results may be observation by a naturalist who may not always know the cause. Results can be data or collections of research by a scientist. These may be interpreted by a writer. Neither the facts nor the causes may connect the result.

THE ROLE OF COMMUNICATION

Communication is an often overlooked player in the ecological mix. Often, advances in communication presage whole new eras. The printing press, an advance in technology, made it possible for English blueprints to be used for an American steam engine which machine resulted in new uses of natural resources. The telegraph, developed in the Civil War period, speeded up communication. It changed government because messages now reached the frontier in hours rather than weeks. It changed social contracts because now news moved faster. Telegraph poles impacted the world of raptors.

One afternoon I was discussing, with a geologist, the history and evolution of right of way law and the impact of roads on natural resources. He was explaining how the building of Interstate 80 across Wyoming had furnished all sorts of new geologic information. New road building technology had dug into formations little documented to that time. The subject of telephone poles along railroads and highways came up. He pointed out that a raptor gets a whole new look on life. They can sit on top of the pole and scan an area of some size simply by sitting there. There are no tree branches to break their view. Hunting now becomes a much easier task. Telephone poles had changed the ecology of the area. History doesn't cover that.

It seems to me, that to more fully understand ecology, you need patience to be able to backtrack to locate all the possible elements. In the political or government arena of "ready-fire-aim" this doesn't always play well. Too often the farmer tells the Congressman, "my God man, we've had a flood" and the Congressman tells the farmer, "my God man, let's build a dam."

Nature, it seems, can absorb or adjust to cyclical or self-balancing events. Time and the ecosystem process blend to evolve at some sort of equilibrium which is a blend of linear sequences and a holistic process. Even over time, it is a complex process. And so, we have a history of the Abenaki and a history of the settlers of Vermont. It is when we try to trace the ecological pattern that we encounter a crucial difference in what we can know. Our model can illustrate the interrelationships of an ecosystem, not so much in terms of time lines but more in terms of results.

The natural ecosystem tends to be a patchwork of diverse communities. Each species in this community seems to bear some relationship to others within a given space. Communities are arranged or connected only in random patterns. There is no apparent organization. There are only relationships, not always evident, or results which can sometimes be linked or connected back to causes.

THE SURVEYOR MAY PUT

. . . a boundary around a national park so that grizzly bears will enjoy the sanctity of the preserve. Putting a boundary around the ecosystems involved is more elusive. Nature's patterns do not always follow the sextant and log book. The patchwork is more likely to hold some surprises. The dogma of an exact science is hard to find. When one probability is determined, another riddle enters the equation. Neither the grizzly nor the wolf will always respect boundaries.

The Native American adapted their way of life to evolving ecological patterns. If the forest burned in places as part of the patchwork design, the Native American adapted hunting and harvesting patterns to fit what nature offered. The new arrivals thought in terms of systems and regular patterns. As they sought to give the landscape a new purpose, they might change what they saw as a tangle of disorder into a more boundaried pattern.

In the European eye, the Native American lived in a landscape that was filled with an unbelievable bounty. But to that same European eye, the Native American lived in what they saw as poverty. Here was a situation they could not understand. Blessed with such great natural wealth how could these people live as what some colonists perceived as savages? The conservation ethics conflicted much as patches on Grandma's quilt.

If nothing else, this sketch of the Great American land story is helpful to understand the almost total disregard of ecological factors and the value of natural resources. If progress was the driving force or ethic behind American expansion, if wealth was the ultimate objective, then it would be safe to say our thinking about the environment has evolved in a long series of bumps and grinds and too many times we applauded the gyrations of the dancing girls rather than considering why they were dancing.

TWO GREAT ISSUES

. . . were never really resolved. The settlers could never decide whether they would loot the land or live with the land. William Cronon, in his book titled *Changes in the Land*, says it this way. "Land in New England became for the colonists a form of capital, a thing consumed for the express purpose of creating wealth. It was the land-capital equation that created the two ecological contradictions of the colonial economy. One of those was the inherent conflict between the red man and the white man. But the second was the ecological contradiction in the colonial economy as well. They assumed the limitless availability of more land to exploit, and in the long run that was impossible. Not only colonial agriculture, but lumbering and fur trade as well, were able to ignore the problem of continuous yield because of the temporary gift of nature which fueled their continuous expansion. When that gift was finally exhausted, ecosystems and economies alike were forced into new relationships: expansion could not continue indefinitely."

WESTWARD EXPANSION BECOMES THE OPTION

When people moved into a new environment, the story repeated itself. They brought social contracts, government and economics with them into what to that point had been ecological stability. The measurement of the benefit of the presence of man to that stability will always be the ability of the environment to reproduce itself. If the environment results from the insistence of man to use it for unlimited extraction rather than use it as a sustainable resource, then the land will cease to take care of man.

We tend sometimes to sweep aside the ecological changes in a region in our haste to embrace only the graphs of linear history. It is harder to think of Vermont the colony, settled for agricultural purposes, with that settlement pushing up the rivers into frontier Vermont. To imagine what it must have been like when those first white spires were built into the clean air of the mountains. Then to shift to Vermont the industrial state, as factories began to dot those rivers and dams began to block the free flow of the stream.

Or the dates when the Green Mountain Boys picked up their muskets and marched off for a cause they probably didn't fully understand. But they intended to defend their land. And then to put dates on the early photos that show the smoke and soot of the factories fogging up the landscape. Then stack up the photos and replace them with a log kept by "economic development" people that shows the demise of the industrial life and to imagine Vermont as an almost pastoral state where tourism is now a major industry. Where they still gather maple sugar and in October have County Fairs and oxen pulls.

But it is more difficult to locate the processes or the causes that produced the ecological connections. Capitalism became a driving force as progress became the ideal. We began to transform our environment. We transformed not only the land and the products of that land, we also transformed our human beliefs and the resultant social contracts. As the ecosystems of Vermont were transformed into a segment of a national capitalistic system based upon consumption, the decision was made to elevate consumption to a very high level. The environment was placed on a much lower level.

Cronon concludes his book with a quote from the geographer, Carl Sauer, a more contemporary observer of the American ecological scene. Sauer says, "Americans (have) not yet learned the difference between yield and loot." Sauer had reference to the long standing way of American life. "Ecological abundance and economic prodigality went hand in hand: the people of plenty were a people of waste."

Today, you can trace this legacy of waste historically by using examples and illustrations that are more evident and certainly a part of history. We do some of that in this book. Not to make the people involved look bad. It is not a game of pointing fingers. We hope we have used the historical tracing for the purpose of finding causes for our beliefs about land use and wildlife stewardship.

What comes out of our model is a view of the environment that is more holistic. Quite often, we can find the problem but are not wise enough to suggest what we can do about it. This much is evident. As a nation, we are like the hunters looking at different maps. We aren't quite as arrogant as we were in the past in dealing with natural resources, but we still have a lot of conflicting maps. We are still blessed with very strong, and divided opinions about conservation. And, as a people, we still have a deep seated faith in the extractionist philosophy. Man continues to be the ultimate predator.

REFERENCES

Babcock, H.L. The Beaver As A Factor In The Development Of New England Americana

Branch, E. Douglas. The Hunting Of The Buffalo. (University of Nebraska Press) 1997

Cronan, William. Changes In the Land. (Hill and Wang) 1986

Sandoz, Mari. The Beaver Men. (Hastings House) 1978

Wilson, James. The Earth Shall Weep. (Grove Press) 1998

Wiseman, Frederick Mathew. The Voice Of The Dawn. (University Press of New England) 2001

Ellis, Joseph. His Excellency: George Washington. (Knopf) 2004

Chapter Seven

The People of the Dawn

Someplace between legend and literature, between folklore and fact, between journal recordings and journalistic exaggeration, between forensics and fiction the truth may lurk. Even then, absolute truth is hard to find.

Trout fishermen are like that. They make the good much better than it really was and the worse much more tragic that it really was. The truth is somewhere in the middle. And if what we like to call history is shrouded in oral translation handed down from generation to generation, the real truth may be even more a mystery. For a broad stroke naturalist, this poses less of a problem. He or she can be satisfied with the spirit of the thing. For the scientist, this ambivalence can be uncomfortable.

My premise is both simple and broad stroke. To understand the land, or ecology, you need to go back to the earliest times. If you used 1600 as a reasonable date of transition, most of the chronology before that time would be oral. This would be any legend and folklore related to the ecology of what became the United States. For the next 100 years, information would come from a combination of oral transmission and some letters or journals.

OUR THREE NATIVE AMERICAN TRIBES

. . . were selected first because of their geographic location; an area that represented an early ecology that could be isolated for purposes of examination. Vermont was fairly typical of New England ecology and had fairly typical zones: forest, floodplain, hills and temperature. Second, we chose Vermont because the first concerns about settler activity and what it was doing to the land were raised in Vermont. We believe George Perkins Marsh was the first ecological voice raised in our country. He was born and raised in Vermont. Finally,

63

the choice of Vermont took us to the Abenaki tribe. Tribe is a misnomer. More correctly, they were a related group of triblets or families, never really connected or organized as a nation. They were early farmers and harvesters which makes them a decent study of early environmental practices

In the great grassland areas of the United States, we chose the Ponca people. They were located primarily in northern Nebraska along the Missouri and Niobrara rivers. They were also early farmers. To illustrate the more western Indian culture, we chose the Shoshone people who in 1600 were located in the Rocky Mountains in Idaho, Northern Utah and western Wyoming. They were fairly typical of the nomadic Indian life style, primarily meat eaters in the early times. They moved from that life style almost directly into a reservation lifestyle.

The combination of these three people captures the basic life styles present on this continent at the times of settlement beginning in the 1600s. The transformation of trapping, for instance, from a part of a sustaining life style to an item of trade or barter, was true for the Abenaki in the early 1600s and was equally true for the Shoshone in the 1840s. All three people were somewhat mobile, adjusting to the seasonal cycles and food sources. All three people were essentially peaceful. They were basically family oriented. They were all spiritual, almost religious. The Abenaki were recruited by the Catholics and the Shoshone by the Episcopalians.

PART OF THE ORAL ABENAKI HISTORY

In the book, *THE VOICE OF THE DAWN, An Autohistory of the Abenaki Nation,* Frederick Matthew Wiseman says this. "I would like to tell you a story, one about my people and their land. The story is a sash woven of many strands of language. The first strand is the remembered wisdom and the Abenaki community. The second strand is our history and that of our relatives, written down by European, Native American, and Euroamerican observers. The third strand is what our Mother the Earth has revealed to us through the studies and writings of those who delve in her, the archaeologists and paleoecologists. The fourth strand is my own family history and its stories. The fifth strand is, of course, that which has come to me alone, stories that I create with my own beliefs and visions . . . The sash is many generations long, sometimes with many strands and simple pattern and sometimes complex, woven of but a few strands."

Dr. Wiseman told me many stories as we sat in his living room in the old Vermont town of Swanton. He lives in the family home, built three generations ago by his grandfather. The Missisquoi river flows at the north edge of the town. Many years ago, his grandfather had taken him to the river and had told him ". . . this is the Missisquoi and its waters flow in my veins, and when I die, I will be buried where I can always see the river. The River is in your blood too, and you will come back to it." He has.

His book divides the very early times of Abenaki history into four periods: The Coming of the Great Animals, (13,000 to 10,000), The Years of the Moose, (10,000 to 6,500), The Years of the Log Ships, (6,500 to 1,000), and The Years of the Corn, (1,000 to 400 years ago which would be 1000 AD to 1600 AD). For our purposes, we will use the date of 1600 to track both the Abenaki culture and the arrival of the European. This is not entirely accurate because it avoids an earlier period of French influence and intermarraige of natives and French trappers in Canada, northern Vermont and Maine. But, it will serve our purpose well enough. That purpose is to challenge the cherished myth that the Euro American came to an untamed "wilderness" and proceeded to conquer a hostile and forbidding natural environment. Actually, Native Americans had been practicing an environmentally friendly type of landscape conservation for some time. The second point we hope to make is that you cannot understand grassroots ecology unless you consider the roots. Too often, in the discussion of ecology, the story has used the arrival of the first colonists as the benchmark for the study of American ecology. That may serve the purposes of patriotism but it omits the first part of the story. To understand what the Euro American did to the environment, you need to consider the way it was when they got to this land. In fact, early arrivals to this country had very little idea of how to reach any accommodation with the land either in New England or Virginia.

HISTORY HAS IT THAT IN 1621

. . . an Abenaki Indian walked into the Pilgrim's village and greeted them with "Welcome Englishmen." He had learned some English, evidently, from previous contracts with fishermen and traders. He told the Pilgrims he was from Maine. His name was Samoset.

He introduced the Pilgrims to another Native American named Squanto. Some sources say he was an Abenaki from Maine. Some say he was a Wampanoag, taken captive by members of Captain John Smith's crew in New England in 1614. In 1619, he was part of an expedition captained by Thomas Dermer. He had become a guide and an interpreter. The English were not strangers to him. The Wampanoag were organized into a simple government, had a tribal religion and were able hunters and farmers. They often traded and hunted with the Abenaki. The Abenaki were never a nation. They were divided into two general groups, Eastern and Western Abenaki. In the early 1500s, some sources say there were 40,000 people. Dr. Wiseman told me he thought that was a good number. They would pay the price of exposure to the white man's disease.

European diseases struck Indian villages with horrible ferocity. Mortality rates in initial onslaughts were rarely less than 80 or 90 percent, and it was not unheard of for an entire village to be wiped out.

THREE WEAPONS DICTATED ECOLOGICAL TRANSITION

The first was the arrival of the bow and arrow. What had been a very peaceful lifestyle between tribes now became more warlike. Improved technology would lead to changing social customs. There was intermittent warfare. The second weapon that changed lifestyles was an improved trap. The beaver population was used for food on special occasions. It became a fur commodity, a means of trade and barter with the Euro Americans. The third weapon that would change the environment was the gun. That change began in Vermont when Samuel De Champlain fired his wheel-lock arquebus, a gun invented in Germany in 1515, into a group of Iroquios. His journal related this event took place "in 43 degrees and some minutes of latitude and was named Lake Champlain." Champlain claimed the land and the lake for the French. 1609 was the beginning of 150 years of turmoil and bloodshed over the right to land that became Vermont.

Until the Pilgrims arrived, or the first 20 years of the 1600s, most of the French or English activity was based on economic considerations. Trade, barter, settlement were the goals. Part of the Mayflower mantra has been the Pilgrim search for religious freedom. Actually, the Pilgrims were only about a third of the people on the Mayflower voyage. The majority of the people had been hired by the backers of the trip to watch out for the economic interests of the investors. This may have been the first American symbol taken out of context to serve a national purpose. When you dig into the early ecology of this country, you do find symbols used that are not absolutely true.

William Bradford, and a minority of the arrivals, crafted what came to be known as the Mayflower Compact. This was a social contract that expressed the purpose of their settlement. It is based on the Pilgrim doctrine that man has the divine right to conquer anything that is hostile, land or human beings. And if the humans are seen as hostile or savage, then like the environment, they are justifiable subjects for submission.

THIS VIEWPOINT BECOMES THE
FIRST CONSERVATION ETHIC

Nature is hostile and an enemy and must be subdued and conquered if man is to survive and prosper. Only mastery of natural forces will release man from the restraints of nature and allow for use of physical necessities. Now, add the epilogue to this ethic. It is not just a social right or ethic, it is a divine right!

One of the roots of ecology that goes deep is that all cultures have the ability, if not the right, to change their ecological stability. They will never leave it exactly as it was because any intrusion by man, however benign, changes the environment. Ecological stability and cultural stability are related and intertwined. In early day Vermont, two very different ecological relationships collide. This confrontation blended ideas about land use, spiritual connection to that land, who owned it or could own it, and even deeper . . . what was the purpose of the land.

This is Dr. Wiseman writing as Medawas, about the Abenaki purpose.

"Great Spirit, let me see the land as it was when our old ones first came to Wobanakik.

Call forth the Soubagwa, the Great Sea from the east to let it fill our valleys with teeming life.

Awaken the Ktsiawaasak, the great animals from their resting places beneath our hills.

Let me see all of this so I may share it with others."

As the Abenaki turned more to farming and harvesting, they also learned from their neighbors to the south how to build "long houses." Built of wood and bark construction, these houses were larger, could accommodate entire families, often 30 to 60 people, and were permanent in nature. This style of housing, together with crops planted in the area, required that some persons stay behind on a year round basis. From a people who roamed to some degree over the northern New England area, the Abenaki would become more bound to one place. This land, generally along rivers and in the valleys, would be the land most coveted by the incoming settlers.

THE ABENAKI YEAR

. . . was arranged in moons. These moons tend to reflect their culture. Harvesting and agriculture were an indelible part of their lifestyle. The year would begin with the Sugar Making Moon. This month began with the ice break up and the run of maple sap. Maple sap would be collected and boiled into sugar and syrup.

This was followed by the Bark Collecting month which was a period the bark could easily be stripped from the trees. The Abenaki would first ask permission of the tree and when finished would leave a small gift, sometimes tobacco. Bark would be used for canoes and wigwams and in the case of large pieces would be used for the longhouses. Bark would be bound into the building using willow branches. Rawhide was too vulnerable to the mice.

Then came the Fishing Moon when the salmon, walleye and sucker were running. Then came the Planting Moon. There is journal evidence the Abenaki had corn fields as large as four miles long. Fields would be cleared by the men using slash and burn. Women did the planting. They would place 3 or 4 grains of corn in a small hole, add 3 or 4 bean seeds. The beans would grow around the corn stalk. Later, pumpkin, squash, and sometimes tobacco would be planted. It was a more balanced horticulture.

During The Blueberry Moon wild fruit in season would be harvested. To influence good crops, there would be dances . . . the Sun Dance, the Deer Dance, the Eagle Dance. Good harvests in a short growing season depended upon help from Mother Earth.

The Harvest Moon ushered in a time for green corn and fresh beans. In the early fall, there was the Corn Making Moon. Harvested corn would be ground into meal to be stored for use in the winter. The silence of the land would be pocked by the boom of wooden grinders pounding the grain. Oral history has it that in the Burlington area, young men would mount their horses, follow the sounds to locate the Abenaki women in the woods and would proceed to rape them on the spot. Soon, the booming in the woods ended. The Abenaki would melt into the land to escape persecution.

In the Moon of Falling Leaves, the people would scatter for hunting. The hunting of the Abenaki people was highly organized. Land was divided into quadrants. Only one of the areas would be hunted each year. The other three areas were rested. This is one of the first conservation principles we have been able to locate in the Indian culture.

Next came the Freezing Moon, when game was smoked or cured. Then the Winter Making Moon, when the people returned to "home" and hides would be processed. The Greeting Moon would be a time for socializing and winter games, the Bough Sledding Moon and the Moose Hunting Moon, a time when heavy snow made hunting the animal easier. The Abenaki calendar reflected the natural cycles. Their lifestyle, their culture revolved around an ecological pattern. It was a pattern that made a minimal impact on natural resources.

THE FIRST ECOLOGICAL ISSUE

Both the Native American and the Euroamerican were affected by the exhaustibility of thin soil. Other than valley land, most soil in Vermont is thin. Both the Native American and the newcomers put maximum pressure on this soil by planting corn. For the Indian, is was less a monoculture crop. They planted some legumes and did limited fertilizing. For the settler, corn was the primary crop.

The Indians had virtually no animals. The Euroamerican quickly added both swine and cattle to the ecological blend. By pasturing corn fields after the harvest they further removed what might have been beneficial tillage. Corn or maize had been developed by the Indians who used essentially a hoe agriculture. The newcomers introduced the plow. The early resultant soil exhaustion was predictable. The sorry condition of the cropland began to be noted in farmer's journals as early as 1637.

The farmers followed the practice of mowing grass in the fall and plowing in the fall. Both practices left the shallow soil vulnerable to erosion in the winter and spring. Any shortage of grass or hay would force farmers to turn cattle into pastures in early spring. New grass would be trampled. The settlers had no ready supply of fertilizer except what they gathered in the winter in the barns. They were forced to use fish for fertilizer.

The Abenaki way of life was slowly ground into the earth. Some families would drift into Canada, some into Northern Maine. In 1787, some who had banded together with the Iroquois at St. Regis crossed the Mississippi and drifted into what was then Spanish Arkansas, settling on the White River. After 1803, and the Louisiana purchase, this group is believed to have joined or merged with the Delaware and Shawnee and later moved into Kansas and eventually to Oklahoma.

Frederick Wiseman says the Abenaki had five choices in what came to be regarded as the "dark years." The five options were, 1 - exile; 2 - fade into the forests and marshes; 3 - live the Gypsy, Pirate, River Rat lives between the Native and American cultures; 4 - merge with the French community; 5 - "pass into the English-American society." Different families or groups chose different routes. Essentially, the Abenaki was finished as a farmer. The men would continue to hunt in sparsely settled areas. They would fish when and where they could. They call this period "The Years of the Fox."

If an Abenaki spokesperson were to part the curtain in our weary saga, they might say some words that would reflect the defiance of the Indian and the will to survive. Again, this is Wiseman writing as Medawas:

"Let the visitors believe they have conquered, that they have the land and its bounty. Let them believe that we are gone, Indian Joe is dead. The forests keep our secrets. The unseen fox has kits in its den. The drums and rattles are not stilled. They are heard in the far places, they are heard in the air of night. The visitors think they have won. Yet the scent of sweet grass troubles their dreams."

Talk to a long time member of the Vermont Fish and Wildlife Department and he will tell you the relations between the department and Abenaki were not always smooth. Angelo Incerpi, former Director of Vermont Fisheries, puts it this way, "Many of them didn't think they needed a license to fish. Many of them thought they could hunt deer or moose anytime they needed some meat. Some of them were pretty confrontational."

ALTHOUGH VERMONT HAS RECOGNIZED

. . . some tribal status, the United States has never recognized the land claims, the tribal status nor the citizenship of the Abenaki. At this time, there are about 2500 "Vermont" Abenaki in Vermont and New Hampshire. Most are concentrated in Northwest Vermont near Lake Champlain.

In her book, *Hands On The Land*, Jan Albers makes this observation. "It is a peculiarity of human consciousness that we experience nature as being outside ourselves, and yet we cannot look at it without projecting ourselves into its center." The statement reflects the Anglo perception of the relationships between humankind and nature. But we need to remind ourselves we are not the only agents of change in the environment.

And too often, we have chosen to view the Native American and their relationship to nature in an almost romantic perspective that shrouded the truth. The Native American had very simple social contracts with the land. They tended to produce a sustained environment. They had rather simple economics and little desire to accumulate goods or wealth.

Very simply, or perhaps not so simply, the Euro American applied a totally different set of social contracts and economics to the land. And Mother Nature didn't always like the gun, the plow and the cattle.

George Perkins Marsh

Was born on March 15, 1801 in Woodstock, Vermont. In 1820, he graduated from Dartmouth College with highest honors and in 1825 was admitted to the bar. For the next seventeen years he practiced law, published a book, and in 1843 he became a U.S. Representative to Congress.

George Perkins Marsh. Photo courtesy of the Library of Congress, LC-USZ62-109923, LC-USZ62-61104.

While in Congress, he helped found the Smithsonian Institution. By this time, it was clear he was a person of extraordinary intelligence. By the middle of his life he knew twenty languages. He would spend twenty-five years as a foreign minister to Turkey and later Italy. In 1857, he published his report on the fisheries in Vermont and this really began his connection to his conservation contribution.

In 1864, he published *Man and Nature*, or *Physical Geography as Modified By Human Action*. Little noticed when it was published, the book today is considered "the Classic of Conservation." In the book he argued eloquently that the human being was assuming a more important role in natural forces shaping the environment.

It was the first book to challenge the American myth of superabundance and the inexhaustibility of earth's resources. Marsh concluded that man's impact on the cosmos was not only probable but that it was also calculable by future scientific research.

Marsh died in Vallombrosa, Italy on July 11, 1882.

Thomas Jefferson

Was born April 13, 1743 in Albemarle, Virginia. He inherited some 5,000 acres of land from his father. His mother was a Randolph so Jefferson enjoyed high social standing. He studied at William and Mary College and in 1772 married Martha Wayles Shelton and they moved to the partially built home that came to be known as Monticello.

Jefferson was eloquent as a writer but not good as a public speaker. He was regarded as the "silent member" in Congress where, at age 33, he drafted the

Thomas Jefferson. Photo courtesy of the Library of Congress, LC-USZ62-53985.

Declaration of Independence. He began a career in politics serving as minister to France, Secretary of State in Washington's Cabinet, became Vice President in 1796 and finally was elected President and served from 1801 through 1809.

While Jefferson is generally thought of in terms of his political activities, he was an avid naturalist. He made trips to New England to bring trees and plants to Monticello. He experimented with different varieties and grew grapes that produced fine wine.

In 1803, he found a way around the Constitution to purchase the Louisiana Territory from France. While the action may have been political, his resultant letters to Meriwether Lewis on what to look for, and report on, were written by a naturalist. They are very explicit. Why list him as a conservationist? His gift to Americans was the western United States.

Jefferson died at Monticello on July 4th, 1826.

REFERENCES

I have relied heavily in this chapter on two sources:

Albers, Jan. *Hands On The Land*. MIT Press. 2000
Wiseman, Frederick Mathew. *The Voice of the Dawn*. University Press of New England. 2001

Other Sources

Cronan, William, *Changes in the Land*. (Hill & Wong) 1986
Jennings, Frances. *The Invasion of America*. (W.W. Norton and Company) 1976
Johnson, Charles W. *The Nature of Vermont*. (University Press of New England) 1978
Wilson, James. *The Earth Shall Weep*. 1998
Wiseman, Frederick M. *The Abenaki, People and the Bounty of the Land*. (The Lane Press, Grove Press) 2001
In addition, two specific visits to Vermont produced interviews with members of the Vermont Fish and Game Department, Dr. Frederick M. Wiseman, Maggie Gilmore of the Vermont Law School and Frederick D. Carson, Department of Geology, at Norwich University.

Chapter Eight

The Poncas . . . Myths and Messages

From the safe distance of 150 years, it is no easier to sort out some American history, and justify specific events, that it must have been just hours after the event. Time is not always on the side of reason. Our history is full of small events, accidents and coincidences that become pivotal moments, significant patches in the big quilt that is our history.

So at the risk of upsetting a considerable amount of prairie lore, it really should be pointed out that the plains Indians didn't live on the prairies. That was a myth. They lived almost entirely along rivers that ran through prairie country. Only the settlers who replaced them thought they could live on the prairies.

History has tended to focus on the larger tribes of plains Indians. Actually, there were many small tribes. The Poncas were one of those. Recorded history places them along the Niobrara River in Northeastern Nebraska. The land was fertile and lent itself well to Native American farming. It naturally produced wild fruit, berries, herbs and roots. Wildlife was abundant enough it would supplement the farming base. And essentially, the Ponca Nation was a gatherer/harvester type society.

This small tribe, probably no more than 1000 people at any time was generally closer to 800 in number. They were peaceful people. Their neighbors to the east and south were the Omaha and Pawnee nations. They often intermarried with the Omaha. To the west were the Sioux who raided them and drove off their horses. Oral history tells of their living in the country along the Niobrara for so long they had no other home. Other stories relate how their fathers had come up the Missouri river to settle along the Niobrara.

Smallpox was one of their deadliest enemies. In 1837, it swept through many plains tribes reducing the powerful Mandans and Hidatsa; the Mandan population went from 1,600 to 150. Records before 1800 on the patterns of life of the "western Indian farmers are sketchy." The sketches of George Catlin and the drawings of Karl Bodmer, who traveled with Prince Maximilian of Wied, the German explorer, are the best we have until the 1830's. John Ordway of the Lewis and Clark expedition emerged as the best recorder of food and agricultural practices as they met the Indians.

We do know the Indians of Nebraska and South Dakota dug large storage pits, as deep as 8 feet and 2 to 3 feet in diameter. In these pits they stored corn and squash and even sunflower seeds. These, in terms of use, would be similar to the "cellar" later dug and used by the prairie settler. Hunting was usually communal, done by stampeding the animals over a cliff or past a line of archers. What they killed was used very soon or was dried for later use.

Women converted the hides and animal skins into the useful items needed for survival. They were the mistresses of the fields, raising the corn, beans, and squash. Tobacco was raised by the men and considered a sacred plant. If farming gave the women a lot of work, it also gave them authority. It was a part of their mystic power, a demonstration of fertility. There is nothing in Ponca history that would indicate they had cattle until 1850 or later. Until they got horses by trading with Southwestern tribes, they used dogs as their "movers" of goods and supplies. There are no references to pigs. We know they moved cattle to Oklahoma in 1877.

IT WOULD BE FAIR TO SAY

. . . the Abenaki people were a blend of hunters and farmers. The Ponca people were more farmers and less hunters. The Shoshone people tended to be more a hunting culture. All three cultures shared certain ecological or conservation commonalities. While the Native American was a predator, killing was not a sport. It was a necessary part of life that directly related to the security of the people. The relationship was filled with ritual and a certain thanks for what nature had provided. It was a spiritual connection.

THEY USED THE BEST LAND FOR AGRICULTURE

These were lands along water. Of course, these lands were the first targets for occupation by pioneers and settlers. We found a good example of the settler mindset in a little book entitled *The Sod House*. It was written by a doctor, a very early settler in Nebraska. He writes, "Indians and buffaloes were plentiful but never helped develop the country. They were obstacles that had to be removed before the natural resources could be developed. Sentimental people always criticized the so called wrongs done in taking the land from the Indians.

It was an economic necessity because they would not develop the land. The Indian had to go or go to work. The buffalo had to go to make way for improved cattle that the white man produced."

THE STORY OF THE PONCA PEOPLE

. . . is less nebulous after 1858. They would make a treaty with the United States that meant they gave up all their land except a part adjacent to the Niobrara River in Northeastern Nebraska. Keep in mind, that in the first 90 years after independence, the United States made 370 treaties with Indians. Every one of these treaties was violated by U.S. citizens or the U.S. government.

When the Poncas entered into this treaty with the government, they were ceded 25,810,000 acres of hunting lands and were to receive $50,000 payable over 30 years. During the two years it took to ratify this treaty, the Ponca tribe was confined to land that no longer sustained game. They suffered drought, locusts and raids by the Sioux.

The Poncas signed a new treaty in 1865. Their land was reduced again. They were moved to agricultural land which also contained their old burial grounds. They gave up hunting and became essentially an agricultural people. They had found some degree of peace on land that was now on both sides of the Niobrara River.

One year later, in 1866, they learned their land on the north side of the river had been lost to the Sioux. The government had given away 96,000 acres of what the Poncas believed was their land. When the mistake was uncovered, the Department of Interior was unwilling to take it back from the warring Sioux. When the Poncas refused to leave their land, the Sioux raided them, destroyed crops and took horses. It was a time of more droughts. Patches of corn were a failure, the wild plums dried on the trees and the Poncas hunted for wild turnips and ate cornstalks to keep from starving.

In 1873, the Poncas and the Omahas, with the U.S. Government as a party to the treaty, agreed the Omahas would sell Omaha reservation land to the Poncas where they could farm. But the Indian Agent, A.J. Carter, encouraged resettlement of the Poncas. In 1875, the Poncas signed a request to be moved to Indian Territory, but there were language problems. They thought they were moving to Omaha land.

The order was put out to remove the Ponca Indians from their old home to Indian Territory. Strangely worded as it was, it contained the phrase "with their consent." Standing Bear, who acted more like a principle chief at that time, and nine other chiefs went with an agent, Edward C. Kemble, to look at their new home. They didn't like the Oklahoma land. Water was poor.

THE STANDING BEAR STORY

The agent told them to pick out a place for the tribe or he would not take them back to Nebraska. They refused. He left them in Oklahoma, a thousand miles from their Niobrara river home, in the winter, with no money nor food. Standing Bear later is reported to have told this story.

"We started for home on foot. At night we slept in haystacks. We hardly lived until morning it was so cold. We had nothing but our blankets. We took ears of corn that had dried in the fields. We ate it raw. The soles of our moccasins were out. We were barefoot in the snow. We were nearly dead when we reached the Otoe reservation in Nebraska. It had been fifty days. We stayed there ten days to get strong and the Otoes gave each of us a pony. The agent for the Otoes said he had a telegram that the chiefs had run away, not to give us any food or shelter or any help."

The Otoe agent said afterward the Ponca chiefs came into his office and when they left, their footprints were in blood on the floor.

When the Chiefs reached their home along the Niobrara, they found the Indian Agent, who had left them in Indian Territory, waiting for them with soldiers. The soldiers put the women and children into wagons and on May 21, 1877 they started for Indian Territory. On the way to Indian Territory nine people died including Standing Bear's daughter. A tornado upset their wagons. Part of the time they were out of food.

The Ponca tribe was now split. Part wanted to stay in the old lands in northern Nebraska. Part wanted to stay in Indian Territory but wanted other land. At the end of a three month journey they reached Indian Territory. They arrived on the unbroken prairie with nothing but wagons and tents. The water was very bad, silty and salty. All of their cattle and many of their horses died. In the first winter, one hundred and fifty eight out of seven hundred and sixty eight people died.

STANDING BEAR RETURNS TO NEBRASKA

Standing Bear's son had been among those who died. Before his death he had asked that his body be taken to ancestral grounds in Nebraska. Standing Bear and twenty-nine of the tribe broke away from Indian Territory and after a long journey of three months reached the reservation of their friends, the Omahas. Soldiers came and arrested them. The orders were to take them back to Indian Territory.

Dr. George L. Miller, editor of the Omaha World Herald, took up the issue and two attorneys defended the Poncas without pay. Judge Dundy heard the case in the United States court in Omaha. Standing Bear and his group of 29 were considered wards of the court, acting in their own behalf.

Standing Bear's defense is based on the 14th amendment which reads, "nor shall any state deprive any person of life, liberty or property, without due process of law." He made his plea for recognition of his basic humanity. He said, "I am a man. The same God made us both."

The judge ruled in his favor and said, "The question cannot be open to serious doubt. Webster describes a person as 'a living human being . . . an individual of the human race'. This is comprehensive enough, it would seem, to include even an Indian."

Both social contracts and the law had been changed. These changes would modify prairie land use, settlement, create water use problems and in fact, modify the ecology of prairie country. The Standing Bear decision was a "landmark case."

After they were set free, Standing Bear and his party settled on an island in the Niobrara River which was part of their old reservation. In 1890, peace was arranged between the Sioux and the Poncas. The Sioux gave back part of the old Ponca lands on the Niobrara. Gradually some of the Poncas moved to this area until they numbered about 130. The remaining members of the tribe settled in northern Oklahoma. Today, they are modestly successful farmers. Hunting is restricted to finding customers for their casinos. Standing Bear lived to an old age and died at his home on the Niobrara in September of 1908. His statue stands in a park in Ponca City, Oklahoma, just north of the Ponca settlement.

THERE IS A THREAD

. . . that stitches the many Indian Tribes together. History has chosen to ignore the ability of the Native American to adapt to their natural resources. Archaeologists tell us the maize or corn first raised on this hemisphere probably grew in Peru or perhaps the fertile mountain country of Guatemala or Mexico. Corn was grown only by overcoming differences in soil, amount of rainfall and different growing seasons. Corn recovered from excavations near Cottonwood, Colorado, called Basketweaver Corn, averaged 4 and 3/4 inches in length. The kernels were reddish brown. A modern ear of corn measures about 10 inches in length, is white or yellow. Corn seems to link most Native American tribes.

How did the Native American come to know that when they moved further north they needed new strains of corn? Further south, it took longer to mature. The more northern corn grew shorter, survived with less water. Some sage has observed that "one of the miracles of Indian corn is the way it was adapted to so many different climates." The white man would "improve" upon the varieties of Indian corn to increase production. Today, some of the advanced breeders of corn are going back to certain tribes for back crosses to restore the original resistance to disease and insects.

LAND IS THE BOTTOM LINE

The Indian view of land simply did not line up with the view of the white man. It was two different social contracts. The white man liked square boundaries. Rivers were often the boundaries between Indian tribes. River bottoms provided better soil and this is where the bulk of the Indian farming was done. In fact, some journals report that a few Ponca women, armed with hand hoes, would, in some years grow as much as 300 acres of corn in the bottom land producing some 3000 to 6000 bushels of ear corn.

When Americans have sat down at the environmental card game, whether it was poker chips or corn, they really believed they were playing with a deck of fifty-two cards. Nature didn't always deal that way . . . still doesn't. The game is not always stable. When we get bad cards and the logic doesn't work, the government steps in. If you need protection on the Oregon Trail, you call for troops and forts. If you want protection on the river, the Army Corps of Engineers steps in. Sometimes, the Corps will build a dam whether you want one or not.

In 1879, Congress created the Mississippi River Commission whose assignment included protection of land along the Missouri River against flooding. In 1927, Congress passed more laws and with sweeping gestures that seldom worked out as planned, instructed the corps to build more dams, levees and dikes. This bordered Ponca land. In 1944, after disastrous flooding, the effort to tame the river became the Missouri Basin Program.

THIS PROGRAM HAS CREATED MANY CONSEQUENCES

One of the enormous impacts has been on the wetlands and floodplains and the species using those areas. Another impact has been on soil quality. Silting has increased. The benefit of alluvial flooding has been removed.

Over a period of 50 years, the Corps of Engineers managed to spend over $25 billion on the "system" that was to tame the Missouri. The ethic that we could manage and manipulate the environment was invoked by Colonel Pick, the river mastermind. Came the floods of 1993. Near St Louis the river rose to 49 feet above normal, 19 feet above flood level. They estimate that 17,000 square miles of land was flooded. Estimates of flood damage would top $10 million. There was more. Over the years, people had grown careless. Only about 15% of the residents along the river had flood insurance. Many people had moved into flood prone areas.

In retrospect, the reason for the damage was simple enough. The channeled Missouri River had no place to go when the water went up. It flowed faster, hit dikes and levees with more force. It had no floodplain to absorb the torrents of water. Looking backward, the damage from a "tamed" river was probably greater than if the river had been left alone.

Slowly, flashlights began to flicker in high government circles. Bill Dieffenbach, an engineer with the Corps, is quoted by Daniel Botkin in his book, *Our Natural History,* as saying, "People have created the flood problems. We have spent billions on flood control, but it doesn't work with these kinds of floods." Dieffenbach moved to the Missouri Department of Conservation.

As planned, dams such as the Garrison Dam would protect towns like Bismarck, North Dakota from flooding. What the lyrics did not intone, was the inundation of virtually every acre of productive bottomland "owned by the Mandans, Hidatsa and Arikara people." The dams had been in a great song book used by Colonel Lewis Pick. It was titled, The Pick Sloan Compromise. The songbook was published in 1943 and was in use through 1945. It had a lot of sour notes in it.

IF THE PONCA PEOPLE DIDN'T KNOW

. . . exactly where their land was because the government was shifting it back and forth, they haven't experienced much improvement. Fast forward to 2002 for an "instant replay." After 13 years of review, the U.S. Army Corps of Engineers is about to release a study of the preferred alternative for managing the six dams that now control the Missouri River." This quote from the South Dakota Wildlife Federation's newsletter of spring, 2002. "Since 1990, the U.S. Fish and Wildlife Service has repeatedly told the corps that current dam operations have flat lined the Missouri River's 'heartbeat' by eliminating seasonal fluctuations in water levels on the river, threatening three federally listed endangered species and leading to the slow demise of potentially dozens of other native fish."

The arguments of the Wildlife Federation are that a three year study by the National Academy of Sciences concluded that "degradation of the river ecosystem will continue unless the river's natural flow is significantly restored. . . . The barge industry, once the power player on the river, has dwindled to a $7 million industry while recreational fishing, hunting and recreation have become an $85 million industry." The economic frame again.

THE VIEW FROM THE DAM

I stood one morning on the corner of Fort Randall dam in southern South Dakota. The dam was a massive structure, a mighty sentinel of concrete and steel. No question, it was a hellava engineering feat! I looked at the water backing up from the dam and wondered how many acres of land had been flooded? Could enough water ever be sold to pay for the dam? How many treaties with the Sioux were under water?

I could hear the boosters stir and mutter, "but look at the jobs it created." I couldn't help but think what it must have been like to have rowed up the river against the current. To have passed the land of the Omahas, the Poncas, the Sioux, Mandans, Hidatsa. To have seen their villages, their farms, watched the rituals as they celebrated natural cycles. Then I looked at the dam again and saw the other rationalization. Nature is a machine and if it isn't, then man will make it one. The engineer at work.

I thought of prairie country, the natural relationships of that land with rivers and streams. How most people are attracted to rivers as they look for "their place." And how, left to their own meanderings, rivers can serve man so well over a period of time. But reduced to a machine by man, what seems to be immediate economic gain soon becomes the loss of natural balance. There are many messages, if man would only listen.

I glanced at a sign along the road. Buffalo Casino it announced. Funny, the bison escaped the Corps of Engineers. The Corps would have had a plan to drown them. They escaped the lawyers too. I have never heard of a court case about who owned the bison. Arguments about the land they were on, the water they drank, how much a hide was worth? Yes. Nobody seems to have bothered much about who owned them.

That really doesn't matter. For the Poncas the bison were gone. Only the bison spirits plodded through the prairie night skies, dark now that the sun had pulled the covers over Fort Randall Dam. The bison had entered the Tabernacle of Waste.

John Muir

Was born in Dunbar, Scotland in 1838 and spent eleven years there. The next eleven years he spent in the backwoods of Wisconsin. He cleared forest, plowed with oxen, and dug wells. But it was nature that drew his attention and animals his sympathy. Though he had no formal schooling, he was admitted to the University of Wisconsin but left in 1863 after 2½ years.

In 1867, while adjusting some machinery, a file slipped and went into his eye. He lost sight in that eye and nearly went blind. As he slowly recovered his sight in one eye, he later said he felt he had been reborn. That began his

John Muir. Photo courtesy of the Library of Congress, LC-USZ62-55012.

observation of nature, a long 1000 mile walk to Savannah, Georgia. Finally, he arrived in San Francisco, California. He walked to Yosemite. For a time he worked as a shepherd, then became a guide, even guiding one of his heroes, Ralph Waldo Emerson.

Now, he was in and out of the valley, got married and eventually would guide Robert Underwood Johnson, editor of *Century*. A series of articles resulted and Congress responded by creating a National Park from what had been a state park. Out of the friendship with Johnson came the Sierra Club in 1892. Muir was President. He guided an increasing number of members, entertaining, educating, inspiring.

John Muir died in a Los Angeles hospital on January, 1914 and the "ecstasy" he found in the mountains was his legacy. In 1976, the California Historical Society would name him "The Greatest Californian." In the game of conservation in this country, John Muir wins the "most valuable player award."

Henry David Thoreau

Was born in Concord, Massachusetts in 1817. He graduated from Harvard where he was exposed to the writing of Ralph Waldo Emerson. He worked for a time in the Emerson household and Emerson was almost a mentor to him.

Thoreau actually lived in two worlds, literally and figuratively. In 1849, he published an essay called "Civil Disobedience" which resulted from his refusal

Henry David Thoreau. Photo courtesy of the Library of Congress, LC-USZ61-361.

to pay a poll tax. This article had a major impact on political thinking in those times and some say his ideas even influenced such people of later days such as Martin Luther King.

But in environmental circles, Thoreau is better known for Walden Pond. In 1845, he built a small cabin and lived there two years as he said "living deeply and sucking all the marrow of life." He kept a complete diary of his thoughts, reflections and what he observed in his close proximity to nature. He used this journal to produce *Walden*, published in 1854.

Many people believe Thoreau's thoughtful advocacy of nature "up close" became a cornerstone for a 20th century movement to preserve natural areas. Nature, when viewed up close, displays finer lessons and rarer beauty. Thoreau was one of the first naturalists to focus on how each part of nature touches the whole of living. "Above all," he said, "we cannot afford to not live in the present."

Thoreau died in Concord, Massachusetts on July 12, 1862.

REFERENCES

Editions of *South Dakota Wildlife Federation Newsletter*, 2001, 2002.

Atkinson, Brooks. *Walden and Other Writings of Henry David Thoreau.* (Random House) 1950

Barnes, Cass G. *The Sod House.* (University of Nebraska Press) 1930

Cronon, William, Miles, George, Gitlin, Jay. *Under An Open Sky. W. W.* (Norton and Company) 1992

Jackson, Helen Hunt. *A Century of Dishonor.* (Roberts Brothers) 1995

Reiser, Marc. *Cadillac Desert.* (Penguin Books) 1987

Ronda, James. *Lewis and Clark among the Indians.* (University of Nebraska Press) 1984

Wilson, James. *The Earth Shall Weep.* (Grove Press) 1998

Chapter Nine

The Shoshone

If civilization moves from the inside out, from the center of any culture to the border, then to study any mosaic of ecology you would begin with the center. People, as part of the ecosystem, would provide insights and clues into the nature of the ecosystem. Food and water become sidebars.

Our sketches try to narrow the history of our Native American case studies into an era that provides for some cultural inspection. Then we have tried to position various pieces of ecological information like land, burning, food sources, farming, and water supply. Now in the case of the Shoshone, how hunting was woven into their culture and ecological patterns.

The life spans of the native people become the threads that allow us to patch together the connections between the original ecosystems present on this continent. The oldest evidence of people is not human remains but rather stone points like the one found near Folsom, New Mexico. It was embedded in the vertebra of a prehistoric bison. At first, the stone point was considered a local phenomenon. As time passed, more of the laurel leaf shaped points were found. Each had a deep groove running down the center of both faces. The points were not primitive; they were delicate, finely shaped . . . made by expert workmen. They were made of whatever the region supplied. It might be agate, jasper, chert . . . everything except flint.

Whoever made the points knew what they were doing because the deep grooving on each face provided a quick conduit for the blood of a wounded animal. About all we could know about these early people was they made good stone points that helped produce use of meat.

We know less about the early inhabitants of the eastern seacoast than we do about the Indian type people who reached New Mexico some 20,000 years ago. These arrivals were Stone Age people. Very shortly they began to grow food. Now ecological connections emerge. Much as in early middle-eastern civilization, the level of culture is tied to agriculture. The more abundant is food, the higher is the level of civilization. To attain higher levels of food production, middle-eastern people turned to irrigation. Water would control the amount of food which produced the level of civilization. This pattern is present in the archaeology of the early Native Americans in the Southwestern part of the U.S. In the past, the study of older civilizations was focused on soil and culture. Water supply may have been a more important part of history. There are lessons here for today.

There is ample evidence these early people did not live entirely by hunting. While they did use natural foods like roots and bulbs, they soon evolved into farmers who produced beans, pumpkins, squash, tobacco, sweet potatoes and peanuts. These people, then, were creating a civilization. Their farming culture would spread, inside out . . . from their centers or villages to the borders.

SOUTHEASTERN INDIAN CULTURE

The Indians in the southeastern parts of the U.S. also had the common bond of farming. People tended to live in permanent villages. This evolved into a "Mound Builder" culture. They developed art, carving and fished and hunted to supplement farming. The Five Civilized Tribes would evolve out of these early times.

These tribes, the Creeks, Seminoles, Choctaws, Cherokees and Chickasaws all became a mingled society by considerable intermarriage. They lived well, farmed well, wove fine cloth and war was their "beloved occupation." The Indians did not wage war as the Europeans understood the term. They did not, as a rule, seek conquest. They did not try to subjugate another tribe. What they wanted were the risks and rewards that went with war, the ritual of manhood, the coups, the bragging rights. They might fight fiercely one day, even take some slaves, and be trading with the enemy the next day. It was not unlike football today, even down to the trading part.

When the Five Civilized Tribes were moved in the 1830's, they took their agricultural practices with them. But as the Indian either moved, or was moved, into more western areas of the country, they were moved into increasingly marginal landscapes. The Osage deliberately chose Flinthill country because they believed the white man would not want to farm it. Tribes that eventually settled in prairie country included the Cheyenne, the Arikara, Omaha, Pawnee, Missouri, Kaw and the Ponca. Doubtless many of these tribes had traded with other tribes for seeds, may have even traded farming practices. What we are trying to establish is the idea that the Native Americans lived with their environment and had ecosystems that allowed for sustainable production long before the white man arrived

MORE NATIVE AMERICAN CONTRIBUTIONS

The Spanish brought horses to Mexico and then California. The animals would escape and multiply. Some would be captured or stolen. They became the animal of the borders. As horses moved up the California coast, the Cayuse tribe became horse owners. They were especially adept at capturing horses, raising them and trading them. In coming years, "Cayuse" became a cowboy term for horse. The Palouse Tribe developed a breed of roan horse with white quarters or white horses with roan quarters. Palousian horses came to be known as Appalousian, taller than most western horses, fast and strong. They have evolved into cutting horses.

So from roughly 1700 to 1900, the Native American was in a series of transitions. As their environment changed and they were thrust into new ecosystems, they shifted from a walking people to a mounted people, a people with primitive weapons to tribes of well armed warriors, from fixed farming communities to people who were moving or being moved to a new land. The horse changed the way they secured meat. It extended hunting ranges, enabled pursuit of larger game. Grass became important for grazing. The man on horseback and his mobile family now inhabited a vast sea of grass, an environment threaded together by streams and rivers. The ecology of the Indian, which had been based more on a sustainable food growing culture shifted to be more dependent upon meat, especially the bison.

What had been an agricultural society in most tribes would be replaced with a culture captured later in the Wild West Show. This "horse culture" would reach its peak about 1850. It would become a symbol of freedom, of space, of a fancy-free, unfenced existence that would dazzle the Prince Maximilians, the Catlins, the Parkmans . . . provide the images for the Russels and Remingtons . . . it would become almost a shroud pulled over the tragic transition of a people being forced to replace former life styles with a life style not of their choosing. Not only did the arrival of the settler and his plow tear up the prairie, it tore up an entire Native American culture.

THE FUSE THAT LIT THE POWDER KEG

What peace there was on the prairie after the Civil War really came unglued because of a small, unheralded incident in Southeastern Wyoming when Lt.

Gratten's boys, some say smelling of bad whiskey, shot a Sioux Chief who had offered to trade them a horse for a cow that had strayed into the Sioux camp. Food was at the bottom of this skirmish and led to further incidents such as General Harney's attack on Little Thunder's peaceful village in 1855 in which women and children were obliterated. Food was the cause of many Indian troubles.

There was little nor any public reaction to the action in Colorado at Sand Creek when Cheyenne women and children were eliminated. This was in 1864, less than two years after the Emancipation Proclamation. In 1865, a General Connor issued orders to his troops, bound for the Powder River, and an expedition against the Sioux and Cheyenne "to attack and kill every male Indian over 12 years of age."

The military philosophy of the Civil War thinking of General William Tecumseh Sherman had come to the short grass prairie country and the foothills of the Rockies. Sherman believed the principal use of the army was not to occupy territory but to destroy enemy personnel. Sherman rested his argument on what he called "the awful fact" . . . as he put it, "that the present class of men who rule the south must be killed outright." So the general rule of conflict in the Civil War, that extermination was the final answer to the struggle, may have been transferred to the west. Man would become a military predator whose goal was elimination of a species.

The story of the settlement of the west was, in reality, not very pretty. We make a similar point about ecology. It is not all pretty, that there is a raw underside, that the "tree of life" is in a constant state of being pruned.

A SHORT STORY ABOUT THE SHOSHONE

These Indian people resulted from the merging of myriad triblets into what became a tribe. Their composite historical figure was Chief Washakie. Other names would play parts in the Shoshone saga. In the early 1800's there would be Horn Chief, Iron Wristband, Little Chief and Cut Nose. On the spiritual side would be the prophet-chief Ohamagwaya who had linkages with the Sun Dance. Most of this is in the oral history of the tribe. The Shoshone tribe was a melting pot. Some local and family bands trace back to mixed Bannock and Shoshone groups from the Fort Hall region of Idaho. Northwestern Shoshones peopled Idaho and southern Utah. The buffalo eaters group probably had Comanche and Yamparika connections. The people who would eventually inhabit the Wind River Reservation in Wyoming would include Salmon eaters, Sheep Eaters, Bannocks, Piutes and Western Shoshones. They were often called The Snakes.

Osborne Russell, one of the mountain men who kept notes of his travels and used them as sources for his *Journal of a Mountain Man*, tells of wintering in 1840 with a Shoshone encampment that included a mixture of Iroquois, Creek and Seminole Indians who were married to Shoshone women. French traders had Shoshone wives. As in the Five Civilized Tribes, the children of these mixed marriages often became influential tribal leaders, the army scouts or guides for wagon trains, the mediators and the diplomats.

The Shoshone tribe, as early as 1825, reached mutual accord with some of the first white settlers in the Wind River Basin when they accepted the mountain men. A better documented track of their tribal evolution would begin in the 1850's, with the emergence of Chief Washakie as their leader. If there was a nerve center for the tribe in this period, it was Fort Bridger in Southwestern Wyoming. Socially, there were strong ties between mountain man Jim Bridger, his Shoshone wife and the tribe.

So as a tribe of very mixed lineage, the Shoshone moved about in smaller tribal or family units from the eastern buffalo hunting grounds of Wyoming, to the Missouri River in Montana, to the Green River country in western Wyoming and Eastern Idaho, down into the Salt Basin country. A Shoshone woman would guide Lewis and Clark. They accepted the arrivals of the Mormans, the traders, the gold seekers using South Pass, Wyoming. In the meantime, the grass along the rivers which they depended upon for a wheat like flour, would disappear under the grazing pressure of wagon trains.

In a sense, the Shoshone reflected the issues that faced all the plains Indians or those in the foothills of the Rockies. What could they depend upon for natural resources and food? Or an economic base? What would they have, or what could they produce that would provide something of value for trading? What would replace the bison hides? They were advised to farm. Where could they get water? Most of the watershed land had been appropriated by the settlers.

There was another issue. Within the tribe, the social contracts had dissolved. They were often deeply divided among themselves. Alcohol was a problem. Both male and female had lost social status.

BUT IN EVERY ONE OF THESE PROBLEMS

. . . natural resources seemed to be involved. The ecosystem that had sustained them was no longer available. By 1885, with the bison herds gone and their cattle herds still small, the Shoshones were struggling with disease, with encroachment from the Arapahoes and the Crows, with government neglect and inept Indian agents. Chief Washakie assumed the role of diplomat or a negotiator for a people who found themselves in a very marginal ecosystem.

In the year of 1862, the Shoshone tribes on the western side of the Rockies had threatened war with the eastern tribes. The Indian agent of that time, Luther Mann, realized that depletion of game and habitat were the real causes of the unrest. Congress asked him to negotiate a treaty that would protect the overland trade routes, allotted him $20,000 for the tribes but made no distinction of how the money should be divided. Mann became the first official to recommend the Wind River country as the reservation for the Shoshone people. The 1860's were critical times for the Shoshone.

The late summer would be time for the distribution of their treaty goods. Then it was back to the Wind River country for the fall buffalo hunt, then a winter camp somewhere in the region, the spring hunt and then back to Fort Bridger. If the treaty goods did not arrive at the Fort in time for fall distribution, the relationships would become strained. The leadership of Washakie was tested.

THE PRESSURE MOUNTED

. . . on both the leaders and the people. It came from all directions. Washakie and his various chiefs would make many trips to the treaty parlays. Four such trips diminished their original allotment of 2,784,400 acres to the total in 1929 of 524,960. This would be some 2,200,000 less than they had originally been promised. On September 26, 1872, they signed away 601,120 acres, on April 21, 1896, they signed away another 64,000 acres that included the hot springs land around what is today Thermopolis, Wyoming and on April 21, 1904, they signed away 1,346,320 acres. These actions reduced most of their productive land not already acquired in one manner or another by white settlers and cattle ranchers. Today, their reservation comprises some 525,000 acres, located North of Lander, Wyoming. The words of the Cherokee chiefs who had written George Washington 200 years earlier haunted the Shoshone.

Chief Washakie had been promised housing for the people. It took many years to develop. He was promised schooling for the children and it would be many years before, with the prodding of Episcopal ministers such as Father John Roberts, schools became available. He had been promised cattle and food, which provisions were delivered either late or not at all. While not quite the Trail of Tears the Cherokee people had experienced, it was a trail of broken promises and appropriation by the white man.

There are parallels between the great Native American Chiefs, especially those who made an effort to find some reconciliation between their culture and that of the white man. They begin with natural resources which the Indians had and the white man wanted. They could never find a common ground on the use of natural resources. The white man would see natural resources in terms of exploitation, financial gain and progress. His relationship was more hostile or antagonistic. He was determined to conquer nature.

In the 19th century, few western white men found little in nature that was spiritual. With the exception of a few naturalists, the white culture built spires to find spiritual connections. Extermination of species, both wild and human, were a part of the social contract. The westerner did not always practice what he preached. He would leave it to the poet and the painter to put the spiritual in his western space.

In terms of government, the Native American lived in a rather democratic society. If there was a "head chief," that person achieved the position or retained it through the common consent of the lesser chiefs. Families or small groups had the option to "go their own way" if they so chose. In the case of the Shoshone tribe, families might wander during the summer, only to reconvene for security or trade purposes in the fall. A tribe usually voted on important decisions, if only by oral consensus. There might be from 30 to 200 signatures on the treaty. Each family, or unit, had a vote.

In terms of economic areas, the red man could, and often did, make a transition into the world of the white man. The American Indian had a long record of trade and barter with other tribes. As we have mentioned previously, it was not uncommon to be engaged in hostilities one week and be trading the next week. Trade was an accepted part of life so the red man could adjust rather easily to the trading post, the rendezvous, the fur trader and trapper, the stranger who would offer guns or whiskey for pelts and furs. In the red man's world there was no money. Land had a value only when it offered hunting or security. The white man saw the red man as an extension of economic opportunity. The red man saw the white man as the source of goods and supply.

What the red man did not quite understand about the equation of supply and demand, is that once a tribe made a commitment to this economic pattern, to disengage from it was terribly painful. Once they became dependent upon the trading post, they were vulnerable to the trading post. Economics would alter their world, their culture, and the very relationship they had with nature.

THE NATIVE AMERICAN SOCIAL CONTRACTS

It would be safe to say that in all three of our Native American case study tribes it was a man's world. As with the Euro American, women were respected but had little status. In either culture, a close, male/female relationship was rare. The Mormons bought women in England when they needed females in Salt Lake City. The early settlers took Indian women as wives when they needed moccasins mended, the mountain men welcomed the Indian woman when their blankets were cold, and Indian Chiefs often had more than one wife.

Still, the Red Man seems to have had a special place in his world for children. The records of the Poncas, the Abenaki and the Shoshone certainly show this. When the tribes would finally accept a reservation life, when the long journey to find freedom had failed, the chiefs would make a final decision based more on consideration for the women and children than on other factors. In terms of a social contract, there seems to have been some underlying feeling for the preservation of the species. The family unit was the primary focus of preservation.

The paradox of the long struggle between the two races is a crazy quilt all onto its own. The tribes that did the most to accommodate the white man seem to have come out with the shortest stick. To the credit of Washakie, his people did not go through the final travail of genocide. He called himself a man of peace and seemed to be able to walk the contorted and dangerous paths between the two worlds. He possessed a commanding presence and seems to have been a powerful orator. These are his words.

"I shall, indeed, speak to you freely, of the many wrongs we have suffered at the hands of the white man. They are things to be noted and remembered. But I cannot hope to express to you the half that is in our hearts. They are too full for words."

"Disappointment; then a deep sadness; then a grief inexpressible; then, at times, a bitterness that makes us think of the rifle, the knife and the tomahawk, and kindles in our hearts the fires of desperation—-that, sir, is the story of our experience, of our wretched lives . . . I say again, the government does not keep its word! And so, after all we can get by cultivating the land, and by hunting and fishing, we are sometimes very nearly starved, and go half naked, as you see us. Knowing all this, do you wonder, sir, that we have fits of desperation and think to be avenged?" These were notes taken by Wyoming Governor Hoyt, who described his meeting with Chief Washakie as one of politeness, cordial hand shakes and warm welcomes from all those present.

Washakie was married for many years to a Shoshone woman who often accompanied him into skirmishes and on hunts. Six years after her death,

Washakie married a young Crow girl who had been a prisoner of war and was very beautiful. Children of his marriages were part of his family. Agency records show that at one time there were 34 in his immediate family. In later life he became especially good friends with Reverand John Roberts, the Episcopal minister who spent many years at the agency located at what today is Fort Washakie, Wyoming.

IN HER BOOK ON THE HISTORY OF THE OLD CHIEF

. . . Grace Raymond Hebard, who spent many hours interviewing the Shoshone people quotes one of Waskakie's last messages to his people. "One more thing I want to see and my heart will be at peace. I want to see the school and the church built for my tribe by the 'White Robes.'" The chief then donated to the Episcopal Church one hundred and fifty acres of desirable land on which the school and the church he longed to see were eventually built.

That was written in 1930. I located the site in 2001, about three miles west of Highway 287 that now runs through Fort Washakie. My wife and I pulled into the driveway. We drove slowly up the road. A pheasant strolls across the grass surrounding the buildings. Quickly he ducks into the weeds. The buildings are still there but in need of some tender loving care. Corners of the brick boarding house are coming apart, some windows are boarded up. There are signs of age and neglect. But the lawns are neatly mowed and surrounding land is well maintained. It was late afternoon in early October. The shadows in the grass were long and lean. A Wyoming wind purred a slight rustle in the crack of the car window. Or perhaps it was a ghost of another time checking on the visitors.

OCTOBER REFLECTIONS

Chief Washakie was born about the time George Washington had finished his second term as president. He died during the administration of William McKinley. History would condense this time, compress it for a nation and for a tribe on the eastern edges of the Rockies in what would become Wyoming. Washakie would see the wagons he called "the white tops" cross the southern part of his land, across the Sweetwater country and South Pass. He came to know Jim Bridger who would argue with incoming settlers that the Wind River Valley was a better trail to Montana gold fields than the bloody Bozeman trail would be. He would see the first locomotives of the Union Pacific come into Shoshone land near Fort Bridger and he would say to his people, "there is nothing we can do to prevent its coming on, this fire horse of the white man. We cannot stop it. We are compelled to stand back and watch it come into our lands. Squeezed into this short space of time, Washakie juggled the future. He saw his people shoved into reservation life and the ecology of their life style changed completely.

The pheasant ducked out of and back into the weeds. I reflected. Old agrarian social contracts are more likely to endure. When a new idea is introduced in rural culture, it falls into quick disrepute and then passes into delayed implementation. Rural folks tend to "set on new ideas." The focus of such a culture is on preserving whatever has been achieved. Rural culture often chaffs at the pushiness of technology

Once the direction of the United States had been transformed by Civil War technology, the use of natural resources was determined. Liberal immigration resulted in the "melting pot" mentality. The general direction of both cultural and conservation ethics in the United States would be established. These huge events coalesced around the Indian population west of the Mississippi. They could put up no effective resistance. Like natural resources, they would be extracted.

By going back to the Native American and weaving them into the framework of ecology, however patchy our quilt has been, we have attempted to establish the role the Indian played in the evolution of the conservation story in the United States. We have argued that the framework for their views of natural resources, economics, government and social contracts would fit our model. The early dimensions of that model had resulted in more of a sustaining ecology than the Euro American system which replaced it. We could have, and still can, learn from their treatment of natural resources.

The Native American story positions a much broader question we face as we enter the 21st century. How will Americans resolve the balance between consumption and conservation? Will we be faced with the choice between the poverty of exhausted natural resources or can we find the prosperity of a sustaining ecology? What sort of future do we face? Will we become the sullen sitters on a land allotment that provides only marginal hope for the future?

REFERENCES

Devoto, Bernard. *The Year of Decision. (*Truman Tally Books) 1942
Ellison, Robert S. *Fort Bridger: A Brief History. (*University of Wyoming) 1931
Heband, Grace Raymond. *Washakie. (*University of Nebraska Press) 1995

Keller, Robert H. *American Protestantism and United States Indian Policy.* (University of Nebraska Press) 1983

LaFargo, Oliver. *A Pictorial History of the American Indian.* (Crown Publishers) 1956

Roe, Frank Gilbert. *The Indian and the Horse.* (University of Oklahoma Press) 1977

Russell, Osborne. *Journal of a Trapper.* (University of Nebraska Press) 1988

Sandoz, Man. *The Beavermen.* (University of Nebraska Press) 1964

Stamm, Henry E., IV. *People of the Wind River.* (University of Oklahoma Press) 1994

Chapter Ten

How Measurement Got
Us into Trouble

Highway 70, which crosses Kansas, doesn't have the songs written about it that Highway 66 does. But you can still "motor west" along it. The space that stretches into the west gives you plenty of elbow room. It can beguile you. At night it fools you. You think the few distant lights of a small town are a couple of miles away only to discover, as you nod your way along, they are more like 20 miles away.

This feeling of space has been fooling people for a long time. It has fooled the boosters and the bankers, the wheat farmers and corn planters, the railroaders and the right away people, the gamblers and the prostitutes. They all thought it would be better than it turned out to be.

Once it was Tallgrass Prairie where it rained more and Short Grass Country where it didn't rain that much, and that makes sense because grass needs rain in order to grow. Folks who used wagons to travel this country tended to overlook the lesson in the grass. Since the grass looked abundant enough, they figured to plow it up and then grow something. And since they were in a hurry to make a living, they couldn't wait around to learn that experience is something you don't generally get until just after you need it.

Those people were part and parcel of two American dreams. One is the dream that if you work hard enough in this country you can make it. Economists would call that "upward mobility." Moving rocks in Vermont to get some farming land was supposed to produce upward mobility. Some of the folks who got back aches in New England thought it would be easier in prairie country so they moved out west only to find that if they wanted fences in Kansas they built them out of limestone. That experience might be where that phrase "between a rock and hard place" was first used. But the idea of "upward mobility" became part of the American Dream. It was a land of opportunity.

If you stuck with it and worked hard you could end up rich. At the least, when they buried you it was in a suit.

What this romantic dream created wasn't all shabby. Somehow, it turned the "Heartland" of the country into a vast settlement of farmers and in good years it would grow so much food we could boast it was "the breadbasket of the world."

Two dreams then. One . . . upward mobility, the chance to make a life for yourself and be somebody. Two . . . the lure of the value of the land and the space to make it happen. First the trails, then the railroads, hurried off to someplace in the distance. Usually, the dreams would run along the rivers and streams. When it came time to organize the prairie country, the early planners got out the chain. They liked squares in prairie country. They even liked "square shooters."

THE USE OF THE CHAIN

. . . had begun in 1785 when Thomas Hutchins, the first official geographer in the United States had unrolled it along the banks of the Ohio River. The chain was 22 yards long. Equally important, it would have four rods. Each rod was 16½ feet long. This idea had come from sixth century Anglo-Saxon England. It went this way: 4 square rods was measured as a day's work, 40 day's work made an acre, 640 acres made a square mile. A "quarter section" was 160 acres. It was a square deal.

So an American standard came to be 4. The railroads would sell or give land in the prairie country in 40 acre units. Civil War freed slaves were given "40 acres and a mule." The subjects of John Steinbeck's story "gave up on 40 acres." We liked squares so much we invented the term "four square gospel" meaning something close to the absolute truth. So from the very start, the 22 yard chain was not only used to measure boundaries, it also measured morality.

This pattern of squares was largely Thomas Jefferson's idea. First, the shape was simple and that made it democratic. In the ideal society as Jefferson proposed it, the core of that society would be the independent, yeoman farmer who owned 40 acres of land. It was the government's job to make it available. He believed "the proportion which the aggregate of the other classes of citizens bears in any state to that of its husbandmen, is the proportion of its unsound to its healthy parts, and is a good enough barometer to measure any degree of corruption." So from the very start, the 22 yard chain was not only supposed to measure boundaries, it was also intended to shape a society.

The romantic ideas of the farmer or husbandman and his place in our American society, the social value of the farm family, is one of our deep roots. This symbol, probably more than any other, has impacted the ecology of America. It is rural and it is stubborn.

SURVEYORS MUST HAVE BEEN STUBBORN

It often took axmen to open up straight lines through the timber. Under brush and brambles must have torn clothing. It was not a line of work for the weak hearted. But 22 yards at a time, a pattern of land measurement was pushed across the continent, first to the Mississippi River, then westward on the great journey across prairie country.

The end result was that by 1900 most of the country had been squared off, into 40 acre sections, half sections, sections, townships and ranges. Every bit of surveyed land was recorded, registered at the nearest federal land office. The system would guarantee legal ownership of land. It simplified the purchase and sale of land. It would enable a property tax system to be established. Interestingly enough, this system would evolve through the form-storm-norm-perform stages.

The historian, Frederick Jackson Turner, in 1893 would argue that "the frontier would promote the formation of a composite nationality for the American people. . . . that restless nervous energy , that dominant individualism working for good and for evil, and withal that buoyancy and exuberance which comes with freedom—-these are the traits of the frontier."

So today, when we speak of the "farming community" we need to be sensitive to the depth of that statement. Because to that time, land ownership had been an almost vertical thing. The top of the local social class would acquire most of the land. The lower classes would own little or none. Class was based on land ownership. What began as land grants from the crown in early New England, Virginia and Maryland would evolve to the Jeffersonian thinking of agrarian ownership and land grants.

The square pattern of land distribution would enable nearly every free American to be a landowner if they desired to be one. And this fact alone, made this country curious and unique in the eyes of European visitors. In reality, land ownership made us a democracy. With the exception of some properties in the south which had survey problems all of their own, any landowner had a vested interest in a law abiding society that marked out land in 40 acre parcels. Even Texas, when it was actually under Mexican ownership, had established the square pattern in most of that state. And now where there was squared off land, legal surveys, meets and bounds, there was a new form of currency. It could be bought, sold, bartered and banked. Even the towns, for the most part, were squares. Fly into Lincoln, Nebraska, Cheyenne, Wyoming, Salt Lake City, Utah. You are struck at night by the long rows of lights. It is a pattern of blocks and squares. River towns hide an ecological lesson . . . they usually follow the river pattern

AT THIS POINT

. . . a heavy eyed reader might already have left us, pondering as she or he laid the book down, what a 22 yard chain had to do with ecology. You miss that when you look at maps neatly spread on the table. But when you walk the land, when you get down on the grass roots level, you find many places where the straight 22 yard line doesn't make a lot of sense. Now, you find creeks, rivers, lakes and sloughs that don't go straight. Nature isn't square. The second lesson is that all the land in even that 40 acre square is not equal in terms of ecological balance. Some of it barely grows weeds, a lot of it should never have been plowed. A lot of what was once plowed is no longer there. It has washed away! But the social contract said the land should be used. It was almost Messianic. So the people plowed it and then built another spire! They needed a place to pray for rain.

Prairie country stayed pretty much in squares until the early 1930s. It was President Franklin Roosevelt who made the impact on the prairie country environment. He brought in Hugh Hamilton Bennett to head what was the Soil Conservation Service. This resulted in the Shelterbelt programs and the creation of thousands of smaller dams on erosion prone land. Many of those dams were on private property. Many would be used more to water cattle than for conservation purposes and usually the focus was on the square mile and less on the watershed. Today, prairie residents like those in Bryson, Texas need to build a six mile pipeline to another community's water supply because the one small well that supplies Bryson's 550 residents can't get enough water from their drought constricted well. The bell is tolling louder for water each year. It is just a matter of time. And the local attitude? "I believe we can hang on if it is the good Lord's will," their Public Works Director was quoted as saying.

TODAY PRAIRIE COUNTRY

. . . faces a decision. Whether to let these watershed ponds fill on up with silt and lose their ability to stop runoff and hold water or to engage in some very costly restoration. The issue usually comes up for discussion when it rains. This is the joker in the deck in prairie country. This is where the phrase "when it rains it pours" originated. On paper, you can blithely proclaim 18 inches average rainfall. But live awhile on this land and you soon conclude that Mother Nature had it about right with her combination of grass and bison. Small ponds have not been the answer.

Water, then, is the true and elusive wild card in prairie country and the arid west. It shows us the importance of a balance between water and soil in ecology. This wild card was an important part of the thinking of John Wesley Powell and his concepts of water in arid country. Powell really never dealt with prairie country. He seems to have missed the lesson of the grasslands. His focus was on the arid west.

My experience, observation and research has tended to show we are trying to do something ecologically and agronomically with prairie country that the land doesn't care much for. Even with massive federal infusions of money where we have added enough water to available rainfall to make agriculture viable, the economics of the effort do not add up. Our account just doesn't balance. If irrigation was going to be the Wizard of the Prairie that leveled out the irregular water supply, it has been a case of diminishing returns.

The latest figures I could locate from a reliable source come from the New America Foundation in Washington, D.C. From that source, at this time we have some $2 billion in annual irrigation subsidies to western agribusiness. Half of this is used on surplus crops. Add another klinker. The underground lake under Texas, Oklahoma, Kansas, Nebraska, South Dakota, Wyoming, Colorado . . . the Ogalala aquifer is dropping dangerously. It provides about 30% of the annual supply of ground water that is being used for irrigation. Meantime, water waste, riparian restoration and erosion in times of heavy rainfall remain critical problems.

The experiment with agriculture in prairie country, however romantic it may have been for the poets, the boosters, magazine writers and movies was a mistake. It has soaked up billions of dollars of federal money for water projects that were sold to the public as recreational benefits and cheap waterway transportation but have been used essentially to supply water for growing alfalfa. Or, to grow more crops that were in turn subsidized by more federal largesse. Even after the dust bowl sent a fairly clear message to a lot of poor, 40 acre farmers . . ." get the hell out," some folks still hang on. Those that loaded up their trucks and headed to California transferred dry land mentality to California and lowered the understanding of water in that state. In Colorado or California, you need about fifty thousand pounds of water to raise two pounds of cow. If you bought that water from the California Water Project, current figures show that much water would cost between 7 and 8 dollars. If you are a farmer or rancher, especially a big one, that same quantity of water would cost you about forty cents. I heard a range expert say it one time so most folks could get a hold of it. "It ain't a water shortage in the west, it's too damn many cows."

Land measurement in this country is based on the idea that land is a commodity. Scenic value or water availability or water rights are usually added on, especially if the land is to be lived on. Square measurement works reasonably well. Water quality and quantity is usually the result of watershed quality. Water conservation in prairie country is a watershed problem. As a result, water conservation has received much less attention than has soil conservation. We have only recently come to treating water resources as a watershed issue that may involve several states. In fact, "water rights" in the country west of the Mississippi River is growing as the most litigious area in conservation law. Logic and reason may not have residence in water law.

OUR WATER MEASUREMENT SYSTEMS

We have developed a wide variety of systems for the delivery of water and as many systems of measurements, both for water quality and water quantity. Most of these are based on the mentality that water is cheap. Water quality is measured in most states as a perfunctory exercise with states using a variety of methods. Not only that, different state laws declare different ownership of the area along rivers. In some cases, Native American tribes are claiming riverbed ownership. So whether ownership extends to the middle of the stream, stops at the water line, is involved with a high water mark of some type all depends pretty much on local interpretation of what is water, what is watershed and how either can be used. We have used federal law, state law and county law. Nothing has seemed to provide a formula for any control of water quality. If better conservation practices are prescribed and implemented, it is almost on a case by case basis.

When the yeoman farmer or the resolute rancher moved into Great Plains country, both species attacked the watershed. One used a plow; the other used cattle hooves. Both assaulted the grass. It was the grass that protected the watershed. Dams, to have any value as a reservoir, need a watershed where erosion is not a factor. Silt and salt are the enemies of irrigation. When these enemies lurk, as they usually do in prairie country, the death of a dam is certain. A dam in prairie country doesn't have "life everlasting."

The Guernsey Reservoir on the North Platte River in Wyoming, one key to Nebraska irrigation, had a water capacity in 1929 of 74,810 acre feet. In 1957, that capacity was 44,800 acre feet. In 2002, when I visited the area, I was told the capacity was at 45,612 feet at an elevation of 4420 feet. I was also told by Ken Randolph of the Bureau of Reclamation, the dam was now flushed every spring to reduce silting. Of course that silt is flushed downstream to silt another dam. Salt does not seem to be a consideration in management of this dam.

The Colorado River can serve as a current example of the salt problem. By international treaty, the U.S. had promised Mexico a million and a half acre feet of water from the river. Most of this was to be used on irrigated land on the lower Colorado. The best of this river water, up high, is 2 hundred parts per million acre feet salt content. When the river reaches the Mexican border, what little water left has a saline content of 15 hundred parts salt per million acre feet. In a trade off for consideration of the emerging Mexican oil supplies, the U.S. agreed to do something which turned out to be $1 million in engineering solutions. The grass roots part of this example . . . the Yuma River plant removes Colorado River salt at about $300 per acre foot. The upriver irrigators in the U.S. buy the river water from the Bureau of Reclamation for $3.50 per acre foot. This produces more subsidy.

There are some positive examples of water quality cleanup, but these also provide an idea of cost involved. The Chesapeake Bay cleanup has been a conservation success as has the cleanup of the Hudson River and the Connecticut River cleanup. On the other side of the country, the work now being done to clean up the Napa River is under way. Poke around a little and you find it can be done. River water quality can be restored. So can lake water quality. Lake Erie is an example of that. But water restoration requires consideration of a quality watershed and takes lots of money.

EARLY EFFORTS IN WATER CONSERVATION

Step back briefly to 1907. One of several settlement waves was beginning to abate in prairie country. There were four or five government organizations concerned with stream issues, three separate government organizations that dealt with mineral resources, perhaps as many as a dozen that claimed authority over wildlife, and half a dozen with authority over forests. Gifford Pinchot had succeeded in putting his Forest Service Bureau under Civil Service in 1904. He was a public figure who saw forestry and the problems germaine to forestry as being tied to the use of minerals, the control of water-power development, the issues of grazing on private and public land. He argued that irrigation of arid country was germane to the central issue of natural resources. The question was "how would they be used for the benefit of the most people?"

Pinchot and W.J. McGee, a self taught scientist from Iowa, urged President Theodore Roosevelt to promote a new concept of conservation. Roosevelt responded by establishing the Inland Waterway Commission. It operated without funds, but would present a comprehensive plan that included hydroelectric development, irrigation, flood control, water transportation and soil conservation. The plan would be endorsed eventually by 41 states but it would fall upon hard times in Congress where it was opposed and never funded. In 1910, the Roosevelt-Pinchot-McGee efforts to unify a conservation effort in the United States would splinter. Individual users of natural resources would continue to pressure the federal bureaucracy with powerful lobby efforts. Pinchot was not sympathetic to wilderness preservation. He continued to be an advocate of sustainable harvest. As prairie country moved toward the great depression and the dust bowl, his conservation ideas increasingly fell out of public favor.

So this muddle of land conservation practices, or the absence of them . . . the welter of water rights laws and customs and wasteful practices, and a population that would come under economic pressure, all got stitched into a shabby quilt that covered about half of the land in the United States, but contained only about 4% of the population. Put that in perspective. Water hungry California is about ⅕ the size of the ten state area of the prairie land in the U.S. Just the people of California alone, much less the highly developed agribusiness use of water, account for increasing strain on California water supplies.

ADD SOME MORE BIG SCREEN MEASUREMENTS

. . . to a significant shift of population in the United States. Only about 6% of the land in America is urban, suburban or rural residential land use. 6% of our land is used for where people live. About 20% of our land is agricultural, 25% is rangeland. So about 45% of our land is used in some way for growing things, crops or animals. That leaves about 49% of our land for forests, woodlands and wilderness. Since 1950, it is estimated that land has been taken out of agricultural use eight times as fast as it has been taken for urban development. There is no reason to assume this trend will not continue over the next 100 years. Most of the land being taken out of agriculture is going back into forest, woodlands and prairie grass. Water is about all that is retarding this trend. Small wonder, then, that irrigated land becomes more valuable and dry land goes back to native grasses.

This isn't the only trend in the great grassland that is worthy of note. It is more anthropology than ecology although the human being plays a fairly

important part in both scenarios. Michael Lind, in an article in January/
February 2003 *Atlantic Monthly,* contends that by 1997, "39% of the Great
Plains farm owners were already designating their main job as 'other' rather
than 'farmer' on their tax returns. From that same source, he says that in
2001, ninety-nine counties in the United States had populations in which
4% of their residents were 85 or older. "Most of these counties were in rural areas in the Great Plains. Meantime, the coastal areas of the country experience substantial population increases."

In the coming decade, if the political atmosphere remains as it is, the Federal government will put $171 billion into direct farm subsidies. In addition,
there will be another $2 billion in annual irrigation subsidies to western
agribusiness. The real wrinkle in this figure is that about half of that money
is used to produce surplus crops. Agriculture simply does not produce many
jobs any more. The newest Geological Survey says that the same amount of
water that supports a 40 acre alfalfa farm with two workers would support a
semiconductor factory with 2000 workers.

Even if all this numerical evidence was sifted down and a fair amount
turned out to be chaff, there is enough going on in the numbers game to merit
some serious consideration of what is happening in the Great Plains. On the
grass roots level, you have only to get off interstate 70, 80 or 90 to find the
funny games Mother Nature is now playing with soil and water. One is if it
were not for irrigation there would be little nor any agribusiness in the plains
country. If farm subsidy were suddenly withdrawn, that would certainly be
the case. It is a land that is being propped up. Two, it is an old population. The
young people, especially the young business people, have essentially left the
land. Three, if you suspended commercial fertilizers and spraying, crop production would drop even more than it has. Fertilizer, the pesticides and no-till
farming are masking a decline in soil productivity.

The exceptions to these conclusions are along rivers. There, the presence
of water, whether for irrigation or recreation, has provided an economic pier
that has kept the economy out of the basement. These small rural communities are attempting to shift from extraction to eco-tourism. There is a growing
awareness that tourism produces mostly minimum wage jobs. It may not be
the elixir some thought it would be.

PATCH TOGETHER A QUILT

. . . of water quantity and quality and there is no pattern. It is more like
Grandma's crazy quilt. Here is a patch that comes from a coat of an employee

of the Interior Department. It leaks! There is no federal department involved with water that has a shabbier record of mismanagement. It has been a grab bag of functions. Powell's Geographical Survey of 1879 got shuffled to Interior, the Reclamation Service in 1902, Bureau of Mines in 1910, the Park Service in 1916. The department was home to the Bureau of Indian Affairs; even administered public buildings in Washington, D.C. Came the Franklin D. Roosevelt administration and his Interior Secretary, Harold Ickes. Now dozens of conservation and dam building projects are put in the department.

Skip to 1940 and Harold Ickes has 13 storage dams under construction. All are sold to the public as partially conservation, partially recreation and partially creation of jobs. Ickes wanted a National Planning Board to deal with conservation and national planning. His idea had merit, but got lost in the smoke of World War II. In 1950, it was proposed to build dams on the Green and Yampa Rivers. The Echo Park dam would be 525 feet high. Proponents of the dam would argue putting the dam there would save 300,000 acre feet of water. A former Chevrolet dealer from Oregon, Douglas McKay, was then Secretary of the Interior. His figures were challenged. Eventually, the estimate of water to be saved was lowered to 25,000 acre feet. Numbers and people lost their credibility and their dam. So for the last 100 years, planning involving water has been a multihued conglomeration of projects and ideas. Ickes idea to consolidate control of natural resources into one department is still debated. Whether it would be possible to gather that many birds into one cage is doubtful, given the powerful forces aspiring to control of water resources. Special interest groups sing of economic development and jobs. This is a strong song.

THE STUBBORN STREAKS IN THE PRAIRIE

In his book, *Cadillac Desert*, Marc Reisner treats the subject of prairie water in much greater detail than our brief sketch does. He writes, "From a national perspective—-forgetting about the farmer's plight—-whether irrigation of the southern plains ends in thirty years, or in seven, or even fifty years does not matter; the fact is, it will mostly end. The important issue, from that same perspective, is what will happen then—-not just to the farmers and the cost of food and the balance of payment deficits, but to the land." He agrees with the Paul Sears conclusion . . . "So long as there remains the most remote possibility that drier grasslands, whose sod has been destroyed by the plow, can be made to yield crops under cultivation, we may count upon human stubbornness to return again and again to the attack."

The first part of this chapter was written in 2000. Other parts in February, 2003. From the wires of the Associated Press this news story. "The west's persistent drought is forcing many ranchers to quit or send their cattle to greener pastures. For the seventh straight year, the number of cattle in the nation has dropped because of lack of forage—-particularly in the west—-and the sluggish economy according to the National Agricultural Statistics Board. Bill Wilkerson, a rancher near Trinidad in southern Colorado, said he shipped most of his herd to Texas last year because of the drought. He hasn't sold it off because of the animal's valuable genes, but the move pushed him $20,000 over his budget. 'It's getting to be no fun anymore,' he said. 'This lifestyle isn't a huge payoff anyway, and we do it because we love it, but it's getting pretty difficult.'" The article then went on to say cattle in Colorado were down 13% from the previous year, Wyoming had a 12% decline while Nebraska had its smallest inventory since 1995.

Then this classic statement from a USDA livestock economist, Ron Gustafson. "Essentially, once mountain states recover and get more rainfall, the cattle will go back there and the cycle will continue."

SO, IS THE DISTANT HORIZON DREARY?

The Jeffersonian squares still spread into the horizon. More sections you find, as you walk the land, are still marked by a house on the corner, but it is collapsing in a weed filled yard that displays various types of farm machinery of days gone by, maybe a windmill rasping in the wind, still squirting an unsteady stream of water into a rusty, metal tank. Much of it drains on the ground. Some is used for cattle that are looking for their parentage. Mostly, it is wasted.

Only the political clout of the prairie country remains fairly intact. Wyoming, Nebraska, South Dakota have small populations. Wyoming senators would represent a half million people. New York senators represent maybe 60 times that. But that one U.S. Senate vote is worth just as much when it comes time to vote for an irrigation subsidy or another farm bill as the vote from the more populous state. It is a measurement that hasn't changed much since Mr. Jefferson was hustling money for the Lewis and Clark expedition.

It is entirely possible that because we like the simplicity and the security of 22 yards of chain, quarter sections and sections, we are lulled by the permanence of the simple. We tend to avoid or evade the more fluid issues of water. After all, a river meanders, it changes flow with rainfall or runoff, it tends to have a diverse and complex wetland ecosystem. Maybe that is why we tend to focus more on land and less on water.

The space in prairie country can be melancholy. It is easy to see great distances and yet . . . folks have had trouble seeing very far. It could be that "God never intended for prairie land to get plowed up."

Oh, we could get the land measured out into squares. It was the water part we can't quite get the hang of.

George Bird Grinnell

Was a mentor of President Theodore Roosevelt. He was a founding member of the Boone and Crockett Club. He founded the prototype of sporting magazines, *Forest and Stream* which later became *Field and Stream.*

George Bird Grinnell. NPS Photo.

Grinnell founded the first Audubon Society in 1886. He was a leader in founding Yellowstone National Park and also Glacier National Park. His conservation philosophy served as the basis of the American Conservation Program when Roosevelt became President in 1901.

Born in 1849 in Brooklyn, New York, he developed a love of birds at an early age and spent time at the John James Audubon home in Assigning, New York. Later he attended Yale University where he was tutored by O. C. Marsh. He served as a naturalist in the Custer expedition in 1874 to the Black Hills. He came to know many of the Indian tribes well, including the Cheyenne, Pawnee and Gros Ventre and became one of their strongest advocates for fair treaties.

His most notable contributions to environmental protection came as Editor of *Forest and Stream*. He advocated a system of game wardens to be supported by small fees paid by hunters. He pushed for enforcement of game laws on the state level, again supported by financial support from hunters, an idea that would become the template for game management.

Grinnell died in 1938 at the age of 89.

Frederick Law Olmsted

Was born in 1822 in Hartford, Connecticut. While he had no formal college education, he became a highly intelligent person and is generally regarded as the founder of the American landscape architectural movement.

At the age of 18, he tried a career as a scientific farmer. When that failed, and after a hitch as a merchant seaman, he was a newspaper correspondent and published several books. He then became a columnist for the *New Yorker*. He was appointed Superintendent of Central Park in New York City. He would meet Calvert Vaux, who was working on the design for Central Park with another partner, Andrew Jackson Downing. Downing died, and Olmsted joined forces with Vaux. Their design eventually was accepted for Central Park.

In 1863, he took the position of manager at the Mariposa Estate in California. That project failed. He rejoined Vaux and, from 1865 to 1887, produced a series of projects resulted in Chicago, Buffalo, N.Y. and Niagara. His last major effort was the design for the World's Fair of 1893 in Chicago.

His health failed in 1895 and he was confined to the McLean Hospital in Waverly, Massachusetts, a place where he had designed the grounds. He died August 28, 1903. His landscape architectural firm stayed in business under the ownership of his sons and their successors.

Frederick Law Olmsted. Photo courtesy of the Library of Congress, LC-USZ62-36895.

REFERENCES

Brown, Lester R. *Building a Sustainable Society.* (W. W. Norton) 1981

Hollon, W. Eugene. *The Great American Desert: Then and Now.* (University of Nebraska Press) 1966

Reisner, Marc. *Cadillac Desert.* (Penguin Books) 1987

Stegner. Wallace. *Beyond The Hundredth Meridian.* (Houghton Miffllin) 1992

Strong, Douglas H. *Dreamers and Defenders.* (Addison-Wesley Publishing) 1976

The Measurement That Built America. An article in American Heritage Magazine, November / December 2002.

Many visits and interviews with farmers and townspeople, some of whom kindly gave me permission to hunt pheasants.

Chapter Eleven

Considerations of the Prairie Model

A study of the ecological patterns in the United States, could, for the sake of great simplicity, locate one set of patterns east of the Mississippi River and another west of that river. This would hold even more so if you drew a line from Virginia west to the Mississippi. In the upper right quadrant, moisture amounts, river patterns, soil quality and density of species would be fairly similar.

Once settlement pushed across the Mississippi, what could be called adaptations of European land patterns began to change. That change became even more pronounced when settlement pushed across the Missouri River and into the prairie states. Moisture amounts became a guessing game. Rapid seasonal changes from too much rain to not enough muddled patterns. Soil quality shifted as sand content replaced the richer loam of the eastern states.

Space now became a major factor. Where land control in the eastern part of the country rested primarily in the hands of the people on it, the western part of the country would see large amounts of land end up under governmental control or owned by large corporations. This would, and still does, present many conflicts for the people attempting to use the land for whatever purpose. This would present new and challenging questions about the use of land and space.

SPACE AS A VARIABLE

One of the variables in any ecosystem is space. Our prairie model uses soil, fire, bison and plants as basic framing. They were the determinants in the original, virgin prairie. Because of the original great expanse of the prairie, or the abundant forests, space was not an issue in early thinking about the use of natural resources.

The space model is a good example of the statement, "it is best to approach nature with a good deal of humility." Nature can quickly get complex. In our organizational model we used a rectangle to allow room for the progression of the stages of the organization. We used the word "simplistic" . . . not to demean the reader or audience, but to keep a multidimensional condition in a visual or linear sequence. We also emphasized that each of the frames or determinants would have a model that extended the initial definitions.

An examination of space requires more than a linear approach. More correctly, space should be placed in the context of a cube. It is multidimensional. In a later chapter, we refer to the concept of Professor Frederick Clements, one of the early Nebraska professors who studied the prairie. He suggested eight determinants of the health of the prairie. They were: (1.) water content of the soil, (2.) humidity of the air, (3.) light intensity, (4.) air and soil temperature, (5.) wind, (6.) precipitation, (7.) ground physiography (8.) soil class, i.e. sand, clay, gravel. We suggested his approach could have

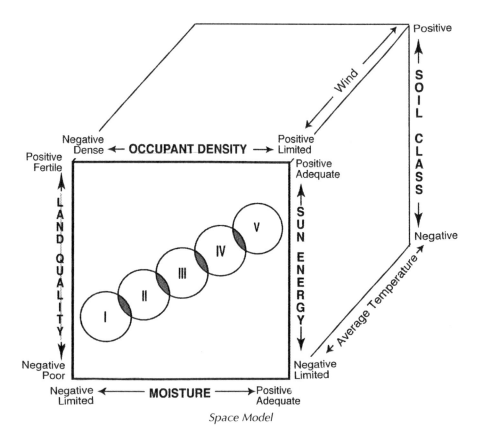

Space Model

a resonance with our model in that he is trying to define the determinants of prairie condition.

To simplify our model, we have used land (soil), moisture, sun (energy) and occupant density. We tried to cover the idea of agreement and how the Nobel Foundation experts had agreed upon the definitions of their determinants when they inspected The Tallgrass Prairie Preserve. The important point is that any ecosystem is a living thing composed of many determinants. Logically then, a space ecosystem is a living thing composed of many elements. In our model, we could have substituted air quality for occupant density. We would then have needed agreement on what we meant by air quality. In turn, it would have four determinants.

Space models can be extended. In our visual model, the frame of sun (energy) might use that frame for one side and then add the frames of air temperature, light intensity and land cover. These, if agreed upon, could be used to produce an evaluation of energy quality in the ecosystem. Cooler temperatures, even at night, less intense light, and land cover such as heavy grass or trees would produce a better quality energy stage than the opposite of these would. During a drought period, soil temperatures are higher on bare land, or heavily grazed land than on land having grass cover. These determinants deserve consideration before you build a prairie pond. All will have a bearing on the evaporation rate of water in the pond.

Small wonder that today, when groups study an ecosystem, be it land, air or water, they put the many variables on the computer in order to consider all the possible determinants.

Another way to view the model, which attempts to simplify a very complex process, is that high performance in any of the determinants is just a step away from either the REFORM or FORM stage. Nature does not stay at the stage of high performance for any great period of time. At some point, the ecosystem will regress and will start over again to find a natural balance.

In a natural state, this is often a long, slow process. A severe, natural disaster will speed up the process. This can be fire, flood, tornado, or hurricane.

WHAT WE HAVE LEARNED

. . . often contrary to common wisdom of the past, is an original condition is hard to replicate, be it prairie, forest or river. And often times, ecosystem space is a factor. Density of human beings is hard to reverse.

With this as background, we come to the many patches in the quilt that make up "Prairie Country." It is a vast panorama of plains, wooded areas along streams, tall grass country, and short grass country. Some folks thought of it as "a Sea of Grass." Many metaphors have been applied to suggest a landscape of wind, sky and grass. It was a land of unsuspected diversity. If the land was divided into 640 acre sections, as it often was, each section could contain as many as 600 to 700 different species of plant, animal, insect and bird life. The grasslands were space, huge space, both above and below ground level. It is hard to imagine so much space has been impacted by the presence of man.

From what we can surmise about this land, from 1000 AD to the mid-1700s, it was essentially in a state of natural evolution. A few Native Americans seemed to have hunted there, there was water supplied by streams and rivers, wildlife was abundant, wetlands and potholes probably dotted the landscape in times of more rainfall. If the part of the ecosystem that was above ground remained in some state of balance, then the other natural world hidden beneath the grass stayed in some state of balance.

Narrow this broad sketch to just one state, Nebraska. It makes a reasonable micro study. I grew up there. Couple this background with limited experience with the restoration of the Tallgrass Prairie Preserve in Oklahoma, and countless journeys through the plains states. Then some Nebraska chronology.

By 1900, the wild turkey, the prairie chicken, the big horn sheep were no longer to be found. Pronghorn antelope had been reduced to an estimated 100 animals. The Nebraska Game and Fish . . . they now have Parks in their mix . . . told me they estimated only about fifty deer at that time.

1875 is probably a reasonable date to begin tracking the story of man's exploitation of the land. By 1880, Nebraska had a fair share of railroads. Some ran somewhere and some just ran. When the settlement of land slowed in eastern Nebraska, the railroad hucksters would get out their drums of plentitude and pound out stories of the "land of plenty" further west. Trot out the canopy of manifest destiny, plenty of "boosterism" that used progress as the holy grail, and the alteration of space continued across Nebraska.

To illustrate that transition, the rise and fall of the prairie chicken population would show how species abundance depends upon space. This proud, little bird is sometimes called "the barometer of the prairies." It is the indicator of the health of tall grass country. If that is true, then tallgrass country is in a world of hurt right now because the strutting of the chicken has become very rare. In Nebraska, chicken habitat was roughly the eastern two thirds of the state. Early settlers regarded them as a reasonable substitute for domestic chickens. Marie Sandoz sketches this idea in her novel, *Old Jules*. Jules was a Swedish immigrant. He would leave home for days at a time to come home with a harvest of chickens. His mail order bride would clean and cook them.

She got tired of cleaning the chickens and made other arrangements for a domestic life. The chickens got scarce so Jules went into politics.

In the early stages of development in Nebraska, small fields of cropland
created quilt like patterns. The chickens seemed to prosper because of the
edges created by small fields. Cover was adequate and there were plenty of
insects for chicks. But population swelled in the 1870s and 1880s. As agriculture moved westward across the state, pushed by economics and abetted
by favorable Federal land granting, prairie chickens first tried to find space
westward and then gave up the battle.

Quail populations would provide another illustration. Again, early settlement would produce abundant edges. The quail prospered. But it was a time
of few game laws and less protection. There were no social contracts that
would place limits on the birds. Where the social contracts of the gentlemen
hunters of South Carolina tended to leave some birds for seed, the prairie
country code was shoot until the shells ran out. In 1917, quail hunting was
eliminated by law. In 1929, all grouse hunting was banned.

EFFORTS TO IMPLEMENT WATER CONSERVATION

Nebraska is on the southern margin of the continental waterfowl production
area. About 1910, the Nebraska social contract for land use was shifting. The
idea was born to drain wetlands and manage water resources. A 19 mile long
ditch was built in Burt County. It carried water from Silver, Mud and
Tehamah Creeks directly to the Missouri River. The first of the Nebraska
floodplains had been converted to crop land.

The process moved westward. The Nemaha river project, the Salt Creek
Project near Lincoln would drain thousands of acres of wetlands. In South-
central Nebraska, 20 to 35 percent of the rainwater basin had been converted
to cropland. The immense flocks of waterbirds no longer stopped on the land
of the huskers. Many of the huskers involved with this work would swipe
away the sweat and rub sore backs. Draining wetlands with mules or horses
and hand swedges is very tough work. So tough that on some Sundays, the
sore backs rested instead of going to church.

Today, along the Platte river near the town of Kearney, Nebraska, efforts
by The Nature Conservancy and private groups, have established a refuge of
sorts along the river. And while the residents of Nebraska and Colorado are
still haggling over water rights and how much water should be allowed to stay
in the Platte, the Sand Hill Cranes have relocated their old space along the
river. In the spring an estimated 600,000 of these birds find rest and refur-
bishment in this location. And that many Cranes can make an immense racket

when they tune up. Tourists are attracted to the river and thousands of bird lovers trek to observation posts where they are able to marvel at the noisy choir of Cranes. The economic impact on the area has been most favorable, not quite like the elk near Jackson Hole, Wyoming, but economically still very measurable.

Back to the land push across Nebraska. The Kincaid Act of 1904 granted 640 acres for settlement in 37 more western Nebraska counties. From 1900 to 1920, population in those counties went from 136,615 to 251,830. Corn production in the state doubled, wheat production jumped four times. The optimism of agriculture and the economic benefits it promised, swept aside any regard for wildlife, the land or natural resources. The economics of agriculture, and the agribusiness community were now positioned in direct conflict with wildlife and natural resources.

By 1920, Stage II of wildlife management in Nebraska was in place. Federal and state laws were formed to protect wildlife to some extent. And while the species could be protected, their need for habitat, space and a species sensitive ecosystem could not be protected.

In Nebraska, the model for wildlife management evolution, first protection and then regulation advanced. What became the Nebraska Department of Wildlife and Parks was first attached to the agricultural department, then was separated into a department that focused primarily on wildlife management and setting hunting seasons. In the early 1950s, under Director Paul Gilbert, it was still struggling for recognition. I was in television in those days and helped Gilbert produce the first weekly program that told their story to a bigger audience. It was the first such TV program in the United States, using what was then an emerging medium.

The model we have positioned for wildlife management, with its stages was certainly in evidence in Nebraska. It reached the biological stage where species substitution, stocked pheasants for prairie chickens, was tried. Experience proved the stages were present. When the habitat was in balance, the species would increase rapidly, then find a balance. Populations would vary depending upon agricultural practices and weather.

World War I was boom time. Farmers expanded acreage and production. Prices fell in the early '20s. Cropland was abandoned. Weeds, native and exotic, thrived. So did the pheasants. The '30s was a time of depression and dust bowls. Game birds and waterfowl suffered. The economic and sociological scars cut deep in the psychological fabric of the people. They also left deep tracks on the ecosystem. Now there was a growing array of federal programs designed to correct past abuses.

Irrigation became a high priority in the state. In 1930, about 1% of Nebraska was irrigated. That was about 532,000 acres. By 1950, that number had nearly doubled. Then for 30 years, first one irrigation project and then another came off the bureaucratic drawing board. Work began on the Missouri River below Sioux City, Iowa to improve navigation and prevent flooding. Waterfowl, wetlands and timber gave way to economics.

THE LEFT HAND VERSUS THE RIGHT HAND

Before World War II, some 250 million acres of land was manipulated out of production by federal programs. During the war, most of those programs vanished. Then following the war, the U.S. Department of Agriculture began an aggressive program of technical assistance and cost sharing to encourage farmers to drain wetlands. The great industrial expansion in the U.S. during the war, the great advances in technology would find their way into land use after the war. Tractors replaced animals, large farms and corporate thinking replaced the small farmer, fertilizers and pesticides changed social contracts. Irrigation was another form of technology that swept the state as pivot watering systems produced broad changes in land use. It also would begin to change underground water supplies.

In the sand hill country of northwest Nebraska, irrigated acres went from 64,500 in 1965 to 535,000 in 1978. In other parts of the state, the bulldozer and the tractor continued the assault on the shelterbelts and were equally destructive to brushy ravines and hedgerows. The voice of Secretary of Agriculture Earl Butz echoed across the state as farmers were exhorted to plant from "hedgerow to hedgerow" because Russia was buying wheat, all that American farmers could produce.

By the early 1980s, either Russia was growing its own wheat or the Russians had stopped eating bread. It was bust time again. Land had been purchased at high prices and high interest rates. Gloom settled on the land. Banks closed and small towns contracted to a post office, a convenience store and maybe a tavern. And sometimes there was just one church left with an unpainted spire poking into the lonely prairie night. And once again, the federal government rushed to the rescue.

1985 saw the emergence of the Conservation Reserve Program and The Farm Bill. The bill didn't have wide public support and not much in Congress. The farmers needed some allies so wildlife benefits were added. A "sodbuster" provision discouraged plowing erosion prone grasslands, a "swampbuster" provision discouraged wetland plowing. By now these programs applied more to big or corporate farmers because the small farmer had joined the plight of the wild things and if not an endangered species, the small farmer had become a rare species.

THE COLD REALITY OF PRAIRIE SPACE

There is something cold and calculating about a litany like this. The churn and vacillation of state and federal programs. The lists of dates and figures. The heartaches on the human side of the equation are missing. There is a hollow feeling about leaving the land, just as there is a gauntness about the picture of the deserted farm house or the barn, roof half gone, that leans into the prairie winds. There are hopes and aspirations in the cracks of weather-beaten boards and whispers of times when the mortgage would be paid off. America has been a land of squatters, searchers, settlers and stayers. And those who stayed on the land, battled the land, used up the space, fought heat and wind, the blowing dust, and then had to give up and leave the land . . . they knew that hollow feeling. Gnarled hands . . . tough. Work . . . God, how they worked! And how most of them prayed. For they really believed their destination was the kingdom of plenty.

This metaphor was not limited to Nebraska. It fit much of the plains country. And for words that describe the prairie in breathless prose, there is a story of giving up and leaving. Sure, there were times when Mother Nature dressed up in her Sunday pants but there were just as many times she served as the acolyte for the grim reaper.

As the small farmer and the small fields with the edges friendly to wildlife disappeared from the Nebraska terrain, the Nebraska Game, Fish and Park Department, as it was by now known, beat the drums for the story of the value of the pheasant hunter as an economic resource to the state. But the CRP program had been rather disappointing to the wildlife officials. Although some 1.4 million acres of former cropland was enrolled, and looked like excellent cover for the pheasant, these lands lacked the cover diversity necessary for optimum habitat. Man could rip it up with great proficiency but has trouble putting it back together like it was.

That brings this brief chronology to 1995 and the 1995 Farm Bill. There were a number of agendas by now. There was a political sense in Congress that the Federal Government should get out of farm management. There was public demand to reduce federal subsidies and reduce the tax consequences. Farmers wanted freedom from complex and conflicting regulation and a growing conservation and environmental lobby were pressing for recognition of environmental issues.

What came out of this mesh of different agendas was a pattern of 10 and 15 year land retirement programs with cost sharing provisions that were designed to protect soil and water resources and provide for better wildlife habitat. Filter strips, permanent woody cover, grass waterways and issues involving native grasses were addressed. Wetlands qualified for inclusion and funds for wetlands restoration were included. It was almost as though the Wizard of OZ was pulling the strings on a landscape of smoke and noise. There just was no L. Frank Baum to get the story right.

SIMPLE LESSONS IN A COMPLEX QUILT

In the expanse of the country, there would be simple lessons. It would be years before the folly of straight rows of plowing proved to be a contributor to the ecosystem. For the big winds could sweep down the long rows of loose soil and carry silt and dust into angry, broiling skies. Strong rains would wash the loose soil down the rows creating ditches.

Later students of the prairie would come to understand this land really rested in the rain shadows of the Rocky Mountains. Walter Prescott Webb, a noted authority of prairie country used the 98th parallel as the dividing marker for geographic and climatological purposes. Climate and weather actually created three distinct prairies. Rainfall determined the composition of the land. That, and the type of rock under the ground, created the mixture of grasses so that ecosystems are slightly different from an area that might have a sandstone substrata to one with a limestone substrata.

Enter the botanist into this shifting landscape. They will tell you the grasses had adapted to their environment. They were, and are, windborn pollinators. The flowers, especially, have developed an intricate infrastructure for capturing the pollen grains that drift on the wind that is almost always present. These grasslands had also adapted to predators that grazed on them. It was almost a system of predator and prey, resulting in a balance between animals and grass.

Our botanist friends tell us, the grasses developed ways to keep themselves from being eaten to death. James Stubbendieck is a range ecologist and Director of the Center for Great Plains Studies in Nebraska. His theory is that cells of prairie grass are filled with silica. This is a gritty material that will grind away an animals teeth. There is evidence of badly worn teeth in old bison skulls. This condition is present in prairie grazed cattle. Bison would graze an area and then move on. This allowed grass to rest and rejuvenate. Fenced in cattle break up this subtle biological pattern. Grasses with less silica are the first to be grazed. When these grasses expire and are replaced with other herbage, the natural balance of the grassland is upset.

Consider that 97% of the land in Nebraska is in private hands. This is in marked contrast to the western states where smaller percentages of land is in private hands. Clearly, the future of ecological conservation and the resultant role of conservation of wildlife resources depends to some extent on the private sector. And to the credit of Nebraska researchers and farmers, the state has begun to struggle with the problems of restoration. But the time that is required and the dollars involved provide proof of the task involved.

The plain truth is . . . and there are some who believe there is no plain prairie truth . . . the challenge is almost staggering. This story was in the July, 2006 *NebraskaLand* magazine. "In June, the U.S. Fish and Wildlife Service issued a 'biological opinion' that a Platte River Recovery Implementation Program would provide protection for four federally listed endangered species . . . A committee composed of water users, environmental groups, federal agencies and officials from Nebraska, Wyoming and Colorado developed the recovery plan of the last *12 years. (Note the time span.)* The main thrust is to provide 10,000 additional acres of wildlife habitat in central Nebraska and to release water stored upstream in such a manner as to retime river flows in order to improve habitat conditions. Shortages to FWS target flows will be reduced to 130,000–150,000 acre-feet. An acre-foot is about 326,000 gallons, enough to meet the needs of one or two urban families for one year. . . . The plan will be implemented in phases, the first to span 13 years, and is estimated to cost $317 million, with $130 million of that coming from habitat contributions from the states involved." So, while you have some good news, it is accompanied by long time spans and big bucks.

In that same issue is the story of Nebraska's Environmental Trust which was set up by the Nebraska Legislature in 1992. It purpose was "conserving, enhancing and restoring the natural, physical and biological environment in Nebraska . . . The Trust shall complement existing governmental and private efforts by encouraging and leveraging the use of private resources on environment needs with the greatest potential impact on future environmental quality in Nebraska." Money in this trust comes from the Nebraska State Lottery. It provides the seed money that will match other grants to fund public projects.

"By any standard, the Trust must be considered a resounding success. As of April 17, 2006, the trust had received $100,691,424 from state lottery proceeds and given out $100,532,761. The trust is governed by a 15 member board that is appointed by the governor and has nine private citizens in that group. To this time, the trust has received 1,772 applications and has awarded 871 grants. Five of those grants have been to the Aquatic Habitat programs to support 24 rehabilitation programs. The Trust has awarded $700,000 to purchase conservation easements on the Niobrara River. In 14 years, more than $100 million has been distributed to groups and organizations."

THE PRAIRIE STRUGGLE

The struggle with the prairie began with the horse drawn plow. It evolved to giant tractors, bulldozers and machines moving the earth to create more production or moving water around to enhance production. The tough hands gave way to technology and science, imported wheat strains and advanced strains of corn and milo, prodigious use of chemicals, fertilizers and pesticides. It came to include wild rides of commodity prices, the ups and downs of rainfall and drought, conflicting government programs and subsidies, the near eradication of wildlife and in some cases natural resources. It was a replica of grandmother's crazy quilt.

1954 was a bit early for environmental statements, some 15 years before the word was coming into common use. That was the year the book, *North American Prairie* was published, written by J.E. Weaver who became Professor Emeritus of Plant Ecology at the University of Nebraska. These were his words on the prairie. "Grassland soils have been thoroughly protected by the unbroken mantle of prairie vegetation. The vegetation and soil are closely related, intimately mixed, and highly interdependent upon each other and upon the climate. Hence prairie is much more than land covered with grass. It is a slowly evolved, highly complex organic entity, centuries old. It approaches the eternal. Once destroyed, it can never be replaced by man."

Hugh Hammond Bennett

Was born April 15, 1881 in Wadesboro, North Carolina. He graduated from the University of North Carolina and from there went to work for the U.S. Department of Agriculture. He was supervising soil surveys by 1909 in southern states in the U. S. and later studied soils in Alaska, Cuba and Costa Rico.

In the 1920s, he began to draw national attention with articles in scientific journals and popular magazines. Then in 1928 he co-authored the article "Soil Erosion: A National Menace." He was convinced soil erosion was more than a problem for farmers, it would also impact rural economy. The Dust Bowl days made him something of a prophet.

Bennett would be behind the push to establish the Soil Erosion Service in the Department of Interior. He became the director in 1933. Then in 1935, he was scheduled to testify before Congress. He timed his speech to a dust storm arrival in Washington and successfully made his point about the need for soil conservation. This resulted in the Soil Conservation Act of 1935 which created the Soil Conservation Service at USDA.

Among the many honors Bennett received was being named a charter inductee into the USDA Hall of Heroes. But he was also honored by such

Hugh Hammond Bennett. Credit: National Resource Conservation Service

groups as the National Audubon Society, the Garden Club of America and the National Geographic Society. His major contribution to conservation was convincing the country it had a serious soil erosion problem.

He died July 7th, 1960 at the age of 79.

President Theodore Roosevelt

Was born October 27, 1858 in New York City, New York. He graduated from Harvard in 1880, but he had grown up steeped in the lure of western hunting and adventure. His first trip to the west was in 1883 when he rode horseback for ten days in North Dakota searching for a bison.

Roosevelt felt Americans had lost a legacy and seemed to feel the loss of wildlife was a portent for potential demise of the nation. When he became the

Theodore Roosevelt. Photo courtesy of the Library of Congress, LC-USZC2-6201.

26th President in 1901, he pushed for sweeping legislation at a time when the word conservation was little used. When he left office, he had enriched the future of conservation with 230 million acres of designated public forests, had begun a system of wildlife refuges, bird preserves, parks and national monuments.

Roosevelt's face is one of the four American Presidents on Mount Rushmore. He was one of ten founders of the Boone and Crockett Club. The major park named for him is the Theodore National Park located in the badlands of North Dakota.

Roosevelt comes very close to being the "father of conservation" in the United States. Certainly he put conservation on the national screen as no other political figurehead had to that time. He not only provided inspiration, he produced action on a national level.

President Roosevelt died January 6, 1919 in Oyster Bay, New York.

REFERENCES

Barns, Cass G. *The Sod House.* (University of Nebraska Press) 1930

Farrer, Jon. *Wildlife and the Land.* Outdoor Nebraska Magazine. Summer, 1999

Hanley, Wayne. *Natural History in America, From Mark Catesby to Rachel Carson.* (Dementer Press) 1977

Pammel, L. J. *Prominent men I Have Met.* "Dr. *Charles Edwin Bessey.*" Iowa State College. 1926

Sandoz, Mari. *Old Jules.* (University of Nebraska Press)

Sheldon, Robert. *These Great Plains, an Essay.* Nebraska Magazine. Fall 1999

Tobey, Ronald C. *Saving the Prairies. (*University of California Press) 1981

Weaver, John E. *North American Prairie. (*University of Nebraska Press) 1964

Weaver, John E. *Prairie Plants and Their Environment: A Fifty Year Study. (*University of Nebraska Press) 1944

Chapter Twelve

Evolution of Grassland Ecological Thinking

Prairie country provides a reasonable landscape to consider the various passages of thinking about different schools of ecology. To track conservation thinking from Emerson and Thoreau to Pinchot and Powell, to Leopold and Carson is not a single path. There are several paths. At least one goes through the Nebraska prairie.

These differing ideas provide several of our conservation ethics. The transcendental naturalism of Emerson and Thoreau becomes an eastern United States "gentile tradition" that resulted in the urban perception of the values of wilderness, parks and wildlife. This gradually added thinking about protection. Add the esthetics of painters of western landscapes and an ethic that is more spiritual than logical emerges.

Then there is the path of John Muir, John Burroughs and others that becomes a roadmap of preservation. The original condition is the ideal condition. Maximum value is achieved only with maximum protection. This viewpoint, certainly obvious in the establishment of national parks, is the path that sometimes crosses with Pinchot, Powell and even Stephen Mather. National "protection" ideas would often conflict with national use.

As the ethics developed there were modifications. That is true of the sixth ethic that nature is a property, or condition, that can be quantified and defined and therefore can be understood and managed as part of a scientific process. Two systems would eventually emerge. One system would be based on isolation and then observation. The other is based on empirical evidence and conclusions

At the core of grassland and botanical study in Nebraska in the late 1800s were names that put a stamp on range and prairie ecology and extended it to practical application. These would be a first generation of scholars that included C.E. Bessey, Roscoe Pound and Frederic Clements. Charles Edwin

Bessey was a botany professor, and educator who pushed for college level botany at the high school level, almost an evangelist of learning about nature. He literally, by 1887, transposed the University of Nebraska "farm" into an experimental laboratory. Bessey was uncomfortable with the systematic study and classification of plants, with dogmatic theories and popular assumptions. He believed in the laboratory tested facts and the microscope. He involved himself in political moves aimed at forest conservation in the sand hills of Nebraska. He got a donation of land and seedlings in the early 1890s, then sought the help of Gifford Pinchot. In 1902, President Theodore Roosevelt created two forest reserves in the sand hills region of Nebraska.

During Bessey's early term at Nebraska, Roscoe Pound was a student. His close friend in the botany lab was H.J. Webber, son of a farmer just outside Lincoln where the University was located. Webber would later work for the USDA, the University of California and eventually would become director of their Citrus Experiment Station. Pound would go to Harvard Law School and in 1890 pass the Nebraska bar examination. A third member of this unusual group of students was Frederic Clements. Born in Lincoln, the son of a photographer, he too would become a devout follower of Bessey and an intellectual companion of Pound. It was Pound who would read *Deutschlands Pfanzengeographie*, Oscar Drude's concepts of vegetative formations and succession. This German idealism was transposed to Nebraska, first as an ecological theory and later, for Pound, as a Spencerian sociological concept of the formation of an organism.

Frederic Clements began as a student at the University where, prodded and stimulated by his friend Pound, he forged a career as one of the world's brilliant pioneers in the field of plant ecology. Some say he became the first "philosopher of ecology." His postulates on ecology would be first espoused, then challenged and later abandoned.

In 1905, Clements used doctoral work to publish his book, *Research Methods in Ecology*. He had moved science into the grass roots using meter-plot and the quadrant. String bounded, small plot identifications of plant species in a quantified geography were compared to other geographical areas of similar size and similar ecological composition. (A very similar method is being used by Oklahoma State University students on the Tallgrass Prairie Preserve in Oklahoma today.)

Clements felt "habitats differ in all degrees, and it is impossible to institute comparisons of them without an exact measure of each factor." Earlier, Clements had co-authored a book with Pound that was concerned with the geographic distribution of species and the abundance of individual species. For Clements, the two theories were linked. Phytogeography was the plant geography on the landscape. The other was his view of ecology, the numbers, the research and the comparative data.

According to Clements, habitat conditions were controlled by eight major physical factors. His concept is applicable to our model. These were the eight determinants in Clement's theory.

1. water content of the soil.
2. humidity of the air
3. light intensity
4. air and soil temperature
5. wind
6. precipitation
7. ground physiography
8. soil class. (i.e., sand, clay, gravel)

These determinants had long been studied in plant geography but had always been regarded as static or pronounced stages. Clements added the idea of continuum. This comes much closer to the Darwinian idea of evolution in biology and refutes what had been, in the early 1800s, the naturalist concept that sensory impressions were adequate for the study of biological nature. For Clements, it was more than that. The ecologist did not simply study nature, he collected and compared data about nature. He used precise measurement to reveal what he could not see.

SCIENCE MOVES INTO PRAIRIE COUNTRY

For many years, naturalism had been the code for the ecologist. George Perkins Marsh began as a naturalist. Later he would plead for science in his book, *Man and Nature.* In those times, nobody paid great attention to the book, at least not in America. Likewise, the ecological scholars on the Nebraska prairie received little attention.

Clements brought several ideas to the science of ecology. First, the idea that habitat or environment had determinants and those isolated factors could be measured. Second was his idea that a plant, or a specie, could go through stages. Plants didn't just grow, they had stages of growth. Third, he felt you could compare quadrats of habitat and compare one quadrat with another if your specie analysis was consistent. This would result in quantified stages of growth and allow for comparisons.

If a patch of contemporary prairie was plowed or denuded, the soil would not immediately heal itself by filling in with mature prairie. Instead, the patch would progress through different stages of grass, grass or shrub or weed mix. It would go through stages until the composition of mature or climax prairie returned again as nearly as possible. Clement's thinking of this change was typological and provided a tracing or a schematic that analyzed growth as a succession of changes. It was this part of his concept that would later be challenged. The great drought years of the 1930s would provide proof the prairie would not return to an original climax condition.

THE STORM STAGE HITS THE PRAIRIE

In 1880, Nebraska had an estimated 5½ million acres of land that would pass for "improved." That is to say, prairie grassland had been broken and another specie of grain or grass had been substituted. By 1900, or in the period we have under discussion, improved acreage had more than tripled. The initial breaking of the prairie soil would enable new species to emerge. The original virgin prairie didn't like newcomers. It had been tough and stable enough in the original state to resist a newcomer. The new ecosystem of plowed or broken prairie was not able to cope. In the simplicity of our model, when the land was plowed, it went back to "FORM." And in a short time, the second stage of our model, "STORM" would emerge and what would emerge would be hard to predict until it emerged. It was a battle between native seeds and newcomers.

The Russian Thistle, a bushy little bugger, was accidentally imported from Russia to the prairie. This was logical because many of the prairie settlers were coming directly from Russia, Germany, Poland, the Balkans, and Denmark and the Nordic countries. If they hadn't cleaned out their pockets real good, or scraped their shoes clean, they might have smuggled in some seed. The Russian Thistle took up residence in South Dakota, spread to Nebraska and began to create vacant farms. The old timers tell of the spines of the thistle disabling a horse. If the weed was well rooted, it might stop the smaller mechanical harvesters in use. Keep in mind, there were no sprays nor pesticides in those days. The University of Nebraska had plenty of opportunity to translate laboratory theory to actual dry land farming, if they could find any dry land farmers willing to listen.

Still, to the credit of Bessey and his students, they responded to the demands on their time as first one and then another strange newcomer showed up in the Nebraska "improved" fields. In short order, the settlers were faced with Asparagus Rust, a fungus from New Jersey, Downey Mildew went after potatoes, Stinking Smut infected the wheat. In western Nebraska a wild species of larkspur was spreading into grazing land in short grass country. What, in the reality of the grassland seemed to be a sort of biological firefighting, would turn into a statewide framework of applied research. This research activity resulted in a stream of publications that dealt more with practical issues and less with scientific theory. But the point is worth noting that what had been naturalistic excursions in botany had phased into harder science. That science, although limited in application, was being used, at least in the collection of data, to provide some basis for examination of what were contemporary grassland problems.

A STATE BOTANICAL STUDY EMERGES

This early work would evolve. First would be the collection of information about prairie ecology from impressions or observation. The second would be the comparisons of quadrats of prairie plant life. Then, Pound and Clements would put it this way. "After more than ten years of active field work on the prairies, it seemed to the sorters that the mental pictures acquired were approximately sufficient to make reference to the commoner secondary species of prairie formations to their proper grades as an easy task. When actual observation, the season permitting, seemed to confirm the picture already formed, this seemed certain. Closer analysis of the floral covering proved the conclusions formed from looking at the prairie formations, without actual enumeration of individual plants, was actually erroneous." The bottom line . . . they weren't so sure that just looking at nature, without measurement, was the way to do it.

Time stretched from 1900 to 1910. Roscoe Pound turned increasingly from ecology to law. In 1904, Pound published an article that spoke of the "sociological school of Jurists." In 1880, in his *Principles of Sociology*, Herbert Spencer had advanced the idea that "society is an organism." His idea was that society, as a whole, and apart from living units, exhibited growth, structure and function much like those in an individual body. Later that idea would be extended. Society was a compound organism that reflected the results of the individual forces exerted by its members. Those acts obey fixed laws. "If society is to progress, then men must manipulate the laws that apply to it just as they have manipulated the laws of breeding to improve live stock."

This application of the principles of ecology to a relationship with society we identify as the sixth conservation ethic. That nature is a property, or condition that can be quantified and defined and therefore can be understood and be manipulated and managed as part of a scientific process.

These ideas of Pound did two significant things. First, they set the stage for the application of science to any discussion of the use of natural resources. Just as John Wesley Powell was pushing science as part of any discussion of irrigation, so the Nebraska school of prairie ecology was pushing science into grassland ecology and sociology. And at no time in the ensuing 100 years would this idea of the application of science be more evident than from 1990 to the present time. Science and data have become the handmaiden of the lawyer as they use their many witnesses in the wide array of court cases being argued today.

The second result was a merger of the neolamarkian position of direct relationships between the environment and the organism. This would raise the social questions of the importance of the environment. Was it more important than heredity? Did environment impact the biological constitution of the individual? Or even the constitution of the society? How important was pollution? Extend these questions to the role of government. Would government intervention that altered the environment promote the progress of society? Should the government intervene? If so, at what point should government intervene? The ideas of Roscoe Pound extended ecological thinking into sociology and law.

PRAIRIE THINKING GOES DEEPER

Our thinking about prairie grassland had come a long way from the view of the sweat stained settler, only recently dismounted from his wagon, who now followed his team in the search for security and freedom in the new furrows he created in the soil. And while the nature lovers of those times might have stirred emotions with their incantations about black-eyed susans, the buffalo bean with violet flowers, the yellow of the golden parsnip, the rosinweeds and wild bergamot, the settler thought more about how many bushels of wheat he would extract from the soil. And as he squinted into an overbearing sun, he probably had little concern for the root depth of the grass he was turning under. He just prayed the soil would grow wheat or potatoes or corn. This much he was certain of. It was his land, under law, and he could do with it what he wanted.

While the Homestead Act promised the settler free land, what they found when they arrived was the best land along the rivers or the rails, was owned or controlled by the states, the speculators or the railroads, who, by some estimates, controlled half a billion acres. Consequently, most of the newcomers settled on prairie farms. What they quickly learned was how to build a sod house. In 1868, a modified version of the prairie plow was introduced. It had adjustable rods that would cut strips of sod. Fall was the best time to build. The grass roots were toughest in the fall. In 1872, a 14 by 14 "soddy" needed one window, cost $1.25 . . . wood for the door, cost 54 cents . . . a door latch and hinges, cost 50 cents. A stove pipe cost 30 cents. After a week of hard work, the finished house would last about ten years assuming the roof didn't leak in heavy rains, in which case the breakfast mush would have a different taste due to the dripping mud. The dirt floor was hard packed, the walls shaven smooth with a sharp spade, coated with a plaster of clay mixed with ashes and then whitewashed to deter bugs and rodents.

The plains wife would keep a canary for company in order to cope with the loneliness. Many women would suffer deep despair and ill health. Some would lose their mind. Otherwise, they kept the cow, the pigs and a few chickens. They would also care for the children, and if any new ones arrived in the winter, especially during a blizzard, they would pray for the care of a doctor. Until medicine got to sodbuster country, survival depended mostly on good fortune. They were tough people, stoic people, poor people. There is a story, as I remember it, of a young friend of my grandmothers whose husband had to stay in bed while she patched his only pair of overalls.

Death was common. The human system had difficulty adjusting to the ecosystem. And still they kept coming even as many gave up, moved to town, moved east or west or just moved on. There was a land rush in the Red River Valley in Dakota country, a land rush in Oklahoma. There was also a $30 thousand grant of money from Congress to enable farmers to buy seed. It may have been the first of many subsidies in prairie country.

A Reverend Olaf Olsson wrote to his friends in Sweden: "The advantage which America offers is not to make everyone rich at once without toil and trouble, but the advantage is that the poor, who will and are able to work, secure a large piece of good land almost without cost, that they can work up little by little and become after a few years the owners of property, which rival large estates in Sweden. . . . The difficulties at the outset are so great that not every person has the courage to overcome them. The best plan is for several acquaintances to settle in a tract, where they can encourage and help each other." Today, in the social fabrics of the grasslands, you still find enclaves of different nationalities. You will also find strong minds, strong opinions, and stout attitudes. Small wonder when the young man, just graduated from the University of Nebraska shows up and says, "I'm from the Soil Conservation Service and I'm here to help you" . . . he may get only a short handshake from a rough hand.

IN TIME BOTH THE ACTIONS

. . . of the settlers and the theories of Frederick Clements would be challenged, on one hand by unprecedented drought and on the other by an English botanist, Arthur George Tansley. Tansley originally espoused the views of

Clements. But where Clements was working in a temporal world of prairie, Tansley, being very British, was involved in heather, heaths and fens, habitat that was a product of succession. The habitat was hardly developmental. Tansley began to distance his thinking from Clements and finally evolved into an alternative approach that was a "system." He may have been the source for the term, ecosystem. Tansley saw nothing as being totally isolated. The act of perception was, in a sense, an act of isolation. But the abstract "isolates" were parts of a system. Even the specific science, the single organism or the single prairie grass specie was a functional part of the ecosystem of which it was a member.

Further, from Tansley's view, systems, not organisms, underwent evolution. While some systems might be more stable than others, it eventually was a case of the strength of internal elements that would withstand the strain of the external pressures. His key point, as it related to ecology, was stated as "There is in fact a kind of natural selection of incipient systems and those that can attain the most stable equilibrium survive the longest." From this concept, it was but a short stretch of logic to extend the system concept to the idea that no system was ever absolutely isolated from or neutral to the rest of the universe. The parts of the universe were connected.

This thinking moved even closer to our model. If you stop the study of the evolution of ecology at the natural resource determinant, and you do not relate it to the other determinant systems, you have an isolated ecosystem. This is very nearly what happens when a zealot argues their particular ecological point of view and ignores the other determinants. You can have the pure naturalist view with no regard for the economics involved, the social contracts nor the government interventions. But, if you assume that all things are connected, then very simply within our model, any internal or external force can disrupt the equilibrium of any determinant.

So, it was natural that as ecology matured as a scientific concept, questions would arise. To what extent should government support scientific study? Would certain conclusions drawn from isolated studies be considered conservative or liberal? What impact of certain government programs would there be on the pure science of ecology? Would government be justified in using, or misusing, scientific evidence to manipulate public opinion regarding government programs? What if private sector, non-profit conservation organizations demanded certain results? Would you have pure science or science as a tool for public discussion and political action?

Today, in our flooded world of information, it is easy to shrug off how science has become a part of conservation causes. A magazine arrives from a conservation organization . . . in my case, Ducks Unlimited and this one from Canada but it will suffice to serve as an illustration, eh? From the personal page of the Executive vice-president. "In my job, I'm fortunate to be privy to some of the conservation world's most extensive scientific research on wetlands and waterfowl. From this broad diversity of scientific studies one common theme consistently emerges; the more habitat on the landscape that's in productive condition, the healthier our waterfowl populations will be." Is this science being used to stake out a position?

SO MANY OF THE QUESTIONS

. . . raised in studies beginning in 1881 at The University of Nebraska would evolve and grow into both a body of knowledge about grassland ecology and a huge box of opinions that would today be packaged into the environmental cause. As the naturalist views of the 1800s were replaced with the scientific views of the 1900s, the cacophony of ecology views became a new game as feeling and sensory observations gave way or made room for science and numbers. Almost! This begs the question . . . can you strip ecology and the natural resources of all feeling and emotion and make it only a science game? I'm not so certain. The Native American says you cannot.

So what was learned in the drought years when soil temperatures went up to 140 degrees? The pastures were the first thing disseminated by the drought which lasted nearly ten years in most of the prairie country. Scientific efforts at the University of Nebraska were headed by John E. Weaver who, as a student, had dug over 100 pits three feet wide, six to ten feet long and six feet deep to study root systems of the prairie grassland. In 1918 he had concluded as a result of his study and research that in the event of abnormal weather, the prairie could remain fairly intact for a period of at least two years. But his work was gradually leading him to conclude that man, and the intrusion of man, could alter a natural ecosystem permanently.

"Man," he was to write, "so inextricably wove himself into the ecological fabric of nature that he could not be neutral. Either he destroyed or he promoted. The scientist had to discover by experiment or historical inference, what, on the path of succession would be isolated from man and then actively guide the path of succession. This was called conservation." To remedy the destructive coactions of man on environment, it then becomes necessary to "reverse both sets of processes . . . coaction and reaction, and turn them into agents for stabilization and control." My understanding of this conclusion would be that what man had destabilized, man would need to return to something resembling an original condition. Further, in addressing wildlife, Weaver felt that "destructive coactions must be replaced with constructive ones, and natural or artificial propagation must be invoked to restore a dynamic balance within species as well as communities." Scientific management, then, could restore wildlife. Today, that theory has been proven with pheasants, turkey and deer.

When the rains came in 1942, Weaver made some further observations and again modified his theory about grassland ecology. He reported that the old tall grass prairie of eastern Nebraska had been destroyed and replaced by mixed prairie. The mixed prairie further to the west had been replaced by short grass. In the case of the tall grass country, the little bluestems and big bluestems had been eradicated completely. In more scientific terms, the grassland formation that Clements and Weaver had once thought to be the terminal climatic condition, in relative harmony with the environment, had been destroyed and then replaced by a different set of ecosystem values, with different determinants of the ecosystem.

So what was once the belief that nature would return to an original condition had again been modified. When a conditions or a species, (and it can be plant or animal,) reaches a stage of optimum stability or what we call "high performance" . . . if all conditions of the habitat or environment remain relatively stable, any stage of optimum stability will adjust slowly. That is true for nature, it is true for an organization, and it is true for a society. If something internal or external happens to disrupt that stable condition, the entity will evolve into a lower stage. That stage can be extinction or death, or if survival is involved, the entity goes back to a "FORM" stage, reestablishes new or revised determinants in a "STORM" stage, goes through a period of adjustment and emerges as a "NORM." Most of the determinants of the system are then in a relative state of stability. But the former entity has not been exactly replaced as it was. As Weaver said in his article in the *American Scholar* in 1944, "The prairie is a slowly evolved, highly complex entity, centuries old. It approaches the eternal. Once destroyed it can never be replaced by man."

THE SOUL OF THE PRAIRIE

So to address the question I raised. Can you strip natural resources of a naturalist connotation and leave only the scientific ecology? I have stood on the Nebraska prairie, waiting for the sunrise to labor into another day. I have seen what Weaver describes in scientific terms. He describes a piece of ungrazed prairie on the protected side of the hill as having lost only 5% of its grass during the great drought. He tells of an adjacent patch, heavily grazed and unprotected by a hill, that has lost 95% of its grass. When I stood where only 5% of the grass was gone, I could only marvel at the strength, the endurance, the patience Mother Nature can have. The earth can endure a great passage of time if it has care. And when I looked at the grazed patch of 160 acres where 95% of the grass was gone, I first got angry at what man's arrogance had done. Then a lump came into my throat and I was sad. You can't measure and quantify anger nor sadness and that isn't very scientific. It's just a human condition.

You feel it, you know it. As surely as the scientist can know numbers. And that is why I am not so certain you can ever strip ecology of all human feeling. I am now convinced there is a dimension in nature that is the soul, the spirit of all living things.

I believe you call this the compassion of nature. It is treating the earth, the air and the water as you would want to be treated. Granted, the conservation, the repair or the care may be based on science and what man has come to learn about the natural resources. Nature can be very tough and even mean. But nature is not totally a machine. There is also a compassion in nature, a soul perhaps, certainly a spirit.

I believe those things will be discovered as Nebraska continues the challenge of returning the Niobrara watershed closer to a natural state. Perhaps that effort will encourage others to understand what we could have learned so many years ago from The Native American.

REFERENCES

Clements, Frederick E. *The Development and Structure of Vegetation.* (University of Nebraska Press—The Botanical Seminar). 1904
Ducks Unlimited Canada Conservator. Summer 2001 Issue.
Spencer, Herbert. *The Study of Sociology.* (Concordia Publishing House) 2000
Tobey, Ronald C. *Saving the Prairies.* (University of California Press) 1981
Weaver, John E. and Albertson, F. W. *Grasslands of the Great Plains.* (Johnson Publishing Company) 1966
Williams, Raymond. *The Country and the City.* (Paladin) 1975
Worster, Donald. *Nature's Economy: The Roots of Ecology.* (Sierra Club Books) 1977

Chapter Thirteen

Tallgrass Prairie Preserve

You turn at the Triangle Building in Pawhuska, Oklahoma. Most of the windows are boarded up. It now stands as a rather awkward symbol of times past, oil boom time. The street heads north and you see the first sign, "Tallgrass Prairie Preserve." The street heads up a hill.

Shortly, houses have a little more space between them. Then the grass commences. Houses nearly disappear. When pilots fly over this part of the prairie they call it the "dark hole." Not many lights blink below them as they head north and west from Tulsa. Even in the daylight, it is a land of shadows, ghosts, and illusions. Another sign as you drive. Preserve Headquarters . . . 14 miles.

Had you turned left in town and driven up west you would have come to the site of the million dollar elm. Ghosts lurk under the tree. In June of 1921, 14 leases sold for $3.25 million. A few months later, 18 leases sold for $6.25 million. In March of 1922, one 160 acre tract brought a bonus of $1 million. On March 18, 1924 the opening bid on a single tract was $1 million. Colonel Ellsworth E. Walters cryed the auctions. His fee was $10. Later they raised that because they liked his style. He also got $50 in expenses to cover his trip from Skidee, Oklahoma to Pawhuska. Finally, the Osage tribe was so pleased with his work they gave him a diamond studded badge. Trivia and oil mix in Pawhuska. Walter's ghost is up there somewhere around the elm where oil leases were sold.

The Osage tribe got the lease money, plus 1/6th royalty on wells of less than 100 barrels a day, 1/5th on wells that ran over 100 barrels a day. When oil was $3.50 a barrel it was $3,500 daily split between producer and tribe. The ghosts of the oil producers and bidders hunker over the land. You would suspect Frank Phillips, founder of Phillips Petroleum, is around someplace, maybe wearing the special ceremonial headdress the tribe gave him when they made him a blood member of the Osage tribe. He had treated them quite well, respected them, dealt fairly with them, and they liked him. His ghost is around.

Preserve Headquarters . . . 12 miles. Somebody had shot some holes in this sign. Locals who can't hit anything moving shoot the signs.

And the ghosts of William G. Skelly, E. W. Marland and others who often got together for some poker. Of course, Osage county was one huge poker game and sometimes the oil lease got on the table when the poker chips ran out. In the shadows are stories that are interesting if they were true and even if they weren't true, as Mark Twain would say, were still interesting. The Osage lady who spent $12,000 for fur coats and $5,000 for automobiles in one day. An Osage man who bought a car and when it wouldn't start he hitched a team of horses to it, stood in the back seat to ride in a parade. And another Osage who bought a hearse and hired a chauffeur to drive him around because he didn't know how to drive.

Preserve Headquarters . . . 10 miles. You are now fairly close to another ghost because John Joseph Mathews still stalks the wooded area to the east. He was the Osage tribal member who was sent to Oxford in England for an education with the understanding he would write the history of the Osage tribe. He did that, and also wrote *Talking to the Moon*, the spiritual journey of the Osage and a sort of Walden Pond in the Osage hills. He died being too educated for the Osage and too Indian for Oklahomans.

HERE IS THE ENTRY SIGN

. . . to the Tallgrass Prairie Preserve. We pull over. "You stand at the South edge of the largest unplowed, protected tract that remains of the 142 million acres of tall grass prairie grassland that once stretched from Canada to the Gulf of Mexico. Today, less than 10% still exists, found mostly in the Flint Hills and Osage Hills of Kansas and Oklahoma.

"In an increasingly crowded and noisy world, what you see is an oasis of space and silence. Here you can experience the same beautiful vistas that greeted the earliest human hunters and gatherers many thousands of years ago.

"This area is indeed a national treasure. Please treat it with respect."

"For additional information about this preserve and its unique history, visit the preserve headquarters gift shop another 10 miles north on this road. This preserve is owned and managed by The Nature Conservancy, a private non-profit organization."

You drive on. Skeletons of blackjack oak are scattered on the right hand side. Stark twisted black and gray remains, they provide a requiem for spraying with crystals, a page of history provided by a past tenant who wanted more cattle pasture production. He applied the chemical, the trees died, the dozer never made it to the service to clean up the tree bodies.

But there are other bodies in this land. There is the legend of Henry Foster who had the first 10 year lease on all this Osage land, some 1,470,559 acres. He died before the lease was approved by the Secretary of Interior. His brother's name was substituted. Then the Dawes Commission broke up tribal holdings, gave each headright an allotment. They could sell the land, but in the Osage Tribe they could not sell the mineral rights. Those stayed with the tribe thanks to the wisdom of Chief James Bigheart, who helped get a tribal constitution approved in 1896.

There are other ghosts. Chief Baconrind and Chief Fred Lookout were Principle Chiefs in the glory times when the Osage became the nation's richest tribe. They experienced what became known as the "Reign of Terror." That would have been around 1920. The oil conditions produced the hustlers, the hucksters, corrupt lawyers, harlots, drifters and the grafters. The usual collection of unmentionables who are attracted to shady closets. Osage leaders appealed to the Federal Government to help with the situation because it had gotten too wild and wooly. The FBI was sent in. They proved to be less than effective.

Preserve Headquarters . . . 2 miles. You turn right. You are now headed to Sand Creek. This sly, little stream has some ghosts along it too. The story is that Frank Phillips was riding along this steam one day with a friend who had told him of a black substance oozing out of a creek bank. To that time, the substance had been used for greasing axle hubs. Later, Phillips, after hitting five dry holes, would bring in his first well. The Anna Anderson was closer to Bartlesville. Frank Phillips would cast a very large shadow.

THE CHAPMAN BARNARD RANCH

We have arrived at a low ranch style building, tile roof, brick construction, big front porch . . . once headquarters of the Chapman-Barnard Ranch. At one time, 102,000 acres of land, mostly owned, some leased. Mr. Chapman the business man. Mr. Barnard the cowboy. The ranch, one of the finest operations in Osage country. Southwest of here at a small railhead, there were loading pens. They will tell you more cattle were loaded at Blackland than any other single spot in the United States. But the ranch became of a model of the best ranching stewardship. Treated kindly by the Chapman-Barnard combine, it would provide the best of opportunities for restoration when it was purchased in 1989 by the Nature Conservancy.

Since that time, the preserve has had only one Director, Harvey Payne, a lawyer by admission and a much respected wildlife photographer. The young man who manages the day to day operation is Bob Hamilton. His title is Science Director. From this point on the tallgrass story is theirs and while it is Oklahoma, it could as well have been Nebraska prairie.

I have shown Payne the model used in this book. Once the determinant of "plants" had been "grass." I asked him about the shift of terms.

"Well," he said, "that has been one of the pleasant surprises we have had here on the preserve. The return of the prairie flowers has been absolutely amazing. You should be here in May. You've never seen so much color. Many of the long dormant flowers apparently had seeds or root systems in the prairie. With burning and the bison on the land, they have really made a comeback."

"You remember," he continued, "we did no burning the first two years. We let the land lay dormant except for some limited cattle grazing. We have been burning since 1993. We are now approaching 400 burns. The first year, we burned some 23,000 acres. The Caddo, the Wichitas and the Pawnee probably all burned grassland. One cowboy, 'Doc Ramsey', who was here in 1890 talks about the land here burning all the way to the river (Arkansas). He also

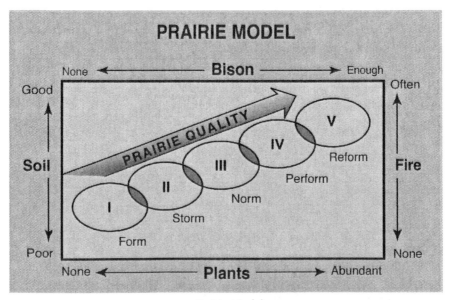

Prairie Model

said he can remember running the antelope with greyhounds. That's the way they hunted them then."

I asked him after the burn, what came back best? He explained the big bluestem grass came back first. Then the rest of the dominant grasses, followed by the Indian Switch grass and the little bluestem.

ROUND UP TIME

It was November. Roundup time for the bison. It's dry. Grass rustles in fall wind. I remember the story an old cowboy told me at the Cowboy Reunion held in early times on the preserve when some of the neighbors believed The Nature Conservancy would turn the land over to the Federal Government. The old timers put their stamp of approval on the preserve. One such oldster told of times when he would ride out into the grass which came up to his saddle. "You could," he said, "get your ass good and wet just checking on the cattle and never getting out of the saddle."

Just south of the preserve, sandstone is the predominant rock beneath the soil. Coming north, this rock formation turns to limestone. The soil is not very deep. Consequently, little of it except in the valleys was ever turned for agricultural use. The limestone provides the nutrient that produces the unusual quality of grass. It is gently sloping country. Some ravines lace the landscape, but essentially it is grass. This is the natural ecosystem.

"One of the things we're doing here," Payne says, "is burning in the summer, fall and the spring. Traditionally, prairie burning had been done only in the spring. That had been the rancher's pattern for many years. They were afraid of off season fires. Since we have been around with modern fire control equipment, we have been able to burn on the preserve in all seasons. Burning in the spring favors warm season growth. The grasses adapt to that. Burning in the fall and summer favors the forbes and broad leaf plants. 98% of a bison's diet is grass. 2% is broad leaf. That explains the profusion of flowers we have now."

THE NATURE CONSERVANCY IS WORKING

. . . closely with two universities, Oklahoma State in nearby Stillwater, Oklahoma and the University of Tulsa in Tulsa. Dr. Michael Palmer from OSU is working on the plant community. He has established more than 250 study plots. He uses a pattern of 10 meter squares. His work is what causes the staff on the prairie to smile and explain they are part of a giant "numbers game."

They have done a plant census. They have located 750 different plant species. Originally, they expected only about 250 different species. Then add another ecosystem in the soil.

Other scientific work being done on the preserve, in conjunction with Tulsa University, involves the cleanup of oil sites. Kerry Sublette, a chemistry pro fessor at the University heads up this effort. Their goal was to halt erosion on old oil well sites and restore vegetation. An early attempt, funded by the E.W. Carter Oil Company, had been disappointing.

The work started in 1997 with two remediation steps. They put in a drainage system. It worked after three years. The site was again covered with grasses. They also learned this method was expensive. They tried a bioremediation system. They tilled hay into the soil. They found it might take several tries at tilling the hay into the soil but eventually it worked. And it was less expensive. Sublette was quoted as saying, "We really think it may be a two step process. First you try the hay and the tilling. If repeated hay applications fail, then you resort to the sub-surface method of drainage."

Both applications may have merit. The processes illustrate again how easy it is to disrupt nature and how difficult it is to heal the wounds. Petroleum reservoirs typically contain a mix of hydrocarbons and salt. During the production or drilling for oil, soil around the well becomes contaminated by salt water. This condition kills the plants around the well. Over time, a scar on the land results. Over 100 wells on the preserve attest to a hope by the Conservancy that drilling and prairie conservation can find accommodation. The same could be said for the entire state where oil well contamination presents cleanup challenges. What have they discovered to date? Spills can be cleaned up. But the effort is pricey.

OVER A TWO YEAR PERIOD OF TIME

. . . I have plied Bob Hamilton, the Science Director on the preserve with dozens of questions. We have talked during the noon hour when he was taking a break from tagging bison or at a quiet lunch in a Pawhuska cafe that featured carrot cake. I asked him what surprises he had encountered in the seventeen some years he had been in charge of the preserve. These were his responses.

"How vigorously the bison graze on a new growth area. That would be one. Then, the effect of late burning. How dramatically it affected the herbage. The impact on the wild flowers. Then the breadth of the interest in what we are doing here on the prairie. Alan Alda was here to film for the Scientific American Frontiers. They did an 18 minute segment, the longest they have ever done. Peter Having, the Dutch photographer, said this was the most exciting filming he had ever done. We shared those segments with storks and otters. They sold rights to National Geographic."

"I guess," he continued, "the discovery of 750 different species of plants, and that doesn't count the mosses nor lichens. Then the response we have had from the scientific community. We have had, or have, over 50 different projects on the preserve. It has become truly an outdoor laboratory."

Hamilton went on to explain how the agronomy people at Oklahoma State University had taken the lead in studies that compared cattle weight gain on patch burn pastures to all-burn pastures. They were now trying to export the idea of patch burning to the ranching community. The patch burning would produce wildlife benefits that whole section burning would not.

People who see bison usually have seen them from a distance. They appear to be rather docile animals when grazing, emitting an occasional soft grunt that is more a grumble. They take on an entirely different character when they are up closer. Then you better have a good iron fence between you and the animal. Hamilton explained their grazing patterns as "you never know." They will sometimes be in what he called "bands." Those might be mostly cows with calves or young bulls. During breeding season, which is July and August, they are generally in two large herds. "Then a thunderstorm comes up and splits them up again. You just never know."

Bison males are generally around 2000 pounds at maturity. They can run up to 35 miles an hour. Calves weigh about 50 pounds at birth. The average grown bison will consume 30 to 50 pounds of grass daily. They prefer fresh green grass that comes up soon after a burn. To date, on the preserve they have experienced about an 80% calving rate. They maintain a herd that has a ratio of 1 bull to 8 to 10 cows. Those are numbers you can come to know from safe observation at a distance. Roundup is less safe.

And it's roundup time. The bison have been rounded up in a large pasture west of the corral. The corral is a maze of steel pipe, most of it donated to the preserve by oil companies and trucked in by Arrow Trucking Company. The maze covers what would be a city block more or less. The corrals narrow into a long chute where the bison are lined up single file. At the end of the chute there is a narrow enclosure that has a collar that pinches the animal so only the head is exposed. Perry Collins, preserve ranch hand, and one of about 15 cowboys working the roundup, takes a long pair of special tongs and goes to the head of the animal which is by this time slightly unhappy in the strange surroundings. Collins grasps the nostril of the bison, the most tender part of the animal, with the tongs. It stops kicking and thrashing. It's now quiet in this location of the chute.

Bob Hamilton takes a wand and waves it past the ear of the beast. This one already has a transponder pinched in so she has been there before. Nearby is a small metal building. Kay Krebs, of The Nature Conservancy staff, sits within earshot of Hamilton. She reads her laptop computer and shouts a number to Hamilton. That number will be the year of the animal's birth. In this case, it was 98. "Three years old," he says, "and she weighs a 1000 pounds." She is about five years ahead of schedule. "She is a big girl," a helper in the shed says and hands Hamilton another transponder to attach to the cow. All this is part of the numbers game of the prairie. It is hi-tech with transponders, computers and special software programs. As little as possible is left to guesswork. The ultimate goal is a herd free of disease.

According to Harvey Payne, the herd is now some 2100 animals. When the original band of bison was released to storm out across the prairie at a special ceremony, there were about 300. They were a gift from the Kenneth Adams family in Bartlesville, Oklahoma. Originally, the plan called for a herd of about 2200 bison. "What we have learned since then," says Payne, "is our land will probably handle about 3300 animals. That is the thinking now." I asked Harvey about brucellosis, a disease of bovines, cattle, elk, and deer. At one time it was more a fear of cattlemen than it is today. It is a form of undulant fever identified by a Scot physician named Sir David Bruce, hence the name brucellosis. "There is no brucellosis in this herd," said Payne. "We inoculate the calves and they are clean from then on." "And the fears of the ranchers?" I asked. "That's pretty much died away," he said. "In fact, the opposition to the preserve has pretty much faded away."

He went on to explain that the three big challenges for the preserve are maintaining a healthy herd, the maintenance of the preserve fences and marketing the culled bison. In the year 2000, the Conservancy sold only 70 bison. But he said that in another six years, that number will probably be more like 1,200 head a year. This year the older cows, those born in 1985 and '86 will go to auction. When the bison reach the age of about 12, their teeth begin to wear from eating prairie grass which is high in silica, the abrasive that grinds on their teeth. Another theory has been proven. A bison can live to be 25 years old, but is not a prime animal in later life.

Some of the young bulls, those from the years '98 and '99 will be sold for slaughter. They will be sold as "grass finished bison." Selling the surplus bison will help to cover an annual budget of $700,000 needed to run the preserve. The Conservancy still has a lease with one cattleman who runs an intensive grazing program for 90 days each summer. Hamilton explained that, "The original plan was to bisonize most of the preserve, eventually eliminating all cattle grazing. Now, with the exciting cattle patch burn results we've had with OSU, the new plan is to cap the bison unit at 24,000 acres and keep 11,000 acres in the cattle burn research."

Back in the chute another bison cow has been introduced to the hi-tech world. She stands quietly, then Collins gets behind a gate, reaches in, pulls a lever and another animal joins the herd. Grunting and kicking and thrashing. She searches for freedom on the prairie. The roundup takes four days. It is not a time for the weak of heart . . . or back.

THE ECONOMICS

As I watched another bison take off from the chute, I couldn't help but recall discussions with Bob Hamilton. He and I shared adjacent offices when I served the Conservancy as Development Director, helping to raise $15 million dollars to get the project off the ground. It was a big shift in philosophy for TNC, from specie focus to a broad ecosystem thinking. Some people had lots of answers. This year, about 11 years later, there was less certainty and more open-mindedness to the options that might be coming down the road. Not all the options involve science. Some involve the economics of a big operation.

The Conservancy continues to pay taxes on the land so each year they make an economic contribution to the county tax base of about $25-30,000. There is little nor any government involved. The county has been cooperative in getting better roads to the preserve. The Osage tribe has become a solid supporter of the project. Local people have accepted the preserve. Visitor counts at the headquarters are hard to come by because some people do not sign the guest register and some are repeat visitors. Each year there will be visitors from nearly every state in the union and oftentimes as many as 20 foreign countries. There has been some economic impact in Pawhuska. There are a number of bus tours that stop to see tall grass and bison.

Focus for a moment on the natural resource. Totally, the preserve probably represents an investment of $25-$27 million. This does not count all the contributions of pipe, road building, and oil well site restoration . . . probably another $2 million conservatively.

STEWARDSHIP IS NOT CHEAP

One of the premises of this book, a point I hope we will prove, is that stewardship of any natural resource is not cheap. It costs money to maintain what is already there, like a national forest or a Yellowstone Park. It costs a lot more money to restore something, even grassland that had survived because of good stewardship. Stewardship is a not a free ride. It has been one of those things we preached a lot about but didn't always put the money where the mouth was. There is a price for conservation. But adopting a sustainable conservation ethic costs much less in the long run than raping natural resources and then trying to put them back.

The operation of The Tallgrass Prairie Preserve is not cheap. To the thousands of individuals, companies and foundations who have donated money, and many still do that, the cost is worth it. It has cost the taxpayer virtually nothing. But the real value of The Tallgrass Prairie Preserve may go much deeper than just the economics. It may be that lessons learned on the vast stretches of grass will unfold new understanding of how to use the land. It could have implications for land use by ranchers, better balance between wildlife and cattle, better methods for recovery of oil stained land. Those values may not be harvested in the immediate future.

I asked Bob Hamilton, who has been so close to this project for nearly 12 years, to tell me what he now envisioned as the purpose of the preserve. It was a tough question. His response was part answer and part reflection. "Our purpose is to create habitat niches," he said. "In the past, bison management could focus on one specie, but you wouldn't have the variety of other species. So we are trying to simulate how it was . . . to create the mix. We are creating a mosaic. It will be as nearly like it was as we can make it."

HOW WAS IT?

. . . way back when bison and prairie chicken shared the space? From George Catlin, chronicler and artist and his notes on display in Gilcrease Museum. They are dated 1841. "I was lucky the other day, with one of the officers of the garrison, to gain the enviable distinction of having brought in together 76 of these fine birds, which we killed in one afternoon, and although I am ashamed to confess how we killed the greater part of them, I am not so great a sportsman as to induce to conceal the fact.

"Seeing the prairie on fire several miles ahead of us, we found these poor birds driven before its long line, which seemed to extend from horizon to horizon and as they were flying in flocks or swarms they would from time to time fill the air. They lit in great numbers in every solitary tree and we placed ourselves under those trees as they settled in them, and shot them, sometimes killing five or six at a shot."

Then words from some of my notes made for a talk given at the fifth anniversary meeting of the Oklahoma Chapter of the Nature Conservancy. It was an evening dinner meeting. "On this weekend, some of you will arise early and venture forth to search for the prairie chicken. And much as the prairie is searching for a new life cycle in this present year of Conservancy ownership, the chickens will be resuming their search for a new life cycle. If the dawn is very still, you will see and hear the little drummer, filled with desire, as he comes to the April grounds. And you will see him strutting, stomping, stepping, pivoting to boom and jig, a little thumping dance. You will marvel at the ethereal sound, the displayed pomposity. And you will watch a ritual that is eons old, the transaction that leads to another generation. And you will say, 'how was it in times past?' And perhaps wonder what right do we have to break this circle? Is it our role that we should save this ritual for future people to see so they might marvel at this dance on the prairie?"

There is documentation of millions of these birds in tall grass country, from northern Oklahoma, across Kansas and into Nebraska. Some sources estimate a prairie population of 2 million. By 1990, when the U.S. Fish and Wildlife Service began their deliberations about declaring the bird an endangered species, the population had swooned to 50,000 birds. In Oklahoma, where birds had once been observed in 16 counties in the 1940s, they could be found in only 7 counties.

In 1999, the Sutton Avian Center located in Bartlesville, Oklahoma started research on prairie chickens. Earlier, they had been very successful in restoring a bald eagle population. The staff would capture the chickens, fit them with radio telemetry and then track their travels. Steve Sherrod, the center's executive director, then made an interesting point. "Nearly all prairie chicken habitats are predominantly on private land, so you've got to have the landowners working with you. Today, political habitat can be as important as environmental habitat." We would call that "the right social contract."

SO THE SEARCH GOES ON

"Where were the prairie chickens?" Tracking, bird counts, blood samples, coordinating work with researchers in other states have all been used to find the answer. To this date, the reasons for the decline are as elusive as the wind that rustles the grass. Theories abound. What has been concluded as of this date is their population has "stabilized" and that's about it. Meantime, the U.S. Fish and Wildlife Service still has any designation under review. Of course, reviewing chickens is tricky because you don't want to count them until they are hatched. As we have said elsewhere in this book, a policy from a government body takes a long time to reach any sort of consensus.

Prairie chickens, and their habitat, may turn out to be as complex as is the salmon story. Daniel B. Botkin traces that cycle in his penetrating book, *Our Natural History*. In the chapter, "Anchovies, Salmon and Single Factor Thinking," he provides an excellent summary of the complex issue the United States faces in restoring the salmon to the rivers of the northwest. It is not a case of "single factor" thinking. It involves habitat, stream quality, the instinct of the salmon, the lack of baseline figures for salmon population, the density of trees along the streams, a myriad of factors that are involved."

You try to place the Tallgrass Prairie Preserve in some perspective against a bigger picture. Many of the answers to any balanced ecosystem are still shadows and ghosts. Tall grass prairie is a land of shadows and ghosts.

We are leaving the preserve now. There are no signs saying "Whoa, you missed it. Buster's Barbeque." . . . or ones so common around resort areas, "Thanks for coming. Drive safely." . . . there is only dust kicking up from the limestone road and now and then a Red Tailed Hawk perched on a post.

Many people have come down this road, with different ideas, motivations or agendas. When you sift through the dust devil swirling across the road ahead of you, about all conservationists have in common is a vague feeling we are responsible for the earth and its occupants.

There is a sense from just being around the prairie it is important to us. The prairie has lessons in space, time and cycles that hold clues to environmental solutions. That is the logical side of how you feel when you leave the Tallgrass Prairie Preserve. But there is an emotional side also.

If you have been around awhile, you carry a little garbage with you for the things that have been done to the earth and maybe even share a guilt trip for things left undone. We sometimes get a little frustrated because what we do takes so long. We try to emulate engineers building bridges but their spans are completed faster than ours. We try to be like the scientist who develops a new serum but they seem to get governmental approval faster than we can. We try to be like a doctor setting a broken leg, but leg bones on a football player seem to mend faster than most of the broken things we try to set right.

Our American arrogance that we can build anything, fix anything, discover anything if we just throw enough money at it does not always seem to fit when it comes to nature. When we reflect on what we do to restore or preserve a diversity of nature, we are somewhat numbed at the scope of the task. We may be much like that dust devil chasing across the road, only to disappear into space. A flurry of activity that lasts only briefly. At our best, we are somewhere between muddlers and meddlers, manipulators and moralists;

well intentioned substitutes for the real thing. Perhaps when we deal with nature, our best posture is one of considerable humility.

We are off the limestone road. It has been replaced with asphalt. Just a little bit ago, it was dirt changing to gravel. The sign by the road says "25 miles per hour." It is a short step from the eternal to another world.

Aldo Leopold

Was born in Burlington, Iowa, January 11, 1887. He attended the Lawrenceville School in New Jersey and in 1909 received his Master's degree in Forestry from the Yale University School of Forestry. He then served nineteen years in the United States Forest Service, ending up in the Forest

Aldo Leopold. Photo courtesy of the Aldo Leopold Foundation Archives

Products Lab in Madison, Wisconsin. In 1928 he left the Forest Service and in 1933 was appointed Professor of Game Management at the University of Wisconsin-Madison.

He was a founder in 1935 of the Wilderness Society and today the Aldo Leopold Wilderness is in the Gila National Forest in New Mexico. He is considered to be the primary thinker behind the modern wilderness-conservation movement. He was the first conservationist to speak in terms of ethics. "A thing is right," he wrote, "when it tends to preserve the integrity, stability and beauty of the biotic community. It is wrong when it tends otherwise."

His book, *A Sand County Almanac* has been read by millions and is one of the most quoted books in conservation. He believed that obligations have no meaning without conscience and the social conscience needed to be extended to land. He elevated conservation to the level of religion and philosophy and argued that without this spiritual dimension conservation had been minimized. He proposed that education was the ultimate solution more than written laws and regulations. "Conservation is a state of harmony between men and land," he wrote. Self interest did not always address the ethical questions. His mark on conservation has been enormous.

Leopold passed away April 21. 1948 in his home in Madison, Wisconsin.

David Brower

Was born July 1, 1912 in Berkeley, California. After a troubled childhood, Brower entered the University of California at the age of 16. He had to leave school in his sophomore year because of emotional and financial problems. He was befriended by Ansel Adams, the great wildlife photographer, who encouraged him to join the Sierra Club in 1933.

What he did. Served from 1952 to 1969 as Executive Director of the Sierra Club. In 1969, he founded Friends of the Earth. In 1981, he founded the Earth Island Institute. He helped create national parks in Alaska, Cape Cod, Fire Island, Kings Canyon, the Redwoods and North Cascades.

What he opposed or stopped. He kept dams out of Dinosaur National Monument, the Yukon and most notably, the Grand Canyon and the Echo Park dam. Some sources use the figure of $7 billion in encroachment construction he stopped.

His use of large, well produced newspaper ads changed the appearance of the Sierra Club from a reactive organization to a proactive group, so much so it lost its tax exempt status and Brower lost his job as Executive Director. He felt neither the National Park Service nor the Forest Service could be depended upon to protect wilderness areas, that both would cave in to political

David Brower. Photo courtesy of Earth Island Institute.

pressure. He saw the forest as a "source of pure water, a perpetual wildlife habitat, a genetic reservoir, a place to learn."

To those who disparaged him as being too emotional, his response was, "After all, people who don't believe in emotion can be thankful their parents didn't share that problem."

Brower lost his battle against cancer on November 5, 2000.

REFERENCES

Notes from many visits with Bob Hamilton, Director of Science and Stewardship.

Notes from Conversations with Harvey Payne, Director of the Tallgrass Preserve.

Personal observation of the preserve and adjacent prairie.

Visits with agronomists from Oklahoma State University and Board Member, Dr. Jerry Crockett.

Visits to the Osage Tribal Museum in Pawhuska, Oklahoma.

Grass, The Publication. (The Yearbook of Agriculture 1948. USDA)

Franks, Kenny. The Osage Oil Boom. (University of Oklahoma Press) 1989

Madsen, John. Where The Sky Began: Land of the Tallgrass Prairie. (Sierra Club Books)

Marlin, James C. History and Ecology. Studies of the Grassland, (University of Nebraska Press) 1984

Mathews, John Joseph. Talking to the Moon. (University of Chicago Press) 1945

Smith, Annick. Big Bluestem. A definitive history of the land that became the Tallgrass Prairie Preserve. Published by the Oklahoma Chapter of The Nature Conservancy. 1996

Personal Articles. Published in news releases and The Nature Conservancy Newsletter.

Chapter Fourteen

Burning

Strangely, one of the ecological connections that stitch our three case study areas together is patterns of burning. Burning was an early ecological problem in Vermont. Prairie fire was a menace to the settler. Today, forest fires are an earmark of the arid west.

To the settler and the pioneer, the forest appeared as a disarray of wood. The newcomers brought European patterns of boundaries, surveys and corners. They would employ burning, the ax and the plow; the claim, survey and deed would be instruments that resulted in the boundaries of ownership. Their goal was to clear the land. This would often cross natural boundaries. What would begin in the New England area would lay the groundwork for years of ecological confusion.

On visits to Vermont, I could locate no validated virgin forests. I was told there are some very old, virgin pines in the Shawangunk Mountains northwest of Manhattan in New York, a large stand of hemlocks and beeches in western Pennsylvania, scattered small vestiges in the Great Smokey Mountains. Pure, virgin forest is just hard to find in the eastern half of the United States.

The areas that are fairly well validated are in New Jersey and North Carolina. Both remained intact more because of a freak than because of any conservation effort. In the case of the Hutcheson Memorial Forest near New Brunswick, the 65 acres of untouched farm land remained in the VanLiew family from the time of settlement in 1701. This is Piedmont land and while there were farms all around the forest, the VanLiew clan kept the forest in the original state from the Revolutionary war to World War II. Finally, in 1950 the family decided they ought to do something about the trees because of their value as timber.

News leaked out they might sell. Conservation organizations got busy. Money was raised in 38 different states. It wasn't enough. Enter a most unusual source to save the trees, the United Brotherhood of Carpenters and Joiners. They bought the property and turned it over to Rutgers University. Rutgers gets the right to patrol and protect the outer edges of the forest, an area of about 75 acres to which they have recently added 150 acres. The Nature Conservancy is trying to add another adjoining 200 acres. To this date, there has been no controlled burning on any of the area. Rutgers can only monitor the virgin forest. If it is invaded by insects or disease, they watch and take notes as they did when the gypsy moth visited the forest. This small bit of forest will remain as sacrosanct as man can keep it.

The other major virgin forest in the eastern part of the country is in North Carolina, north of the famous Biltmore Estate, a site that has a role in the story of ecology in the United States because of the work of Gifford Pinchot. Pinchot inherited the Biltmore experiment begun by Frederick Law Olmsted. Originally it was designed to train local lumbermen to log selectively. The hope was this would produce "sustained yield."

LITTLE SANTEETLAH CREEK

. . . is near Robbinsville, North Carolina. Just north and west is the largest example of remaining virgin forest east of the Mississippi River. The map near the beginning of a figure 8 trail says it is 3800 acres. Many years before it became the Joyce Kilmer Memorial Park and came under the protection of the Park Service, the Babcock Land and Timber Company had timbered along nearby Little Santeetlah Creek. They floated sugar maple and tulip tree logs down the Little Tennessee River. They had worked primarily around Slickrock Creek. A sign says it's still there and the rocks still look slick.

For some strange reason, the lumbermen chose to work west along Citigo Creek. This was lumbered until a major fire swept through. In the early 1930s, what is now Little Santeetlah Forest, sat in the middle of a vast area of timbered country. The lumbermen would look at the valuable wood, but whether it was depression time, too much money to take the wood out . . . whatever, this small bit of virgin forest was just sitting there when the Forest Service came looking for a memorial to honor Kilmer.

There you find a spiritual aura, a quiet . . . no loud voices in this old forest. It is a maze of poplars, hemlock, chestnut and oak. Stand in the forest and you get a kink in your neck looking up . . . way up . . . at majestic, aged trees. Some are 20 feet around the base. They make a canopy of trees. Your attention shifts to the forest floor. Small trees are growing beside a downed, rotting tree. Where there is an opening in the canopy, a younger tree may race against time trying to beat the canopy closing. Fungus and herbs hasten the decay process. Upturned roots, often covered with dirt, tell of wind that felled a whole tree, breaking smaller trees as it fell.

Two thoughts came to me as I scribbled in a notebook. The first was that virgin forest is not unlike virgin prairie. While there is great display and great life above the ground, there is another world on the ground and beneath it. When you become more fully aware of where you are, you begin to pick your way very carefully amongst all that silent life. You come to realize why the term, "virgin forest" . . . it just seems that not much has been violated. It is the way Mother Nature intended it to be.

The second thought that came to me was the value of these 3800 acres of forest land on today's market. You simply can't find much virgin forest in the eastern part of the United States. Wood was too marketable! It was converted to some form of utility. For the oak tree it was barrel staves or queen posts that braced the roofs of the wooden houses of that time.

Small wonder the staunch settler went after the timber with saw, ax and determination. Natural resources would receive little consideration in the face of the economics. The wood in the forest was too valuable.

PATTERNS ARE PUSHED WEST

The native patterns had revolved around permanent villages. But these moved from time to time. So did burning of forest land. And even if this was only a rather crude type of selective burning, it was effective. This same practice was used by the Indian across Ohio, Indiana, Iowa, Missouri and into the prairie country. The first purpose was to open up patches of land for agriculture, for the native corn, beans, squash and other staples. The by-product was the creation of the "edges" that would provide high quality habitat for birds and animals. It was a mosaic of burned areas that created the "edges that attracted wildlife and game."

These very early patterns, from what we can learn from old journals and settler notes, provide evidence of the effect of burning on the ecosystem. What happened in the Vermont ecosystem in terms of settler burning and clearing, with the degradation of the ecosystem and the resulting erosion of soil and runoff into streams and rivers would set a pattern as population moved westward. In the southern part of the U.S., the pattern was burn, clear, plant tobacco or cotton, exhaust the soil . . . move on. In the prairie country, it would be clear, plow, stop the burning, exhaust the soil, move on or abandon the farm. In the western part of the country, it would be reduce the forest to bare ground and move on. Burning was used by Euro Americans as a means to an end, not as a means of ecosystem management.

Burning can impact the land and the attendant ecosystems in several ways. Some are direct alterations, some indirect. First, if the burning is in a forest, it will clean up brush and undergrowth. This will open up the under story. The burning can also remove a canopy of trees and allow sunlight to hit areas that had previously been shade. This impacts water qual ity where the forest shade protects a river. Burning will replenish nutrients in the soil in both the forest and the prairie.

In Vermont, grapes often grew in the trees. They had started as vines, close to the ground. As the trees grew, the grapes grew up with the trees. When the trees were cut, the grapes were cut too. Another source of food was gone, for humans and animals. Grapes are a good example. The forest was not just wood. It was a composite of many things, not the least of which was fruit that provided food. This same situation is true today in bear country when forest fires or drought reduce fruit supplies. The bears then move around to find food. Sometimes they look in garbage cans.

Another result of burning was species substitution. The newly burned areas would attract wild animals. With settler intrusion, these animals were replaced by cattle, in some cases pigs, and in a few cases sheep. Often this resulted in denser population. This created more destruction of soil and the shrubs in the under story. All this created another dramatic change in the dynamics of the ecosystem. This is part of the Vermont story.

TODAY, IN PRAIRIE COUNTRY

. . . the common wisdom is that spring burning will increase the quality of the grass. The quality of the grass is not the issue. The real concern is the nutritional value of the grass. How much weight gain is produced for the cattle? The concern is for the cattle, not the grass and the land.

Weight gain per animal has become the new computer game in prairie country. And most farmers, small feedlot operators or ranchers will be involved in some sort of computer program that enables them to carefully track weight gain. If burning produces more weight gain, there will be burning.

The point is fairly clear. Burning can be good or bad. It was bad for the land in early Vermont. Erosion quickly reduced the quality of the land. Whether burning is good or bad for prairie country depends largely on how future prescribed burning is implemented. Burning in the forest lands of western states awaits some type of divine intervention. Man can't seem to get it right. Burning produces ecosystem benefits. But when economics becomes the driving force behind burning, the impact on the natural resource can be severe.

There were, and are, other subtle ecological results from burning. It does, no question, increase the rate at which nutrients are recycled into the soil. If burning is only in the spring and fall, the non-woody grasses, herbage and shrubs grow more abundantly. But the soil, over time, tends to get drier. This happened in Vermont.

But even more important to the ecology, and a point largely missed in scientific study until just recently, has been the creation of the "edges."

We didn't seem to fully understand that those areas, great or small, whether between woody areas or grassy areas, become the belts of ideal habitat for wildlife. Today, the practice of burning whole sections of land, fence line to fence line, may be one of the reasons for the demise of the prairie chicken or the quail. The jury is still out on that one.

There was one other significant change in our land ethic as the immigrant and settler became the dominant factor. The newcomers tended to use land on a year round schedule. Many times, when crops came out, cattle were moved in. There were fewer periods of rest; there was more crowding of production into the land. Economics were at work. Taken from this viewpoint, the struggle between the Native American and the newcomer was not only a struggle between cultures and their respective social contracts, it was also an ecological struggle. The central conflict would be how people used the environment and natural resources. It would be a struggle between two rather different ways of approaching the relationship with nature, even to how the seasons of the year would be utilized. Indians even burned differently and for different purposes.

An understanding of the differences in ecological culture can lead to a better understanding of the long and painful struggle between the Native American and newcomer. Was wasteful extraction necessary to produce an economic dynasty? John Locke, the English philosopher put it this way. "The lure of gold then becomes a single step to wealth, not encumbered by the labour or any husbandry of the land."

Where we have come from is important only if we are trying to figure out where we are going. If we really are hunters comparing maps in a truck the philosophy of John Locke and Jacque Rousseau may not be relevant. But, our government borrowed heavily from them in choosing how humans would relate with their government and with each other. Neither of these thinkers had a great deal to say about the use of natural resources. Neither did our original constitution. Economics, not nature, is the driving force behind our government. We need to understand this to deal with today's divisiveness. We are back to interests versus ideals.

ECONOMICS RAISES ITS UGLY HEAD

Our government says we have two uses of land. We have private property and we have communal property. Any time we shift from one concept to the other, different agendas for the use of that land will emerge. If land is moved into the realm of private property, then social contracts of the people must evolve. Outside urban areas, there are very few laws governing burning. It is mostly social contracts. But both our social contracts and our government have favored progressive production, extraction, resultant waste for the sake of progress. Pursued to the logical end, burning is determined more by economics than by long term benefits to the land.

This reality is hidden in the Ten Ethics we have suggested. It explains much of the confrontation we have experienced in developing any sort of consensual land ethic. You can blame John Locke and the founding fathers for some of this. Later thinking following Locke was that of Marshall Sahlins. He would propose two ways to get rich. Either the existing social contract would be based on producing more or desiring less. The Native American found happiness in desiring less. The Euro American thought the Indian was lazy. He believed in "The American Dream." That dream was based on consumption, not conservation nor sustainability.

A SHIFTING SCENARIO

It would seem a reasonable assumption that the best measure of ecological stability is the ability of that system to successfully cope with environmental change and still maintain the stability to reproduce itself. Renewal cycles simply must be a consideration of stewardship.

Burning is a determinant that leaves a mark on the land. We assume it did that on "virgin land." We don't know that for sure so we have no absolute benchmarks. Frances Jennings, in his book *The Invasion of America*, refers to land, not so much as virgin, as widowed. It seems to me it is more divorced. It had been in some state of harmony and no longer enjoys that relationship. If early settlement patterns didn't divorce themselves from natural resource patterns, they simply didn't consider a possible state of compatibility.

The forest, however imposing, even hostile and wild as it may seem, is a more fragile ecosystem than the prairie if you consider the natural state as the initial ecosystem. Burn the forest and how it renews itself is a slower process unless species substitution produces a new ecosystem. Experience is now showing benefits of more diversified planting. Burn the prairie, a much tougher ecosystem, and within months, if moisture is present, the prairie produces growth that will soon put the original ecosystem back into some state of reasonable balance.

Put man into either natural equation and you produce a new set of dynamics. In the natural progression, there are no social contracts. But if man is involved, what happens to the burned forest becomes important. The early social contracts that related to forests related more to general agreement on consumptive purposes. Even Thoreau, while he squatted on Walden Pond, did his fair share of sawing and chopping.

This consumptive social contract was the rule until Yellowstone National Park was established. In the early 1900s, more feeling for the western forests began to emerge. And initially, it was a feeling based more on esthetics. The value of "saving" the Yellowstone area began as a grass roots movement. Fires in Yellowstone Park were not part of the discussion.

In fact, how the park would be managed, if it would be, did not play a major part in the early strategy for preserving the park area. The park was saved, with its forests, so others could see it, enjoy it and use it. This was a shift in our social contracts with the forests.

IN LATE AUGUST OF 1910,

. . . fires in Idaho and Montana would cause another shift in forest social contracts. Driven by hurricane force winds, 3 million acres of forest were consumed and 85 people killed. Stephen J. Payne, in his book, *Fire In America : A Cultural History Of Wildland And Rural Fire*, writes "the fire swept 30-50 miles across mountains and rivers, oblivious to natural barriers and terrain." The fire was to become a seminal event for federal forest and wild fire policy. The "Big Blowup" as it came to be known, created an opportunity for forestry people to establish a policy of fire suppression that would remain the cardinal policy of The Forest Service, enshrined in the public consciousness by Smokey the Bear.

The year 2000 ushered in another century and very possibly a new way of thinking in forestry circles. Massive and stubborn fires hit in nearly every western state. "We have a major problem from an ecological standpoint."

That from Ron Dunton, national fire program manager from the Federal Bureau of Land Management. "We need to start work to bring back the natural balance." The Forest Service then goes to work on what it calls "a cohesive strategy" for treating millions of acres of unnaturally crowded forest. They would propose to use mechanical methods and selective burning. They would also burn up a lot of money. They estimate the cost would be $825 million a year for 15 years. But fighting fires also costs millions of dollars.

Millions of board feet of timber are lost. The tradeoffs are enormous. There are no easy, single solution answers.

The reasons for this policy shift stems from two reasons. Frank Carroll, a long time Forest Service employee now working in the private sector is quoted as saying, "My feeling is this could be a watershed year. It's a rude awakening, a real sharp kick in the butt telling us we have to do something. We're like alcoholics taking a drink thinking we're going to get a different outcome."

There was also some different language from some of the environmental groups. A spokesperson for the National Wildlife Federation representing the Rocky Mountain Area admits "thinning has a place, timber harvest has a place." Other conservation groups have admitted the need for different policies. What has changed their view? Could it be that when fire got close to Santa Fe there was more concern for housing? In 2001, when it got close to housing near Jackson, Wyoming . . . expensive housing . . . it certainly changed the strong positions of some of the more vocal conservation groups in that area. Maybe if burning gets close to home it changes perspectives? It's easier to be a zealot from a distance.

One lesson we learn from burning is no two burns are ever exactly the same. No two ecosystems are ever exactly the same. What is true for Vermont is not true for a forest in Brazil, the forests in Mexico that host Monarch butterflies, or forests in Oregon that provide habitat for the spotted owl. Each is a local ecosystem. To conceive that it can be managed from the fifth floor of some building in Washington, stretches the imagination. Burning is not a national pastime. What is good ecosystem management in Vermont is not likely to fit into an ecosystem in Wyoming. The use of controlled burning on a Georgia plantation managed for quail hunting will have entirely different results than a similar management policy applied to the watershed of a Montana trout stream.

PUBLIC INDIFFERENCE

In a culture obsessed with speed and quick results, we have been slow to embrace the lessons of burning as a determinant in any ecological model. We have drifted between a policy of national federal control and policies that are ecologically local. Unless a policy touches the public, personally and locally, the American people are, as a whole, rather apathetic about policies. Policies are usually stimulated by small advocate groups. It is doubtful we would have had wilderness areas set aside by Federal mandate had it not been for the agitation of John Muir and the further advocacy of the Sierra Club. Even though forests account for one third of the land mass that is the United States, only those people directly concerned with the forest as an economic opportunity really cared much about the policies that made up forest management. People who like to hike in the woods are not a very well organized advocacy group.

Many Forest Service people I have visited with are of the opinion that political strength in today's arena of management policy is so evenly divided between the timber interests, the grazing, mineral development and recreation-conservation-general use groups that none of these groups can effectively stop legislation or administrative action it opposes. At the same time, no group is able to push through their agenda if one or more of the other groups find it objectionable. Try running your snowmobile through that last statement if you disagree.

SO WHAT IS THE PURPOSE OF A FOREST?

If you were part of a poll and were asked that question tomorrow, how would you respond? That you would agree with John Muir they should be preserved. This raises the question of what good is the church if there is no congregation? Or, you would vote for conservation. Use the forests so long as that use can maintain some standard of conservation. Then who decides the standards? Will we have snowmobiles or just elk trails? What purpose is the highest and best use for the most people? Will you vote for economic uses or recreational uses? Do you want the Federal Government to run the forests? Or should the states be responsible for management policies? Or, should we simply let the free enterprise system set the policies?

If you believe, as the Swedish people came to understand, the forest is an integral part of the economic system but it is also vital to the needs of the general public, then you would conclude the forests are an important link between population and resources. Morse and Barnett in their writing on the economics of natural resources say, "The exceedingly varied natural resource environment imposes a multitude of constraints—social no less than physical—upon the process of economic growth. This presents expansionist man with a never-ending stream of ever-changing problems." One problem would seem to be that as land use becomes more intensive, land policies become more rigid. As land use is more firmly established, it can be modified only with great difficulty or great expense. Once that specie known as the "blue-winged developer" gets into the forest, it is hard to reverse the process of development.

When Gifford Pinchot set up The Forest Service and established many of its policies, he saw a forest as a sustainable resource. It was a renewable resource and if managed properly, would supply the needs of Americans for wood products for any foreseeable future. In fact, as late as the middle 1960s, the Forest Service Manual in section 2430.1 outlines five purposes of the forest. This was the language:

"1. Be designed to aid in providing a continuous supply of National timber for the use and necessities of the citizens of the United States.
2. Be based on the principle of sustained yield.
3. Provide, as far as feasible, an even flow of National Forest Timber in order to facilitate the stabilization of communities and of opportunities for employment.
4. Provide for coordination of timber production . . . with other uses . . . in accordance with the principles of multiple use management.
5. Establish the allowable cutting rate which is the maximum amount of timber which may be cut from the National Forest lands."

If the Forest Service is committed to this policy, is there a public consensus to support it? First, as citizens, we really haven't sorted out our social contracts regarding the use and management of forests or the role of burning. The timber industry has a long record of "cutting and running" when economics are not favorable. There are strong arguments for selective cutting as the way the forests should be managed by the government. You find strong arguments about replanting and whether that is really efficient for the timber industry. You hear arguments about replanting after a burn and whether that is good economics. Another expert says you let nature take its course and the forest will restore itself. Of course, that idea may be a 100 year wait if Yellowstone is an example.

We have come to understand that quality of forest protection for streams can dictate water quality. We know The Forest Service can protect a forest with fire breaks, clearing and selective removal of underbrush. At this time, we think this might be a partial solution to forest fires. We have also found this is a very expensive process and takes a bigger budget than The Forest Service has at this time. In short, there is no consensus.

On the more optimistic side, we seem to slowly be coming to some realization that each ecosystem has its own determinants. Certainly, we have come to understand the single entity in the ecosystem relates to the other elements in the ecosystem. We seem to have that part of it connected; that a species does not stand alone. What we may not have patched together is the idea that national policies are fine, but one management policy for one forest does not fit all forests. The principles of forestry may have some resonance, but all the singers may not be on the same page. Forests may need decentralized management.

One day I was standing along a river watching the construction of a dam in the company of a chap from the Corps of Engineers. I was mumbling something about sandy soil, good river bottom land, the hardwood trees that had been growing along the river and were now dozed into huge piles of tangled roots and branches and were being burned. That was before some fisherperson had asked why they couldn't leave the piles of wood for fish shelters when the lake filled? I asked my host what they would do with all the displaced squirrels. Once again, I learned you never joke with an engineer. Brushing aside my feeble attempt at humor, he said . . . " man, looka those new bulldozers move that yardage!" We just had different criteria.

And quite possibly, there may come a time when a visitor riding in a Yellowstone Park van will say to the driver, "And you say they just let this whole valley burn. And the driver will say, "Well in those days there was a lot of public pressure for natural forests and the forest managers were convinced they would come back in a few years. They just didn't understand how long 'a few years' would be."

At the conclusion to his *Principles of Forest Policy*, Yale University Professor Albert C. Worrell writes this. "If our population and average income continue to grow as now appears probable, there is bound to be an increasing and changing demand for benefits from forest resources. Many of our present policies will have to be modified or even changed completely to remain relevant to new conditions. At the same time, our average educational level will continue to rise, and the mass media will become even better at communicating and interchanging ideas. It appears quite reasonable, therefore, to expect basic nature of policy formation to remain much as it is now but to anticipate substantial future improvement in the efficiency and equity with which the process operates."

That was written in 1970. The professor may have been a little optimistic.

Robert Marshall

Was born on January 2, 1901 in New York City, the son of German immigrants. Marshall decided in his teens he wanted to be a forester. He earned three college degrees, one a Ph.D in forestry from Johns Hopkins University. In 1929, he took the first of many trips to the Central Brooks Range in Alaska. He was one of the first persons to explore this area of Alaska.

Marshall began writing about his experiences. His first book, *Arctic Village* was published in 1933. He was a strong socialist and believed private interests would eventually destroy American forests. His views were expressed in

Robert Marshall. Photo courtesy of The Bancroft Library, University of California, Berkeley

an article, "The Problem of the Wilderness" published In *Scientific Monthly* in 1930. In 1935, he and six other men founded The Wilderness Society. Marshall supported the group initially and when he died, he left one quarter of his $1.5 million estate to the society.

The Wilderness Society was small at first, partially because of the strong positions embraced by Marshall. Later it would grow into an organization of

250,000 members. In his public life, he was Director of Forestry in the Interior Department's Office of Indian Affairs and later became head of Recreation and Lands in the Forest Service. A wilderness area in western Montana, south of Kalispell bears his name. His legacy is captured in his words, "To us the enjoyment of solitude, complete independence and the beauty of undefiled panoramas is absolutely essential to happiness."

Bob Marshall died in November of 1939 on an overnight train.

George Catlin

Was born in 1796 in Wilkes-Barre, Pennsylvania. His father was an attorney who wanted his son to follow in his footsteps. His mother was taken captive in the Wyoming Massacre of 1778. From her, he would hear the stories of Indians, the frontier and "the wild west."

During the 1830s, Catlin made five trips to an area that ranged from present day Oklahoma to North Dakota. In this period, according to sources in

George Catlin. Source: Gilcrease Museum

Gilcrease Museum in Tulsa, Oklahoma, he visited 50 Indian tribes and made 500 paintings. He wanted to set up this work in a gallery "unique and imperishable for future ages." But his checkered career and the morality of his motives has always been in question.

It is said that when he was traveling in North Dakota he had a vision of sorts of what might happen to the beauty of these resources in some future time. When he returned to Washington D. C., he would include in his lectures the idea of national parks where the preservation of natural resources and spaces could be preserved. His advocacy of this idea spread to the idea of natural parks.

Catlin has been called "a cultural P.T. Barnum" and even some Native Americans have "a profound resentment toward Catlin because of some of the treatments in his portraits. Catlin had money problems, and he took his portraits and some Indians to Europe where he presented a vaudeville show, which some called a "Wild West Show." This failed and he was put in debtor's prison. Eventually he returned to America and the Smithsonian gave him a studio in which to paint. His portraits and notes had preserved part of American history and he had sown the seed of national parks.

George Catlin died in 1872 in Washington, D.C.

REFERENCES

Barns, Cass G. *The Sod House*. (University of Nebraska Press) 1930

Jennings, Francis. *The Invasion of America*. (W. W. Norton) 1976

Locke, John. *Two Treatises of Government*. (Hafher Publishing) 1948

McPhee, John. *Irons in the Fire*. (The Noonday Press) 1997

Pyne, Stephen J. *Fire in America: Cultural History of Wildlife and Rural Fire*. (W. W. Norton) 1997

Worrell, Prof. Albert C. *Principles of Forest Policy*. (McGraw-Hill) 1970

Denver Post. August 2001

Interview with Dr. Fred Wiseman. Swanton, Vermont. October 2002

Nature Conservancy. *"Students of Fire."* Fall 2002

Tulsa World. March 20

Visit to Joyce Kilmer Memorial Forest. October 1992.

Chapter Fifteen

Symbols and Symbolism

Meandering across south-central Wyoming is a river that has played two roles in the history of the state. The Sweetwater River was probably named by Gen. William H. Ashley in 1823 because the water tasted sweeter to his trappers than the water they had been drinking. Sweetwater County, through which the river flows, is 10,495 square miles of big country.

Sweetwater country was to host a trail that became the symbol of westward migration. Actually there were several trails. One that continued west from Nebraska to South Pass City in southwestern Wyoming was generally known as the Oregon Trail. At times, this trail was taken by the Mormons, although they generally stayed on the north side of the Platte River. The trails became symbols. They supplied movie material that portrayed the indominatable American pioneering spirit.

One summer a Philadelphia writer came to Sweetwater country. He wrote a book called *The Virginian*. It stoked the imagination of the American public. Later, the story Owen Wister penned of the strong, soft spoken, masculine cowboy became a prototype cast in a variety of Class B motion pictures and would be made into at least a dozen major pictures.

More than any single happening in the folklore of the west, *The Virginian* provided the symbol of the American cowboy. In time it was enlarged to become black hats and white hats, bad versus good, Tom Mix jumping his horse, Tony, across the chasm, the solitary figure on a horse seeking independence and freedom. That cowboy symbol of freedom is still rooted in our folklore in spite of the fact that there is no character more dependent upon environmental conditions. The cowboy needs grass and water for his horse, water for cattle, a strong horse for transportation and the list goes on.

But we have made him a symbol of independence. Painters and writers picked up on the romantic view as the art world joined the book world in creating a symbol which lives on today on cigarette packages.

THE REAL COWBOY WASN'T THAT ROMANTIC.

The original Texas cowboys before the Civil War were mostly Mexican vaqueros. Later, post Civil War era, they were a collection of war veterans, some black cowboys seeking freedom and a better life, and no small number of boys and young men who wanted excitement and adventure. What they got was hot, dry cattle drives, raging prairie blizzards, fifteen hour days in the saddle and not many baths.

Fiction warped the truth. Popular dime novels by writers like Ned Buntline and the traveling Wild West shows forged more symbols. The newer trails took on names and Jesse Chisholm, a half-Cherokee, Oklahoma trader who never did drive any cattle, found a niche in frontier lore with his Chisholm trail. Cattle have a smell. So does money. Where there is money there are usually the lawless. Soon there were names like Billy the Kid, the Dalton Gang and the James Brothers. There were the settlers versus the Cattle Barons and range wars. The Cheyenne Club in Wyoming, in 1880, was a Wyoming symbol of gentility even if some members got ousted for profanity or hitting a bartender.

All of these dissonant events changed the ecology of the land. If the symbols lived on, and many have, the scope of symbols in the creation of American ecology cannot be dismissed. A symbol has been called a mnemonic device. In advertising parlance, it has the visual ability to grab the public imagination. It can build brand awareness. It can be as simple as a rock. You can "own a piece of the rock." As a public relations tool, it replaces an actual item with an image. It can replace thinking with emotion. It can distort the real issues. It can create unbelievable public awareness words alone cannot.

Symbols are neither white hats nor black hats, to use a common symbol. To illustrate how complex issues use a symbol is not difficult. Environmental and ecological issues often use them. Often, the symbol becomes the "apostle of the single, simple solution."

WILD HORSES AND ANNIE

In 1952, in Storey County, Nevada, a petition was signed by 147 citizens protesting an application for a permit that would have allowed the use of airplanes to chase down wild horses destined for the slaughter yard. A lady by the name of Velma B. Johnston showed photographs of an earlier roundup. They depicted wild horses, many bleeding and half dead, being crushed into a truck. The Storey County Board of Commissioners banned the use of airplanes to spot and pursue wild horses. While the case was being argued in 1955 in the Nevada State Senate Chamber, a heckler referred to Ms. Johnson as "Wild Horse Annie." A symbol had been born.

Nevada passed a law in 1959 that prohibited chasing and capturing wild horses on public lands using vehicles and aircraft. Wild Horse Annie pushed the issue to the United States Congress. Her battle was to prevent inhumane capture of wild horses. She also wanted some policy that would allow the wild horses and wild burros to roam public lands.

In December of 1971, nearly ten years later, Congress passed the Wild Free-Roaming Horse and Burro Act. In 1976, the BLM took over an adoption program that had been started by Wild Horse Organized Assistance (WHOA) under the direction of Wild Horse Annie. To that time, the group had adopted out approximately 10,000 horses. In 1978, the Public Rangelands Improvement Act was passed by the U.S. government giving the government the right to transfer ownership of up to four animals each year to individual adopters. This was 19 years after the original petition had been presented in the Storey County courthouse.

These are excerpts from a July 16, 2000 story in the Desert News by Amy Joi Bryson.

"A dry spring, a hotter than normal summer and the infestation of grasshoppers and crickets have combined to deplete the vegetation the horses need" . . . in the area near Skull Valley, Utah.

"Crouched on a rocky hillside, a group of people waited. The storm clouds rolled by, spilling large drops of rain, and still they waited, hugging jackets closer."

"In the distance, you could see the horses, all seventeen of them, running bunched together, tails stretched out, manes flying. . . . Two 'Judas' horses waited for them . . . When the wild horses get to just the right spot inside a winged, camouflaged fence, the 'Judas' horses are turned loose to gallop into the corrals, with the wild horses following."

"When the wild horses thundered past, with all their mystique symbolizing the Wild West, the watchers were awestruck by the feeling they were witness to something few people see."

The story ends. . . . "The tails will stretch out behind them, the manes will fly and they will thunder past."

Now, with no bias, there is romance, the old West, wild horses, and no small amount of symbolism involved in this story. It does capture your imagination. "Wild Horse Annie" captured your imagination.

My search for "the rest of the story" on wild horses took me to the front of the Bureau of Land Management building in Elko, Nevada. I passed a display of brochures. Smokey the Bear smiled at me. He is one of the more widely recognized nature symbols. I explained (to a live person), I was trying to re-search wild horses for a book. I needed help.

THE REST OF THE WILD HORSE STORY

I met Kathy McKinstry, the Wild Horse Specialist in that office. She had graduated from the University of Wyoming with a degree in wild horse man-agement. We talked Wyoming, found common interests and then talked wild horses.

She sketched briefly for me more than thirty years of BLM dealing with the wild horse and burro issue. In Nevada, they have established Herd Manage-ment areas. They have some financial constraints. The annual BLM budget for wild horses is $18 million. Nevada gets about 10 to 15% of this money al-though they have about 50% of all the wild horses in the U.S. Another inter-esting example of government accounting.

In 1992, they developed a strategic plan. Ninety people or organizations were part of the planning. They began to contract out the roundup of the an-imals. They would pay from $100 to $300 per animal, depending upon the size of the roundup. Animals 3 years or younger, go up for adoption. All oth-ers go back to the range. Later, to get a better balance, they raised the age to 5 of all animals that went back. Animals that are judged as unhealthy are dis-posed of.

I asked Kathy if she felt they were making progress with the program. "It's frustrating, at times," she said. "I've seen what wild horses can do to the range. If we had no cattle grazing, we might be able to reach some accom-modation." An example of economic determinant versus natural resources. The problems stem in part with the adoption program. Sometimes, the horses are not bought at the auctions. Other times, they are purchased and the buyer does not comply with the rules. Enforcement of the program is difficult. "It just seems sometimes," she said, "we can never get the job done. Some-times," she said very quietly, "I'm not sure there is an ecological niche for the wild horse. They are prolific. The stallions do their job well. If the mares are healthy, the herd can expand in a very short time."

Buried in the 10th and 11th report to Congress, in a booklet I picked up in the Elko office, are some numbers. If you totaled the wild horse population as they present it, you would get about 43,000 animals. This does not include burros. If the budget for BLM for the horse keeping is $18 million, then divide the horses into the budget and you get about $4200 per horse. To keep wild horses in some humane program is very expensive. There may be no answer to a balance of the wild horse and the burro with the reality of the range, no simple, single solution.

I took the brochure of Smokey the Bear with me. Here is another symbol. The original purpose of Smokey the Bear was to create an awareness of the danger of fire in the forests. Smokey was a badly burned cub when found. His birth as a symbol was equally painful. There was much discussion within the BLM of the value of the little critter as a public relations tool. As it turned out, he did his job better than expected. So well that he may have tilted the management of forest land from some controlled burning and management of undergrowth to what became the pattern of benign neglect.

The symbols to which the American public have gravitated have not always depicted true condition. But they certainly served to grab the public attention. The much publicized snail darter was not the real issue in the Tennessee Valley. The real issue was the question of how many rivers would be dammed up for economic purposes at the expense of fertile river bottom land where people lived. But the little critter did a fine job of damming up public opinion until the overflow inundated objectivity.

THE SPOTTED OWL

. . . would be another example. On one side of this more recent issue, you have the timber side. Their slogan is "Save a logger, eat an owl." On the other side, you have the environmentalist slogan, "Save an owl, ban logging." It would be safe to say the two "simple solution" sides in this issue were some distance apart.

Strangely, there never was an organized "Save the Owl Society." The city of Sweet Home, Oregon became the epicenter for this conflict. One day, the Sweet Home Online Web Site got a request wanting to know if there was such an organization. To find the answer to the question, the SHOL (Sweet Home on Line) sent a request to various major Oregon newspapers asking, "Is there an organization devoted to the conservation of the Spotted Owl?" They also sent the same request to any known "hotbeds" of opinion and the most divergent public views that represented extremism on either side. They got no response.

Eventually, the owl made it to the United States Supreme Court. Along the way, the little bird would ruffle a lot of feathers. For many years the Willamette National Forest had produced and sold more saw logs than any other national forest. Then things changed. The cost of high tech equipment for the mill, logging restrictions and spotted owl closures combined to cut off the steady stream of logs. In the 1980s, the two hundred thousand acre Sweet Home Ranger District was annually producing eighty six million board feet of timber. By 1992 the height of the spotted owl controversy, the district produced one hundred thousand board feet. Willamette Industries laid off 410 employees. The forest industry had paid 60% of the tax that went to support schools. In 1997, the State of Oregon paid 62% of this cost. The economic impact of the owl was severe.

A HOMESPUN SYMBOL

Ruth resides in a senior citizen home. Ruth decided one day she needed to do something for her "home" so she did some investigating. She then purchased and planted an apple sapling. To honor the "planting," the home held a special service. There was a reading from Psalm 104, a simple prayer and then the oldsters sang an old song named in honor of this symbol.

> "Oh, the Lord is good to me
> And so I thank the Lord.
> For giving me the things I need.
> The sun and the rain and the appleseed.
> The Lord is good to me."

Ruth hopes others will catch her idea and they will plant these special saplings, partially for the sake of the next generation, but also for the sake of history. She calls it a "conservation conscience." This is the rest of that story. John Chapman, born in Leominster, Massachusetts on September 26, 1774, conceived a plan, almost altruistic, for the betterment of man. So he followed his religious beliefs, seeking to bring harmony between man and nature, believing God manifested himself in the plants, animals and outdoor surroundings. His mission was confirmed with a vision of fruit trees lining the streets of Heaven. The vision came after a near fatal accident.

In the late 1780s, he set out in a canoe to plant orchards of apple trees. The trees were intended to provide for settlers and pioneers as they traveled westward during the expansion era. Each tree was to provide another tree or two for the farmers to take home and plant. They would pay as they could in any fashion affordable. He also planted an orchard of his favorite Rambo apple trees at the Harvey farm in Nova, Ohio. One of the trees is still alive and produces fruit. In the spring of 1996, a storm nearly toppled the old survivor. Years before, the old apple tree was visited by American Forest representatives. Seeds and cuttings were taken. The direct offsprings of this ancient Johnny Appleseed tree were grown into saplings. Today, John Chapman rests at Archer Cemetery in Fort Wayne, Indiana. His symbol, Johnny Appleseed, continues his journey.

This story is about a symbol that has remained fairly true to its actual purpose. Too often, the lines between fact and fiction become blurred. The symbol becomes a mix of entertainment, education, marketing and pure hokum. Today, in our electronic pop culture, showmanship is entwined with the subject to make a better story and even conservation organizations blend facts and high hopes to produce a story that sells better. The bugle call for the effort can be as basic as "do it for the ducks." The ducks are not the real issue, but harvesting ducks is easier to sell than the wise use, or restoration of wetlands, a more complex story. Real issues are hard to sell.

SOME SYMBOLS HAVE IRONIC TWISTS

When he was hired as a meat hunter for the railroads pushing westward in the plains, he was probably Bill Cody. Reports say that in those times he could dispatch 100 bison a day with his rifle. In 1867-68 he earned his nickname, "Buffalo Bill," by shooting 4,280 bison in 18 months. He became a guide for the "sports" wanting to shoot bison and the slaughter would be escalated to even higher levels. The Earl of Dunraven visited Wyoming looking for excitement. He is quoted in the book, *The Story of Man in Yellowstone* by Merrill D. Beal this way.

"We killed elk, white-tail and black-tail deer, antelope, swans, immense geese, ducks and small game without count. This elk running is perfectly magnificent. We ride among the wild sand hills till we find a herd and then gallop after them like maniacs, cutting them off, till we get in the midst of them, when we shoot all that we can. Our chief hunter is a very famous man out west, one Buffalo Bill. To see his face flush, and his eyes shoot out courage is a sight to see, and he cheers us on till he makes us as mad as himself."

Later, Buffalo Bill would produce a Wild West Show. One of the highlights of his show was the staging of the Battle of Little Bighorn. The popularity of this production prompted brewer Adolphus Busch to commission a print which became the lithograph that hung in saloons across the country and proceeded to rewrite history. Although there were no known American survivors of this fracas, there have been reports of well lubricated men emerging from Montana saloons with vivid stories of how they escaped the "injuns at the Bighorn." So symbols were born and history rewritten.

The evolution of a symbol, Buffalo Bill, was considered to be perfectly acceptable. It was the social contract of those times. Today, if you mentioned the facts in Cody, Wyoming, you would be careful with your talk. I mentioned this one day in that Wyoming setting of beautiful art galleries and museums. I was severely upbraided by a Wyoming lady who proceeded to advise me that William F. Cody "had done so many wonderful things for this town . . . he is held in the highest respect." Symbols can lose their rough bark with enough time.

THE END OF THE TRAIL

James Earl Frazier created a sculpture, perhaps the single best known work of western art, regardless of medium. It is known as "The End of the Trail." It depicts an Indian, huddled over his horse. It is dejection, rejection and resignation. It was intended to be a symbolic appeal for spiritual help. Frazier also created the images that were used for the "buffalo nickel," the coin that has the bison image on one side and the Native American face on the other. This was an effort by Frazier to link the culture of the Indian people with the source of their sustenance.

Less well known is the sculpture Frazier titled "Buffalo Prayer." This magnificent piece is on display in the Whitney Gallery of Western Art in the Buffalo Bill Historical Center in Cody, Wyoming. Frazier had planned that a bison skull be placed at the foot of the praying Indian and would symbolize the destruction of the bison. It would have completed the spiritual message of the art. The artist wanted to connect the bison and the Native American. For some reason, the skull was never included in the piece so the intention of the artist, or the exact meaning has never been conveyed. It is an incomplete symbol in the total irony of the bison story.

Symbols play a significant role in the accumulation of conservation, environmental or ecological thinking in our culture. Certainly, they have played a major role in our social contracts. They have been used for educational purposes, for pure economic purposes and in some cases have been pure marketing inventions by media or even government. Often the symbol becomes a marketing tool. It may obscure the real ecological issue. It can create the image of the "right answer." Americans love "right answers." We much prefer "right answers" to the "right question." Ecology is difficult to reduce to "right answers." It almost demands the "right question." There are no single, simple solutions. On the other hand, if you want a crowd at your show, you probably feature Buffalo Bill and not William F. Cody.

Franklin D. Roosevelt

Was born in 1882 in Hyde Park, New York. He attended Harvard University and Columbia Law School. He entered public service through politics as a Democrat. He gained election to the New York Senate in 1910 and was later appointed Assistant Secretary of the Navy. In 1920, he was the Democratic nominee for Vice President.

It is difficult to separate the FDR social programs from what were also considered "conservation." The Civilian Conservation Corps is a good example. It had two goals according to Roosevelt: to create jobs for unemployed young men and to restore land to productivity. Actually, the program

Franklin D. Roosevelt. Photo courtesy of the Library of Congress, LC-USZ62-93456.

was used to provide water in arid regions, to construct irrigation projects, to restore grasslands and to build roads, trails and other improvements in national parks.

Refuges were improved, new ones added as were the national parks, Kings Canyon in California and Olympic in the state of Washington. In 1936, he signed the Soil Conservation and Domestic Allotment Act. At that time he said, "The history of every nation is written in the way in which it cares for its soil. The United States . . . is now emerging from its youthful stage of heedless exploitation and is beginning to realize the importance of treating the soil well."

Roosevelt was searching for the best use of land, or "the highest and best use for the most people." This was also his position on the TVA. Conservation . . . yes. But the goal was best use.

President Franklin D. Roosevelt died April 12, 1945 in Warm Springs, Georgia.

Charles Edwin Bessey

Was born in Milton, Ohio, May 21, 1845. He received his B. S. in 1869 from Michigan State University and later an honorary M. S. from that school. In 1879 he received an honorary Ph. D. from the University of Iowa and in 1898 a L. L. D. from Grinnell College.

Bessey has been described as a "protagonist" by academic writers. He was a forerunner in his insistence that scientific programs developed by colleges be made available to the agricultural community. This was especially true in Nebraska. He strongly felt scientific research should result in practical application. Much of the research or study produced under his auspices eventually became the basis for grassland management today.

He was President of the American Association for Advancement of Science from 1910 to 1912, was President of the Botanical Society of America in 1895 and 1896 and was inducted into the Nebraska Hall of Agricultural Achievement in 1918. He authored many articles including The Essentials of Botany in 1884. He lectured extensively at Iowa and Iowa State Universities.

He is best known in conservation for his initiation of a tree planting program. This idea resulted in the Nebraska National Forest in Halsey, Nebraska. This was the first man made forest in the world. That story is sketched in this text. He was also active in getting state and federal protection for the Calaveras sequoia trees in California.

Bessey passed away on February 25, 1915.

Charles Edwin Bessey. Photo: Charles E. Bessey, Papers (RG 12/07/10). Archives and Special Collections, University of Nebraska-Lincoln Libraries

REFERENCES

Beal, Merrill D. *The Story of Man in Yellowstone. Yellowstone Library and Museum Association.* 1972

Buskirk, Amy. Charles Russell. (Barnes and Noble Publishing)

Kirkpatrick, James J. Article in the Tulsa World. 1992

McDerrnott, John Frances. *Audubon in the West.* (University of Oklahoma Press) 1989

Abundance and Population Characteristics of Northern Spotted Owls. USGS Biological Resources Division. Port Angeles, Washington.

Interview with BLM personnel. 2000.

National Wildlife and Burro News. U. S. Department of Interior. Summer 2002

Northeastern Nevada Public Lands Outdoor Recreation Coalition. A pamphlet handed out by BLM.

Quill of the Hill. Volume 35, Issue 10, October 2000.

"The Wind River Rendezvous." Vol. XII, April/May/June 1983

USA Today. September 12, 2000. pp.9-A

Various Authors. *How Myth and Reality Converge*. Gilcrease Museum Journal Spring 1996.

Chapter Sixteen

The Myths that Hover Over the Land

Myths and symbols that have mingled with our American history are not limited to bears, owls, wild horses and Little Big Horns. Sometimes they are political allegories that in time become children's stories. The myth of the small freeholder can be found in L. Frank Baum's best known work, *The Wonderful Wizard of Oz.*

His story is based more on economics and social contracts. The real story from 1865 to 1880 in the Great Plains country of South Dakota, Nebraska, Kansas, North Dakota and Oklahoma had been the extraction of native wildlife and a massive species substitution. Cattle replacing bison and cereal grains replacing grasslands. Coupled with this ecological shift would be cycles of depression and drought from 1880 to 1900.

Land giveaways of many types had encouraged settlement of the land by landowners who took deeds or filed claims on prairie land. There was no early technology for clearing land or breaking soil. This rode on the backs of people or animals. Even people who hung on to settle the land, if not broken in spirit, were not always certain they had found the Promised Land. As is still true today, they were willing listeners to promises of quick fixes.

A lady emerged from the Kansas plains and became known as the "Populist Firebrand from Kansas." Her real name was Mary Elizabeth Lease. She was a powerful orator, according to some, better than William Jennings Bryan. She espoused the populist cause which was a political effort to form a coalition of farmers and urban workers. Lease ended most of her speeches with the exhortation, "Raise less corn and more hell."

L. Frank Baum's heroine, Dorothy was based upon the real life Mary Elizabeth. She becomes an idealistic girl from Kansas who gets swept up in a tornado. She is better known from the movie that starred Judy Garland than she is from the original book. For most people it is a fairy tale that captures the imagination and provides wonderful music about rainbows. In reality, the story offers valuable insight into the culture, politics and prairie ecology in the late 1800s.

THE PLAYERS

On her journey, Dorothy picks up some friends. Her good natured but dim-witted scarecrow buddy is the farmer caught in an economic system that is destroying him. He is searching for a brain so he can survive. He is a hick, a stereotype often tacked on the farmer by the urban elite.

The Tin Man is the American worker or laborer in the late 1800s. Working conditions in those times were less than humane and the Tin Man was looking for a heart that would restore the humanity he had lost working in the factories. He is also the allegory of the worker joining forces with the farmer in a political coalition.

The cowardly lion was a jab at William Jennings Bryan. He was the fair haired boy of the populist movement. Baum thought Bryan had backed down on supporting certain key populist issues. The Wizard depicted the powerful men of the east who lived in the city of money, The Emerald City. Baum wanted to show such politicians or power brokers as bumbling, ineffective persons hiding behind screens and pulling strings.

The Flying Monkeys were the allegory for the Plains Indians. Dorothy and her friends needed to defeat these fierce creatures to get to the Emerald City. Of course, the settlers had to resolve the Indian problem.

Then there is the Wicked Witch of the West. She is the sum total of all the conditions that plagued the farmer; drought, pestilence, windstorms, floods. She meets her demise and is destroyed by water. This was the allegory for irrigation and water control, the technology that would conquer nature and provide the quick fix for more prosperous farming.

The Jefferson ideal of the small freeholder who would farm 160 acres of land was struggling in prairie country. Grain and livestock prices rode the string of the yoyo. The myth of the small, freeholder farmer badly needed a friend. The *Wizard of Oz* was really about this condition.

A COMPARISON

A brief comparison to other systems can be useful. Soviet agriculture, since the 1930s, has been collective. Soviet agriculture is subsidized. Today, Russian farmers fear privatization. They do not see it as the myth of the freeholder but rather as another form of collectivation. They see it as a power shift from the hands of the government to the hands of a few powerbrokers.

Of course, agriculture, like media, or retailing or petroleum, can end up in the hands of a few. So a nation that has never had private ownership of their farm land is saying "thanks, but no thanks." A news reporter in Russia would find their few, small farms in private ownership struggling. There would be some evidence of American "vertical integration," where land operation is incorporated with manufacturing and then with retailing in a commonly controlled line. This is achieving only marginal results.

Switch back now to this country with the Soviet experience as background. There was, in the U.S. in the late 1800s very little "environmental concern." That is still true in the Soviet Union. Economically, in this country, U.S. farmers have learned that "vertical integration" is limited help to the smaller farmer. Farm Coops have been only marginally helpful. The farmer still faces a tough supply and demand market, and operates with very few anti-trust laws; has very little collective ability to bid down the cost of seed, diesel, fertilizer or spray. Profit margins get very thin. The American prairie farmer is still looking for the Wizard of Oz.

THE PRICE OF RAPID DEPLETION

More recent stories coming out of Kansas have less to do with allegory and deal more in reality. In Salina, Kansas, the Land Institute's Prairie Writers Circle has run a series of articles dealing with prairie conditions. Donald Worster, a historian and author writes, "The United States is a nation with incomparable natural advantages, but we are depleting them rapidly. . . . Each year we have to pound our land harder to extract its wealth. A hard-pounded land soon becomes a degraded land. Environmental degradation at home is the true cost of empire abroad . . . In the pursuit of global supremacy, we have polluted every part of our nation. And now we are told we must live with that degradation as the price of glory."

He then cites a long list of environmental conditions he feels the United States is not addressing. These include cleaning up rivers, preserving forest ecosystems, saving native prairie, drilling or mining regardless of resulting damage, losing millions of tons of topsoil every year and continuing to drench soils with chemicals. These are all costs, according to Worster, we are paying because we do not want to slip back as a world power. Then he adds a conclusion. "Eventually, all empires falter and collapse. They don't deliver on visionary promises they make. They fall because they bankrupt themselves and, more often than not, they bankrupt the environment that supports them."

From that same source, another author, Charles Francis who is a professor of agronomy at the University of Nebraska, writes, "We must build a sustainable, more natural agriculture. We already have some idea of what must be done. Agro ecology, an emerging study of food production as ecology, can help . . . this new agriculture would employ genetic diversity over space and time. In place of today's huge fields of single crops, it would plant diverse mixtures of food crops and supporting species that can better adjust to weather changes. . . . And it would replace many of our major annual crops with perennial versions that need not be seeded every year."

He then lists a number of viable options that could be exercised to provide a more sustainable type of agriculture. "We must," he writes, "redirect federal farm programs to move agriculture down a more sustainable path. By emphasizing production of a few, key crops, today's government subsidies encourage farmers to continue farming the old way. Taxpayer dollars should focus on how farmers care for the land, not how much they produce."

But the old myths of the prairie die hard. Begin with our ethic that nature is hostile and an enemy and must be conquered. Only mastery of natural forces will release man from the restraints of physical necessities. Doesn't massive spraying in the grasslands fit into this idea? Then move to the second ethic. Nature is a resource that should be used for man's needs and if extraction is necessary, then it is justified for the social and economic gain of man. This thinking is still a strong part of the social contract in prairie country. It would also seem our fourth ethic, that nature is property and can be used as any other material property would also apply. Some of our conservation ethics have very deep roots. Many of those roots can be found in the prairie country.

THE HARD LINES OF OPPOSITES

Of course, causes can create myths. Using South Dakota as an example, the Farm Bureau claims to have nearly 10,000 members. This announced number gives any member of the South Dakota legislature reason to pause when a bill that concerns the farmer or rancher is in committee or on the floor. What is not so well known is just exactly how many of those 10,000 "members" are actually farmers or ranchers? How many just buy insurance from the Farm Bureau? Because, as in most farm belt states, anybody who buys Farm Bureau insurance is counted as a member. The Farm Bureau becomes a formidable opponent unless you discount the mythical numbers.

The South Dakota Wildlife Federation would tell you that "almost any wildlife issue that benefits the sportsman/woman has been fought by the Farm Bureau." Here are some excerpts taken from the South Dakota Farm Bureau Policies Handbook for the year 2001.

"The Game, Fish and Parks Department's total budget should be approved by the Joint Appropriations Committee of the South Dakota Legislature. All revenue collected by Game, Fish and Parks should go into the general fund. The Secretary of the GF&P should be appointed by the commission and be directly responsible to that board. The landowner positions on the GF&P Commission should be required to earn at least 60% of their income from agricultural production."

All of these positions by the Farm Bureau are political in nature. If the Legislature is controlled by the rural members, and it is in most prairie states, then moving financial control and putting restrictions on who is on the GF&P Commission will put the control of wildlife issues in the hands of the legislature. Indirectly agricultural interests control wildlife.

But read on in this booklet. "Farm Bureau opposes any expansion of present wilderness areas or the establishment of any new wilderness areas in the state of South Dakota. We oppose additional acquisition of land by GF&P. Before any property is acquired by GF&P, public hearings should be held in the vicinity of the proposed acquisition. Public oral or written testimony should seriously be considered in any decision. At least two GF&P Commissioners should be in attendance. Current Game Fish & Parks policy allows 48 hours for action on a depredation complaint. We favor shortening the time period to 8 hours. Because of the perception of prairie dogs as being comparable to poodles or other small canines by those people unfamiliar with them, we would like to see their name changed to 'prairie rats' as a more descriptive and fitting label for this rodent."

There is more. Keeping in mind that one of the concerns of prairie country biologists is the disappearance of weed seed that supports bird populations such as quail, pheasant or even wild birds, the Farm Bureau policy toward weed control would cause conflict, if not head knocking. The Farm Bureau policy states "If the GF&P is not controlling noxious weeds and pests on its property, it should be accountable if these weeds or pests spread to adjoining property. Boundary lines must receive proper maintenance."

One hundred years later, there are still strong echoes of the Wicked Witch, still strong positions, still strong talk. The "battle lines" drawn by the Farm Bureau in South Dakota illustrate how often conflict collides with conservation. "Collision" takes us to another story, not as well known, but worth relating. Rachel Carson and her part of the story is better known. Let's begin with Fred.

THE RACHEL AND FRED STORY

The story begins in Switzerland. A chemist with the J.R. Geigy Company began experimenting with an odorless, white, crystal-like compound called dichloro-diphenyl-trichloroethane. He accidentally discovered that even the compound's residue would kill houseflies. He sent the chemical, which became known as DDT, to the New York office of J.R. Geigy. It lay there until a chemist translated the claims about the substance into English. He passed a sample on to the Department of Agriculture. They passed it on to their entomology research station in Orlando, Florida.

By now, it was World War II and the U.S. military was searching for a better way to protect troops against typhus, the fever spread by lice, and protection against malaria caused by mosquitoes. In 1942, the First Marine Division was pulled from combat. Of their 17,000 men, 10,000 were suffering from fever, chills and headaches, malaria symptoms.

Extensive tests were run in Florida. DDT was the spray of choice. The military began spraying beachheads in advance of invasion of the South Pacific islands. In one instance, nine thousand gallons were spread in the area around Saipan to abate dengue fever. The Marines were successful in taking the island thanks to DDT. The compound would be used world wide to fight disease carried by insects.

Fred Soper was born in Kansas about the same time Dorothy came to life, 1893. He received a doctorate from Johns Hopkins School of Public Health, worked for the Rockefeller Foundation. He became the lightning rod of that organization in a world wide effort to treat lethal diseases. Then, beginning in 1958, the U.S. government pledged the "equivalent of billions of today's dollars for malaria eradication, one of the biggest commitments a single country has ever made to international health."

In 1963, the money allocated by Congress began to run out. Some countries had shifted money to other health budgets. Research was also beginning to show that insects could develop immunity to DDT. It required repeated spraying, often in even larger doses. The dream of Fred Soper to create a world free of malaria and mosquitoes was coming apart. He took the turn of events hard. He was convinced his mission was to save human lives. He had an intense focus on only his cause.

It was in this time frame "Silent Spring" was published and Rachel Carson stepped into the environmental spotlight. She would write, "The world has heard much of the triumphant war against disease through the control of insect vectors of infection, but it has heard little of the other side of the story — the defeats, the short-lived triumphs that now strongly support the alarming view that the insect enemy has been made actually stronger by our efforts."

Soper felt that Carson missed the point. The issue was whether the mosquitoes were gone long enough to disrupt the cycle of malaria transmission. He felt that spraying was so effective that it did produce a condition where you could stop spraying and not fear the consequences.

Today, DDT has become a symbol . . . of all that is dangerous about the efforts of man to interfere with nature. Only two countries continue to manufacture the substance, India and China. Some 90 countries have signed on a treaty that places DDT on the restricted-use list and most users have developed plans to phase out such use entirely.

The thesis of "Silent Spring," rested more with the indiscriminate spraying of DDT for agricultural purposes. There was a time it was being sprayed with great enthusiasm around the Midwestern and western countryside in efforts to control Dutch Elm Disease, the gypsy moth and the spruce budworm. And Carson did not exactly separate her misgiving about rural agricultural spraying and using DDT for urban or humane purposes.

She had done her research rather well and was on solid ground. The chemical industry led the assault on her credibility. She survived that corporate assault. But she had some powerful symbols going for her. Consider this quote she used from a bird lover in Alabama. "There was not a sound of the song of a bird. It was eerie, terrifying. What was man doing to our perfect and beautiful world?" Of course, Soper would counter this by pointing out the world was neither perfect nor was the human condition that beautiful in the gambiae invasion of Egypt.

Fred Soper ran into a great truth we are beginning to appreciate more in the early part of 2000. Even the very best of intentions can produce adverse consequences. There may be immediate benefits but these are offset by longer term risks. In nature, there may be no simple, single solution to an ecosystem problem. In fact, in her book, Carson quotes Professor Carl P. Swanson who said, "Any science may be likened to a river. It has its obscure and unpretentious beginning; its quiet stretches as well as its rapids; its periods of drought as well as of fullness. It gathers momentum with the work of many investigators and as it is fed by other streams of thought, it is deepened and broadened by the concepts and generalizations that are generally evolved."

CONNECT THIS STORY

. . . to the present issue of spraying grassland by ranchers who want more grass for cattle. What Carson had to say on that subject has stood the test of scientific time. "Spraying kills off the weaklings. The only survivors are insects that have some inherent quality that allows them to escape harm. These are the parents of the new generation, which, by simple inheritance, possesses all the qualities of 'toughness' inherent in its forbearers. Inevitably, it follows that intensive spraying with powerful chemicals only makes worse the problem it is designed to solve. After a few generations, instead of a mixed population of strong and weak insects, there results a population entirely of tough, resistant strains."

Today, there are many voices that echo the admonitions of Rachel Carson. As this chapter is being revised, Cornell University ornithologists, reporting in Proceedings of the National Academy of Sciences, reveal the population of the wood thrush, a migrant bird "that breeds along the acid-rain-prone slopes of Eastern mountains whose population is slowly declining" . . . may be the second hand victim of acid rain. It seems the rain damage goes beyond the trees. It also may hinder the breeding of the birds by depleting "snails, slugs, and other calcium-rich foods vital for birds to lay eggs and for their babies to thrive." And again, we may have evidence "that science is like a river."

MEANTIME ON THE GRASSROOTS LEVEL

The pattern has developed promises and performance. Congress will allocate money but when farmers apply for help there is no money available. Some farmers turned to the Agriculture Department's Environmental Quality Incentives Program (EQIP) for help. However, in the year 2000, the department was able to approve only 16,443 requests of the 76,168 applications it received. Eighty percent of the applications were turned down because of no funds.

This brings us to the Wildlife Habitat Incentives Program, (WHIP), a voluntary, incentive based conservation/farm policy designed to encourage farmers to become better stewards of habitat for wildlife. This information came from the March 2001 issue of the South Dakota Wildlife Federation Magazine. "This year, under current federal funding levels, WHIP will receive 20 million dollars. By comparison, the Federal Crop Insurance Program is funded at $2.2 billion annually. Of this amount, 23%, or $500 million annually, goes directly to the insurance companies. In other words, the insurance companies receive over 25 times the amount of money that is earmarked for WHIP. Despite the fact that the crop insurance program has been highly criticized by farmers for these windfall payments to insurance companies, Congress continues to ratchet up the funding. The American Farm Bureau Federation lobbies heavily for such increases and currently owns or controls about one quarter of the companies approved by the USDA to provide crop insurance." Perhaps somebody has found the way to the Emerald City!

On the grassroots level, the soil under what was once a sea of rich prairie grass slowly diminishes in quality under the duress of economics and production. Drought conditions cause thoughtful people to check the slowly dropping levels of underground water supplies. Water wells are drilled deeper. Soil erosion continues. Great dams continue to fill with silt and salt.

A strange paradox blows in the wind. Under the cover of promoting their cause the organizations involved enhance a polar position rather than a search for an accommodation. The care of the cause or position seems more important than the care of the land. The care of the land may be the greatest myth of them all!

REFERENCES

I have relied heavily in this chapter on two sources. One is an article published in *The New Yorker* magazine, July 2, 2001, authored by Malcolm Gladwell, entitled, *"The Mosquito Killer."* From this, and internet sources, I got the Fred Soper information.

The second major source was *Silent Spring,* published in 1962, written by Rachel Carson. Houghton Mifflin Company.

An article in Nebraska Land Magazine, written 1999, provided information on L. Frank Baun. Author was Amy Cyphers.

"Small Farms in a Vice Grip" An opinion column by George P. Pyle, Director, The Prairie Writers Circle.

South Dakota Wildlife Federation Magazine. June, 2001

"Field and Stream" Magazine. March 2001

"Plowing up a Storm" was both a pamphlet and an ETV documentary.

Chapter Seventeen

Social Contracts

By now, our two mythical hunters have been on the road for many hours. It's been a long drive. They have had plenty of time to talk. Marvin says to Charley, "Charley, what do you think about this ecology stuff. You heard about it?"

Charley shifts to the other cheek resting on the seat and grunts, "Sort of . . . saw something about it in *USA Today*. Jeez . . . they are talking about getting rid of low mileage SUVs, outlawing snowmobiles because they make too much smoke . . . telling you to cut down on air conditioning. How the hell will people in Phoenix live without air conditioning?"

"Beats me," says Marvin. "And those environmental people are really going after corporations. Claim they are more interested in stock price and don't pay any attention to natural resources. I've got a little Harley stock and I like those gains. Hell, people form corporations so they won't have any personal responsibility for things like that."

"On the other hand," chips in Charley, "I agree with them we ought to do something about forest fires. Makes no sense to keep burning up all that wood the way it's been lately. Pretty soon there won't be a place to hunt."

They pass a sign that says they are now on Indian Reservation Land. Charley lights a cigarette. "Ya know," he takes a deep drag, "ya think about the bufflow. Millions of 'em. Covered this country where cattle can't make it. Those dudes just went out and shot 'em by the thousands . . . for hides, bones and just plain sport, like shooting prairie dogs. That was a lot of meat wasted. I was always taught you ate what you shot. Like we do."

He butts out his cigarette. "Where the hell are we," he asks, reaching for the map on the dashboard. He found the ever present map.

They had found their way off the gravel road at daylight and are headed down a long strip of asphalt in their late model, extended cab diesel truck that now has a full tank. They have rearranged gear, some of which they had bought earlier at Cabelas. They are eager to meet their $500 a day guide. To kill time, they have talked wildlife, wildlife management and no small number of "social contracts."

The odds would be good if you opened their glove compartment you would not find a small book written in 1762 by a Frenchman, Jean Jacques Rousseau. He wrote about social contracts as he grappled with the idea of how much "right" the individual should have and how much power the government should have.

THE IDEAS OF ROUSSEAU

Rousseau suggests cooperation among consensual equals is a mechanism for social action as long as it permits a quality of life to be enjoyed for its own sake. What is morally right for the individual is balanced against what the common interest prescribes. The closer the right in the social contract is to consensus, the more power that right has. Today, we see this idea of attempting to locate consensus when the Forest Service mails out thousands of requests for comments that sample what people think about a proposed plan to control elk on the Jackson Elk Refuge. Or, a highway department asks for comments on their latest Environmental Impact Study that involves a highway relocation. They are trying to find a consensus.

Rousseau had several ideas on the subject. One was that a moral choice and freedom were matters of acute and constant concern. He thought man's social arrangements were the products of human choice. Man must bear the responsibility for the kind of society they construct and accept. He thought as long as several men united and considered themselves as one body, they have but one will which is to promote the common safety and general well being. He believed government is simply the servant of the people.

By natural rights, Rousseau meant rights only in and by any society. They were "natural" because they represented the fuller realization of human nature. Society was not an "aggregation of individuals, but an association." A social contract was "a triumph of interests" distinctively different and more completely human than were the individual interests. All of these ideas fall pretty well in line with our use of "social contract" and our use of that term in our model for ecology. We just had to unpeel a little bark.

Rousseau also felt the basic individual instinct was self-preservation and survival. As a person passes from that state of nature to a civil state, they went through a social change. Justice is substituted for instinct. Actions took on a moral character which they did not have before. He seems to say that as a single individual searching for a pelt, the mountain man did not have a lot of human morality. But as he become part of a community, the social contract becomes more the code of behavior. Now, only in the community is the human morality definable.

THE SOCIAL CONTRACT

If the social contract is the power to define morality, then that power may rest with a particular group within the community, but that power is not absolute. If that power rests with the majority, then it rests to some extent upon the consent of the minority. This concept is present today in much of our conservation thinking. If U.S. Fish and Wildlife Service declares a specie to be endangered, the ultimate power of that ruling will prevail as long as it has a recognized moral majority. But, the minority, which might be some environmental group, has the right to challenge that power.

But social contracts can evolve. Our first ethic states that "nature is hostile and an enemy and must be subdued." There is little impartial judgment, little concern for the rights of nature. Self interest is the paramount consideration. This social contract is individual and survival of the individual is the central theme. Independence and self command becomes a moral code. In the "law of the west," the rugged individual "took his gun and shot the sonofabitchen' wolf."

If we believe the human animal is not all logic and thinking, then most human beings would have feelings. Now, a larger question bearing on ecology emerges. Do all living things have feelings? Because if they do the human animal will experience empathy. If humans have physical sensations then other living things must experience physical sensations. We know the human can bond . . . to horses, dogs, birds and in the world of Disney to fawns. What were once regarded as "wild things" with no feelings now have a different moral position.

The end view of this progression of empathy now emerges in ecology as what has been labeled "the Leopold land ethic." This is our seventh ethic. Nature is an ethic and subject to a moral code. If the subscribed code is not a party to a formal social contract, "then the individual should accept the moral ethic of personal stewardship." Personal stewardship becomes a balance between self interest, awareness and empathy. It is the responsibility of the human individual to take care of all living things.

THE ECOLOGICAL QUESTION

Extend the idea of social contracts to animals. Do they exist in animals? Begin with the single animal. The basic instinct is survival. The most natural community or society is the family. When the needs of the family no longer prevail, single members become independent agents or they become members of a union. This results in one of four arrangements. A. The Herd B. The Pack C. The Flock D. The Extended Family

The herd is a gathering of social animals. Affinity is the commonality. Horses provide a good example. A single horse seldom leaves the herd. The mares and foals form strong bonds. Herds seem to have codes of behavior. There will be a lead stallion. Horses will bond with humans. When separated from humans for long periods of time, they will usually revert to being "wild horses," but still members of the herd.

The Pack is a predator organization. Members band together for hunting, security and breeding. Wolves will hunt cooperatively. Lions, coyotes, jackals form packs to hunt more efficiently. They seem to band together for practical purposes, have an organization and a pecking order. You can find the human version of the pack in urban areas. They are street gangs and prowl the urban jungle.

Rocky Mountain Big Horn Sheep, domestic sheep and geese are examples of a flock mentality. They are generally rather docile. They tend to live and let live. Individuals may be wanderers. Eating is their main focus. The society is based on huddling or banding together. While on occasion, the ewe may wander off and leave the lamb, they usually reunite. The "good shepherd" analogies in the Bible were not accidental. All three of these animal or bird communities have social contracts.

You also have the concept of the Extended Family present in animals and the social contracts surrounding that ethic provide both a social study and ecological study. In the human condition, the term can mean family, tradition or necessity caused by unusual events. Self interest and empathy can be factors. Perhaps the most usual case would the cow that will adopt an abandoned calf. The bison cow will also adopt an orphaned calf. Our premise would be that social contracts are present in the animal world.

A FRENCHMAN VISITS

A young Frenchman in his mid-20s arrived in the United States in 1830. He stayed nine months, traveling to major cities like New York, Boston, New

Orleans and Pittsburg talking to presidents and former presidents, leading people of those times. His goal was to study democracy. His name was Alexis de Tocqueville.

His conclusions and thoughts were published in the book, *Democracy in America* in two volumes. This from his book: "Among the new objects that attracted my attention during my stay in the United States, none struck my eye more vividly than the equality of conditions. I discovered without difficulty the enormous influence that this exerts on society; it gives a certain direction to public spirit, a certain turn to the laws . . . it creates opinions . . . and modifies everything it does not produce."

What he meant by "equality of conditions" was probably what we would call equality of opportunity. This would be Old World conditions versus the New World. In the Old World, economic, social and political standing resulted from birth. This American feeling or sensibility he found, resulted in different political relations, altered roles of the worker, the woman and the church. America simply had different social contracts. He felt Americans were ambitious and acquisitive with an appetite for materialism. Far from satisfying the people, it seemed to make them more anxious. "It is a strange thing to see with what sort of feverish ardor Americans pursue well-being and will often show a vague fear of having chosen the shortest route that can lead to it . . . In addition to the goods (they) possess, at each instance (they) imagine a thousand others that death will prevent (them) from enjoying if they do not hasten."

Many of his observations relevant to those times are relevant today in any study of sociology or ecology. He also picked up on the importance of land in the American social fabric. Land was at the core of American social contracts. What law there was concerning appropriation and acquisition of land more often than not evolved as an aftermath of a social contract. De Tocqueville commented on the abundance of land, how it was the driving force that would encourage movement of population. He was aware of the freedom of the new landowner to use the land as they might choose. He located our second ethic.

Rousseau saw the use of land as much more a social contract than it was governmental. Of course, the exception to this would be the large tracts of land acquired by the Federal government for national parks, the building of huge dams for power and irrigation and channel works which came after his visit. Otherwise, the social contract of land use rested in the hands of the people, not in the hands of government. As Rousseau suggests, morality, ethics and "right" are all parts of the social contract that deTocqueville would discover.

As our society pushed west, these views of land use would be modified. East of the Mississippi River, land use was more a case of individual use becoming the social contract. This resulted in a different view of natural resources in the eastern half of the country. As the power of government grew, and land use pushed west, new national attitudes toward land and land use would emerge. Still, the philosophy of Rousseau is applicable.

EXTENSIONS OF THE SOCIAL CONTRACTS

Social contracts are hardly limited to land use. For instance, the treatment of predators was more nearly a social contract than it was a law. From the early times until the middle 1900s, the coyote, wolf, skunk and rattlesnake were commonly regarded as enemies of the person who settled the land. Their eradication was generally seen as beneficial. Today, some people have changed their view of the wolf.

What had been a social contract for eradication has been changed into a law of protection. However, the place of the wolf in the natural ecosystem remains a burning issue in some parts of the country. And when the economics of the rancher are threatened by the presence of the animal, the social contract in some areas is "shoot, shovel and shut-up." It is a case of the government, economics and new social contracts all being challenged.

In the late 1800s, market hunting was an accepted way to make a living. In many cases, the market hunters lived on the margin so ethics and morality came behind self interest and survival. Hunters eliminated some species such as shore birds on the Atlantic coast and the passenger pigeon in Ohio and Wisconsin. This heavy hunting was the social contract of the time. As we moved into the 1900s, laws were passed to control this activity and even international treaties were negotiated.

By the early 1930s, another type of social contract was becoming a force in the world of natural resources. No law was passed to create Ducks Unlimited. They registered as a non-profit organization. It was, as are most conservation organizations, a social contract between like thinking individuals. Their stated purpose was to protect, restore and create wetland habitat. Their funding came entirely from private sources.

Early thinking was that ducks were produced in northern areas. The breeding and nesting of ducks happened largely in Canada. But by law, the United States could not send money to Canada for wetland restoration. But a private source could. The program that resulted in Canada was initially focused on restoration of breeding habitat that would grow more ducks. At the time this was happening, the Canadian government was encouraging the farmer to drain wetlands and put that land into growing wheat. The same was true in the United States. There were, in fact, financial incentives for the farmer to do this. One hand is trying to restore wetlands, the other hand is draining it. But those were the conflicting social contracts of the time.

THE WATERHEN MARSH STORY

We were standing by a sign on a DU project west of Melfort, Saskatchewan, Canada. It was one of the first Ducks Unlimited projects in Canada. We were with Bob Santo, the DU fellow from Melfort. "This is Waterhen Marsh," he said, "and it is an interesting story. Early on, farmers weren't interested in ducks"

He was encouraged to tell his story. "Well," he said, "the Canadian government had incentives for farmers to drain land, especially if the section was part dry land and part wetland. So this farmer proceeds to drain about a quarter section. When he got it drained, he found peat . . . a lot of peat. What to do, eh? So he set it on fire. Well, peat burns with a good, black smoke. Before long, depending upon wind direction, he began to get lawsuits from his neighbors. They didn't like the heavy smoke. Well, what to do?"

You could visualize the heavy clouds of dark smoke drifting across the Canadian prairie. Santo went on. "About the only way you could put out the fire was to reflood the wetland. There was no government subsidy for this. So, he asks Ducks Unlimited, which was just getting started, if they would flood his peat field. And Waterhen Marsh became one of our Canadian projects. The other early one was Big Grass Marsh."

The story of Waterhen Marsh is not an unusual example of the frame our model provides on display. You have natural resources, government, individual social contracts and economics all interrelated. Travel the country that stretches from Canada to West Texas and you will uncover many such stories. The subsidy from government for draining land. The subsequent discovery the drained land has only limited agricultural value. Only lately have we come to discover that prairie potholes provide values for wildlife other than just ducks. Water levels in the area are better sustained. Pollution is reduced. Even humans benefit from an improved quality of life.

In more current times, global warming becomes an issue as the value of ground water and surface water takes on new meaning. In this book, we have tried to remain true to the stated purpose. We were aware of global ecology issues, but we wanted to focus on the grassroots conditions in this country. Of course, there is now strong evidence that if the U.S. cleaned up some of our messy contributions, such as CO_2 emissions, we would make a major contribution to world conditions. For more on this, we strongly encourage the reader of this book to also read the Al Gore book, *The Inconvenient Truth.* It is a story well told and well documented.

NEW SOCIAL CONTRACTS

There is encouraging evidence that new social contracts are emerging in spite of a rather negative government posture on the national level. State wildlife departments are working more in partnerships with private land owners and other conservation organizations. Ducks Unlimited has fewer big projects in Canada and is focused more on problems in this country. New agricultural practices have lessened the impact of erosion. More area potholes are protected or restored.

New social contracts are also emerging in fish management. Early chronicles or notes taken from the journals of sportsmen in the Rocky Mountain area tell of the unbelievable harvest of trout. One story has it that one of the generals who was supposed to support Custer in his Little Bighorn escapade, had trouble getting his men up to Montana because they were so busy catching trout in Wyoming.

Fish, like game, could be found in great abundance. The initial thinking was catch all you could . . . there would always be plenty. In fact, I have a family picture of a boat full of walleye pike where the smaller fish had been tossed on the ground and only the larger ones kept. "There would always be plenty." But this attitude shifted. Usually state government took the lead and limits were placed on fish. Initially, these were quite liberal. After World War II the idea grew that catching was more important than the meat and there was a place for resource conservation. The idea of "catch and release" came to be a code of sorts on certain streams. It was a social contract that slowly spread.

The idea was subsequently picked up by state game departments and first one stream and then another was designated as "catch and release" water. Economics became a part of the equation. Trout fishing became part recreation and part business as big dollars flowed into a river area. Guides and local tackle shops pushed the idea because more trout meant more customers. States established "Blue Ribbon Streams." And what began as a social contract in many areas evolved into law. In essence, trout fishing was being converted to a sustaining resource.

The idea of a sustaining resource went well until Mother Nature put on her waders. First, the game departments found that certain species of hatchery raised trout would not reproduce in the wild. Then, primarily in Colorado, an ugly little pest called Whirling Disease was discovered in trout. Fish mortality escalated. The only recourse seemed to be closing fish hatcheries. Once again, man and the scientist discovered they had not learned all there was to know about trout management.

By this time, another social contract was in the mix. Trout Unlimited had been doing research on the value of stocking trout. They became convinced conservation money would be better spent on repairing riparian areas than on stocking trout, that it provided greater value. Today, state game departments in Montana and Wyoming have begun to do less stocking and are shifting those funds into more stream repair often in partnerships with private landowners.

The grassroots characters of Charley and Marvin may or may not be stereotyped. They add to the jumble of ideas about ecology, social contracts, and common practices as seen from the bottom up. They represent the churn and evolution of grassroots ecology. There is no shortage of stories about self interest, greed and arrogance. It isn't only Charley and Marvin, there is also plenty of that on the higher levels. In reality we may have reached up on the dashboard many times to find a map that would tell us "where the hell we are," only to find many conflicting roads.

OCTOBER REFLECTION

It was a moonlight night in a small South Dakota town. A long walk in crisp October air had taken me to the edge of town. It was a town of many churches, including ones with the ever present spire. As I scanned the wooden building, showing faded white paint, mauled by too many Dakota winter winds, a wedge of geese honked their way across the sky. They passed over the spire. Their "V" was briefly silhouetted against the moon as they moved on their way. They honked an affirmation of a conservation achievement, the restoration of a species. But they were mute on the subject of human social contracts with wild things.

I wondered, if Alexis de Tocqueville had been visiting this church in this town on this night, what might he have written. This land had seen so much, the evolution of so many social contracts. Not the least of which were the ideas that you worked hard, you prayed hard and you hoped you would be rewarded with some measure of security and economic well being.

What sort of map would he have used in these times? Would he observe the land had taken care of the occupant rather well most of the time? That perhaps it was time for folks to take care of the earth and the wild things on it? After all, he was rather perceptive.

Barry Commoner

Was born in May 28, 1917 in Brooklyn, New York. He received his bachelor's degree from Columbia University in 1937 and Master's and Ph. D. From

Barry Commoner. Photo courtesy of Barry Commoner.

Harvard, the latter awarded in 1941. After World War II, he moved to St. Louis and became a professor of plant physiology at Washington University.

Commoner was an important figure in the environmental movement that emerged in the late 60s and early 70s. Where older environment or conservationist efforts had leaned heavily on federal action, the new effort attempted to reach the grass roots. Commoner would help lay the ground work for Earth Day which began in 1970. His comments about citizen knowledge of the environment were often acerbic. He tried to connect the social world to the environmental condition.

He used Lake Erie as an example of human and technological intervention. His conclusion that the economic system should be based on social goals rather than private gain, that ecological consideration had to guide economic and political decisions. (Now we are very close to the implications of our model.) He was a strong advocate of solar energy and believed Americans could come to depend on renewable resources if the policy decisions were not based on private profit.

Whether a student of ecology agrees or disagrees with Commoner, he is an important study. He ran for President in an effort to call attention to the plight of the environment. As a citizen willing to enter politics, he contributed to raising environmental awareness of Americans.

Gifford Pinchot

Was born August 11, 1865 in Simsbury, Connecticut, into a family of upper class merchants and land owners. Pinchot made several trips to Europe and

Gifford Pinchot. Photo courtesy of the Library of Congress,
LC-DIG-ggbain-00800.

entered Yale in 1885 with a desire to go into forestry work. Yale had no such degree at that time, so he graduated and went to Nancy, France to study.

Pinchot came back from Europe convinced trees were a crop. They could be managed and harvested as a sustaining resource. He accepted an offer to supervise the 5000 acre Biltmore Estate. He wrote a pamphlet on the experience claiming his management produced profitable production, a nearly constant annual yield and an improvement in forest condition. These principles marked his forest career.

He would commit himself to forestry in 1894, would become the Head of Forestry in the Agriculture Department in 1898. By 1901, this division had grown into the Bureau of Forestry. Pinchot proved to be an exceptional administrator, very self confident and very opinionated. Yale men came to dominate the department. Pinchot family money enabled Yale to open the first graduate school of American forestry.

President Roosevelt, while governor of New York, had formed almost a conservation partnership with Pinchot. Together, they moved forest conservation toward rational, scientific, federal control of public forests. They got civil service laws in 1904 and the Forest Service became almost an elite and very efficient organization. Pinchot's legacy was a concept of regulated use of national resources with "the most good for the most people."

Pinchot lived a long life, passing away in 1946 at the age of 81.

REFERENCES

Beal, Merrill D. *The Story of Man in Yellowstone*. (Yellowstone Library and Museum). 1960

de Tocqueville, Alexis. *Democracy in America*. Penguin Classics; New Ed edition 2003

McPhee, John. *The Control of Nature*. (Farrar, Straus and Giroux) 1990

Rousseau, Jean Jacques. *The Social Contract*. Translated by Charles Frankel. (Penguin Classics) 1968

Wickstrom, Gordon M. *Notes From an Old Fly Book*. (University Press of Colorado) 2001

Articles from Duck Unlimited Magazines.

Interview with Bob Santo. Ducks Unlimited, Saskatchewan, Canada.

Chapter Eighteen

Why Do We Hunt?

On Saturday evening, September 8, 1804, the Lewis and Clark expedition stopped near Big Cedar Island along the Missouri River. The next morning, they found themselves in a "Sportsman's Paradise." There were buffalo everywhere. Deer abounded. "The men were wild to hunt but took only what they could preserve," wrote Lewis in his journal.

On Tuesday of that week, they camped near what is now the county line between Brule and Charles Mix counties in south central South Dakota. The people they found there were mostly Indians and a few fur traders, many of whom were married to Indian women. The Indians looked upon the traders as a market for their goods and a way to barter for firearms, supplies and alcohol. In 1855, fifty one years later, Fort Randall had been established by General Haney. By a special act of Congress, all settlers on Fort Randall military reservation as of January 1 of that year were given homestead rights to the land they occupied. Charles Mix County was created by the First Territorial legislature in May of 1862.

THE PHEASANT HUNT

We were standing at the end of a South Dakota cornfield in Charles Mix County some 140 years later. There were 20, perhaps 24 people assembled. Each wore some type of orange garb, vest, hat, a game carrier. The leader was a retired Army Corps of Engineers Colonel. The members of the group ranged from 12 years old to old timers, a mixture of local family and visitors. It was nearly noon.

"I want," said the Colonel, "two flankers down each side of our drive through this field. About 100 yards ahead. Then two more, one on each side, about 50 yards down. You six, you-you-you two-you and you . . . you guys, take the pickup truck and go to the far end and block. Don't shoot low. Call on the radio when you are set. Leave now and don't slam car doors. Be quiet. There are a lot of birds in this field. We scouted it this morning."

The pickup left. The remaining members of the group spread out along the end of the field. Guns were loaded. A few hunters checked the fit of their guns to their shoulders. Some knelt and squinted, wind in their face. The sun glistened now and then on gun barrels.

We waited. And I thought to myself, "my God, it must have been like this, without a radio, long ago in Charles Mix County. This is communal hunting. We aren't Native Americans and these aren't bison we are driving, just pheasants. But the idea is pretty much the same. We have formed a human chute. If everybody cooperates, the birds will be driven down to the end of the field. Penned in at the end, they will be like the bison before the time of Lewis and Clark.

Our technology is much better. Are the reasons we hunt any different or better? Certainly, we don't need the meat. We have an ATV for chasing cripples, radios for communication, polarized sun glasses to help with sun glare. The latest Gortex apparel to protect against changes in the weather. Science has advanced in 140 years. Has the hunter mindset changed?

My ruminating is interrupted. The radio in the hands of the Colonel crackles. He glances at his watch. In a distant field there is a boom of guns. Watches set ahead of ours? He waves to his line. The hunt and the season has begun. We advance and disappear into the corn. The older hands zig and zag down the rows. The young studs plunge ahead. "Easy there," says the Colonel.

When we have finished the day, or gotten a limit for everybody, we will lay the pheasants in a row, or two rows. Some hunters will kneel in front of the display, some will stand behind it. Cameras will click. The hunt has been recorded in a Kodak moment. Is it a hunt or a human machine?

I remember the first pheasant I ever shot in a Nebraska cornfield. It was a transforming moment, someplace between laughter and tears. The gaudy gentleman from China is a beautiful bird. I felt a twinge of sadness for the pheasant as I held him. I also felt a small twinge of pride for what I had accomplished. It had been a good shot, a clean kill.

Years later, I would read the story of The Old Man and the Sea. I could relate to the words Hemingway put in the old man's mouth. "I love you, fish, and I respect you. But when night comes, you will be dead." In time I was to learn that hunting is a combination of many things . . . pursuit, challenge, camaraderie, peer group pressure, social contracts . . . and sometimes twinges of apology and regret.

THERE IS A DISSONANCE

. . . in hunting, an inconsistency between one's beliefs or between what one does and what one believes. There is a cognitive discord. There really is no way any hunter or fisherman can deny that what we do is any less than a blood sport. "Catch and Release" tends to sanitize that idea a little, provide a little elbow room. Even then, with the best of intentions, there are times when a hook goes too deep or the fish struggles too long and death is the end result. We reflect that death is part of an eternal struggle. We are party to that struggle. To hunt, we must accept this and make do the best we can.

I would not presume to defend either hunting or fishing. Finer minds than mine have pondered that conundrum. Ortega talks about social evolution. He tells how aristocrats found liberation from daily humdrum by turning to "The Joy of Life." Hunting was the primary diversion of the noble gentry. Fishing, for some people, also reached that lofty plateau. Hunting has traditionally been a male activity. For the Native American, it was a way the male gathered "coup," got bragging rights and manhood. But in that culture, it was also a necessity for gathering food as well as a social activity. That was the social contract.

The Euro American has mucked it up. We have added guides, packers and land leasing. We have established game departments that depend upon licenses to meet budgets. We have added millions of dollars in sporting magazines and advertising. We have added the economic infrastructure of conservation organizations. We have provided passionate positions for addicts who hunt and equally passionate positions for those opposed to it. Hunting provides a huge gym where folks get a lot of jaw exercise.

Hunting and fishing involve all four frames of our model. Because of the connections to natural resources, hunting, as part of our national fabric, simply must be considered in any study of ecology. It is a significant use of natural resources.

HUNTING MANIA

The urge to be a hunter can be powerful. In some cases, it replaces the sex drive. This specie will desert family, church, business and social responsibility for the fields, marsh or mountains. Dental appointments are rescheduled. The only doctor in small towns in hunting country seems to have other plans. Groups get together and observe long held rituals. Times past are recalled. As more women join hunting and conservation groups, they bring new social customs, some involving single parent families.

Blend economics into the social aspects and it has become a complex world. Recreation is big business. Add up equipment, guns, gas, motels and licenses and it is hard to justify the cost based upon the meat brought back. A chicken from Tyson costs far less than a pheasant from prairie country.

Time has become the critical element in hunting economics. Success is still a driving factor for many hunters. For many of this breed, they simply must be able to respond to that time honored question, "How'd ya do? . . . by responding "we got our limit" or "I got my animal." So in the case of hunting big game, that usually means having a lease lined up or locating a good guide. There usually isn't time to scout, time on the hunt will be limited. A good guide is essential. Time is the dominant word.

For the guide, there is now a social contract. It was inscribed on the tablet of Boone and Crockett, (evidently Moses didn't hunt), that every hunter will shoot their own animal or kill their own pheasants. This means it is incumbent upon the guide to get the client into a favorable position, at the right time, and hope the client's physical condition is such they pull the trigger. And that the quarry is a prime specimen.

If the client happens to succumb to a serious case of the shakes and misses the quarry, the guide is obligated to repeat the process until the client is successful. Because the guide has a social contract with other guides and the hunting peer group. After all, the guide has advertised "100% success" in the ad they ran in the sporting magazine. The economic dimension is now a tangle of related complexities. When a well meaning environmentalist suggests closing the season to protect animals or birds, they simply do not understand this is not just a natural resource decision, it is also a very serious economic decision.

The paradox of this scenario is the person who pulls the trigger that resulted in killing the quarry can now claim "to be a hunter." Although in reality, they have been taken to an area already scouted, put in a position to shoot, had been taken care of, fed, provided accommodations and even had the quarry pointed out to them. They now have a "trophy" to prove their prowess and prestige that a hunter once had to earn. In deference to the hunter, this scenario is also quite common for fishing guides.

THERE IS NO QUESTION

. . . but that the value of time and the affluence of the population have impacted hunting and fishing. Many of the old social contracts have fallen by the side. The hunting fraternity has become hard pressed to defend what they believe is a "constitutional right" to own guns. These rights are under pressure because technology, space and purpose of the hunt have changed.

Weaponry for the hunter has advanced to the extent that hunter deficiencies have been minimized. Guns have continued to get more powerful, shells have been improved, lead is being replaced by alloys, and sights are stronger. Radio equipment allows for better communication. The ATV and snowmobile allow for access to more remote areas. Just as technology shifted the balance of survival from the Native American to the intruder, so has the advance of technology shifted the odds of survival for wildlife.

The second great change in hunting in the last fifty years has been space. More hunting pressure is being compressed into less space. Less private land is available for hunting and even acquisition of land by game departments for public hunting or "walk in" areas has not kept pace with increased hunting pressure. Membership in conservation organizations has grown. This has recruited more new hunters. The new faces are generally from urban areas.

In some states, game farms have emerged and this has become a maelstrom of controversy. In Michigan, private land is limited because there is so much state owned land. The game farm operator has been forced to defend their operation. While it may be their land, they may not be able to use it as they please. A number of coalitions and animal rights groups are challenging this old ethic.

THE SHIFT IN PURPOSE

One of the most significant evolutions in hunting tradition has been a shift in purpose. Except in a few isolated cases, hunting is no longer pursued for food. Most hunters probably still abide by the ethic or social contract that you do not waste game. It is still regarded more as a harvest of a sustainable natural resource. Hunting still provides a challenge, but more what Ortega characterized as "the human being at maximum alert in his contest with a worthy animal adversary only a little less worthy than himself." The challenge is the thing. The human animal still looks for challenge, in sports, the job, in daily life. When the challenge of life is gone, part of life is gone. A day of hunting or fishing will renew that reality.

I would argue that most healthy males still look for challenge. If they hunt, they feel some primordial urge, and wait patiently, or not so patiently, for the hunting season. Deep down, old genes lurk. They say to modern man, you can outsmart the opponent. A distant voice calls and says, "come on, let's go play the game, let's confirm that you can do it. We respect the quarry, maybe even love the quarry, but it's a game. You can win if you try."

Only now and then does the average hunter enjoy "one of those days" when everything is perfect. I would argue that hunting and fishing both provide a maximum opportunity for exposure to failure. Failure is a yardstick by which eventual success is measured. The odds for failure only enhance the eventual self-esteem. It is more like the Indian coup'.

Now and then I am assaulted at a social function by a nice, older lady who admonishes me because I hunt. She takes me to task and asks me how I can stand to look into the eyes of a dead animal or game bird. I have done that. I have seen the eyes of the mallard drake or crippled pheasant. I think of Aldo Leopold's classic prose describing how he looked into the eyes of the dying wolf. I reflect on predators and prey, about the reality of nature and try to justify my actions by shooting well and respecting the spirit of the quarry. Yes, I have seen the spirit in their eyes. I reconcile this because of the small, even selfish satisfaction, I have won. I view it as a harvest. I renew a commitment that tomorrow I'll plant something in a conservation cause to replace or renew what I have harvested. In a bigger equation, there is self justification in renewal of natural resources.

On the other hand, tomorrow the well intentioned lady will address her slot machine or bingo card. She will be there because she likes the challenge. She pulls a handle for the same reason I pull a trigger. Or she hollers, "Bingo" for the same reason I holler "good dog" for a stout retrieve. It means she has experienced something special. She has chosen to compete, sometimes against poor odds, knowing she won't always win, but she likes the feeling of just being there and hitting three bells in a row. Of course, a jackpot is even more exciting! And her buddies will gather around and congratulate her on winning!

I can relate to her success. I only wish she could relate a little more to mine. Most humans like to win.

WHY PEOPLE DO IT?

I have looked at many sources for information about this "why" of hunting. I located material that came from The Archery Manufacturers and Merchants (AMO). They asked the archery hunters why they hunted with a bow. People who responded were allowed multiple responses.

27% responded "to take an animal."
46% responded "being with family and friends."
53% responded "for the meat."
78% responded "for the challenge."
78% responded "for relaxation."
86% responded "for being close to nature."

A July of 2003, *Field and Stream* Magazine had an article that related to what their readers thought was the future of hunting. It also included some statistics about why they hunted. Their base of response was 2,897 replies. They found the average age of hunters was going up and the number of hunters was going down. "Today, only 6% of Americans 16 and over, hunt." The article observed, "In the language of wildlife biology, the sport of hunting is having 'recruitment failure'—-a condition that, in the wild, ultimately leads to extinction."

Of those polled, only 13% considered themselves trophy hunters. 69% replied they would hunt on a preserve where pen-raised pheasants and quail were added to the wild population. On the subject of hunter ethics, 41% thought hunters acted about the same as they had 10 years ago, 33% thought ethics were worse and 25% thought they were better. As to hunting opportunity, 64% thought they had fewer places close to where they lived where they could hunt, 14% felt they had more places, and for 22% the opportunity was about the same. 80% had never paid for guide services, 91% said they would report hunters they discovered breaking the law and 66% said they would support a tax increase to enable the state to buy more land that would afford hunting. These figures parallel my informal cards.

THE HUNTING IMAGE

An interesting question involved how hunters felt about how the media treated them. 90% felt hunters were presented less favorably than they should be. When asked what they felt were the bigger threats against hunting, 41% felt it was a case of shrinking habitat, 22 % felt it was the animal rights movement, 25% said it was anti-gun legislation and 5% thought it was hunting laws and regulations. Then, depending upon your perspective, a very telling question about membership in conservation organizations. Only 50% of the respondents personally belonged to a conservation organization. Tie this to the question about the cost of hunting licenses. 50% felt they paid too much for hunting licenses, 37% thought the price was about right, and 13% said it was too low.

So you can argue this on both sides of the fence. There has been a long standing argument that without the support of the hunter and fisherman, there would have been no conservation effort in this country. If this survey has any validity, 50% of those supporters feel they are paying too much. Half of them buy their license but belong to no conservation cause. From this information, it seems there are still a lot of hunters who have their hand out. Only about half of them care enough about a sustaining resource to support causes that make their sport possible.

One other aspect of this subject relates to ecology. That is the question of "what is the impact of recreational consumption on the natural resource?" Fishing has at least three considerations. The first would be people who need fish for subsistence. This would involve existing agreements with the Native American. This issue is the source of more conflict in the Northwestern United States than in other areas. Salmon fishing, the decline of salmon runs, irrigation versus the salmon hatcheries . . . these are all polarized areas of contention.

The second consideration would be commercial fishing which is still widespread globally as well as along American sea coasts. This is another area of substantial conflict. The third consideration would be recreational or sport fishing which is less consumptive.

This much about fishing we do know. Put too much pressure on the resource and you can seriously damage the fish population if not entirely deplete it. Secondly, if you consider fish a crop, some species can be grown, produced or farmed much like any commercial crop. But even in these artificial circumstances, erosion, waste disposal, pesticides and dams can impact fish management. Still, from a very broad perspective, we may know more about managing and manipulating fish populations than we know about management efforts for game and bird populations.

HUNTING IS A SLIGHTLY DIFFERENT GAME

. . . and research on the impact of hunting on populations is less specific than for fishing. We do know, from experience, that without protection and regulation, human predators can eliminate an entire specie. It took Federal Regulation to stop poaching and market hunting. Even then, the general public was not concerned because the perception was one of abundance. When the general media exposed the idea of extinction, social contracts that people needed jobs shifted to some concern for the wildlife involved. Slowly another idea emerged. Maybe protection and regulation was not enough. Maybe it was time to think about restoration.

THE BEGINNING OF THE REFUGE

It was a German immigrant, one Paul Kroegel, who generated a battle between conservationists of that time and the poachers who had been slaughtering brown pelicans for feathers on a tiny island called Pelican Island. It was the bird's last breeding place on the Florida east coast.

This four acre speck of land on the Indian River drew the attention of Theodore Roosevelt while he was President. He asked if there was any law that would prevent him from declaring this a Federal Bird Reservation. He was told there was not. "Very well," he is reported to have said, "then I so declare it." And so was born a system of wildlife refuges in the United States that now numbers 540 and protects some 95 million acres of land. This does not include some 3000 Waterfowl Protection areas.

For all that we have learned about waterfowl management, there is much we do not know. The duck population was blessed with abundant rain, snow and water in the early 1900s. Government subsidy had provided a benevolent canopy of habitat with the CRP program. The prairie pothole country responded with a record number of ducks. Federal authorities responded with longer seasons and increased bag limits. The hunters rejoiced. So did game departments because the sale of licenses went up, the number of duck stamps went up, the demand for shells, equipment and decoys went up. As the natural resource improved, the economics improved.

Things were ducky! Except for a Delta Waterfowl Report issued by this highly respected conservation organization. It warned of a coming decline in the fall forecast for mallard numbers based on their research. "Mallard numbers should be showing growth, not serious declines. Two areas of primary concern exist. First, the impact of predation continues to take its toll of mallards . . . Secondly, we are troubled by the dramatic increase in hen harvest as the result of doubling the hen limit." When they tabulated the hen harvest in 1997 and 1998 seasons, 63% was from additional hens in the bag. Total hen harvest was approaching 2 million of the approximately 5.5 million total mallard harvest.

Now the research of a private organization does not square with figures from the U.S. Fish and Wildlife people. Not only is the hunter a little bewildered, the confusion contains economic implications like the possibility of falling license sales. Delta introduces a new song for the hunter. It is called "Don't shoot the brown mallards." Or, "Shoot only drakes." It is a new social contract for the hunter. Once again there is no consensus of what is best for the natural resource.

ON THE AMERICAN CONTINENT

. . . what are the odds people will continue to hunt and fish in the near future? Where and how will increasingly be a matter of how well we can align the economics, the social contracts, government and natural resources. How much pressure those natural resources can absorb will continue to produce conflicting views. As the population shifts from rural to urban areas, more people are farther from the grass roots. I have had guides of urban recreational users tell me people they have exposed to the sport or recreation come back from that experience with a much deeper respect for nature and the environment. That's a slow learning curve, one or two converts at a time.

As we have learned more recreational uses of about nature, the questions have gotten larger, not smaller. It is almost like people concerned about the environment are busy swatting mosquitoes . . . you swat one and three more are buzzing around. Common wisdom would seem to suggest there are big issues. Water quality and is our water supply endangered? Global warming and how much time do we have to address that problem? Is this related in any way to drought conditions in prairie country? How important is the aging of the population in rural America and how many people can urban America absorb?

Hunting is a small part of a much bigger equation. It may be like the canary in the coal mine or the prairie chicken in the grass lands. We can swat at snippet issues like ATV machines in the wilderness areas, how much smoke and pollution will we tolerate from snowmobiles, how many fishing derby promotions can we stand, what do we want to do about guys who climb into their boat in raincoats under which they have hidden two stringers of fish to dump into their live well? And in South Dakota they will protest the mowing of roadside ditches because this takes away pheasant cover but the farmers need the grass for cattle feed.

Perhaps it is just much easier to swat the mosquitoes on the grassroots level than it is to grapple with the bigger, more distant problems. But the reality is that if we cannot resolve the smaller issues, there is less hope for finding solutions to the bigger environmental problems. There may come a time when we put hunters and anglers on the endangered specie list. Who knows? We may already be endangered and just don't realize it.

How much patience will nature have with people who insist on turning sport into a machine, a TV event, a promotion or a fine can of worms?

Rachel Carson

Was born May 27, 1907 in Springdale, Pennsylvania. In 1929 she graduated with honors from Johns Hopkins University before going back to Johns Hopkins to get her M. A. in zoology.

She had grown up in the rugged environment of a simple farm house, where her mother infused her with a love of nature. Rachel joined the U. S. Bureau of Fisheries and was a writer on a radio show, "Romance Under the

Rachel Carson. Courtesy of the Lear Carson Collection, Connecticut College.

Waters." In 1936 she passed the civil service test and the Bureau hired her as a full time biologist. Over the next 15 years, she went up in the ranks and became the Chief Editor of all publications for the U. S. Fish and Wildlife Service. So, a love of nature, a background in zoology and professional writing experience provided her credentials to write what became her legacy.

She started work on "Silent Spring in 1957" and it was published in 1962. Her first successful book had been "The Sea Around Us." In early 1959, now retired from government work, she still retained valuable contacts with scientists and researchers. She began to focus on DDT. That part of her story is covered in our text.

By April of 1963, her health had begun to fail. Still she continued a heavy schedule of appearances on television. She would pose this question. "Who are we to say that those who come after us may never see some of today's rare and endangered species?" She continued to wonder why so many people failed to recognize they were a part of the earth's ecosystems. She lost her battle with cancer at the age of 56. She has been called "the mother of the modern environmental movement."

Rachel Carson died on April 14, 1964 at her home in Silver Spring, Maryland.

Stewart Udall

Was born in St. Johns, Arizona on January 31, 1920. His public school years were spent in St. Johns, a Mormon community founded by his grandfather. He attended the University of Arizona, took time off for two years as a Mormon missionary, was a World War II gunner in the Air Force in Europe and finally graduated in 1948 with a law degree. It is interesting to note that Udall is one of only three persons in our All-American Team of Conservationists who was born in what we would call "the West."

In 1960, after terms in Congress, he persuaded the Arizona Democrats to back (then) Senator John F. Kennedy. Later, President Kennedy would

Stewart Udall. National Park Service Photo

appoint him Secretary of the Interior. As a member of the Cabinet, he achieved notable success. During his tenure the Wild and Scenic Rivers Act and the Wilderness Bill were passed. The Land and Water Conservation Fund was established. Four new parks were added to the National Park System, fifty-six new wildlife refuges were established, twenty historic sites and six new national monuments.

Udall sees conservation as part of a political process. He says, "If the forester and the reclamation engineer symbolized the conservation effort during Theodore Roosevelt's time, and the TVA planner and the CCC tree planter typified the New Deal, the swift ascendancy of technology has made the scientist the symbol of the sixties. I think a historical look at conservation would reveal an ebb and flow, with two high tides of conservation, one under Theodore Roosevelt and one under his cousin Franklin." In his own right, Udall can justly be regarded as one of the great conservation minds in our American history.

He is still living in Santa Fe, New Mexico.

REFERENCES

Audry, Robert. *The Hunting Hypothesis.* (Macmillan Pub Co) 1976

Beal, Merrill D. *The Story of Man in Yellowstone.* (Yellowstone Library and Museum Association) 1956

Hemingway, Ernest. *The Old Man and the Sea.* (Saunders College Publishing) 1961

HAPET (Habitat and Population Evaluation Team) *Delta Waterfowl Report.* Vol. 2002 Issue IV.

Leopold, Aldo. *The Fierce Green Fire.* (Falcon) Paperback 2005

Michener, James A. *Chesapeake.* (Fawcett) 1986

Sports Illustrated. September 2, 2002.

Chapter Nineteen

The Model Applied To Wildlife Management

Since 1900, when we began a long trip to unsnarl environmental problems, we have surrounded ourselves with friends who told us the solutions to environmental balance were simply a matter of accumulating enough data. The scarecrow became the scientist and brains would replace bumbling. Scientific inquiry could reveal solutions to environmental problems.

The asphalt roads led increasingly inward. We evolved into a more urban culture. We had less and less contact with the land. We also fell in love with the predictability of technology. Only in times of great drought, flood, tornados, hurricanes or forest fires did we marvel at the unpredictability of nature. As a culture, we lost much of our heritage of personal contact with the land and natural things.

The sportsman or outdoors person would cling to the love of unpredictability. They were often a small minority fighting a rearguard action. Money, machines and science became our mantra. We lost touch with the human in humanity and the life in wildlife.

THE VALUE OF MODELS

The simple model we have proposed can show shifts of stages or passages. It can explain many of the evolutions involved with natural resources if you allow for some reasonable margin of error. In the prairie, this is often referred to as "Osage windage." The stages are helpful to visualize slow evolution. Agreement on semantic differentials allows for consensus.

As we understand more about ecosystems, we know they are not stable. Always there is evolution. In his book, *Our Natural History*, Daniel B. Botkin

talks about developing computer models to produce calculations that involve
several variants in the forest ecosystem. The news reports the costs of forest
fires. "The costs are enormous," says Jeremy Fried, team leader on the study
for the Pacific Northwest Research Station's forest inventory analysis program. "This study is to determine how much biomass might be available for
an energy-producing plant proposed for southwest Oregon." At issue is how
much brush and small timber could be removed as a way to lower the risk of
wildfire.

Fried and his associates are building a computer model, extending bits and
pieces of half a dozen other programs. These allow silviculturalists to experiment with numerous variables such as the species of trees involved, tree density and diameter, roads, topography and distances from timber mills.

We tend to overlook the point that scientists talk best to other scientists.
Many people are suspicious of scientists. The scientific world does not always relate well to the media, who tend to regard data and facts as less ap-
pealing to their audiences. In fact, if Ben Franklin had taken his key and kite
outside in a bad storm today, the news would probably report it more as an
entertainment event and less for the scientific importance.

Back to Fried. "Economics is where problems develop," he says. "If a logger takes only small trees, it costs an arm and a leg." He then points out when
the land is flat, there are good densities of timber, there is good transportation
available, brush and small trees can be eliminated at the same time the logger
is taking profitable timber. Then comes his conclusion. "Only on one-eighth
of the landscape were we able to get positive net return. It seems the biggest
fire dangers are on steep mountains, the same areas it is expensive to get good
timber logging returns."

Fried's model becomes a way to visualize relationships. It can be used to
explain interdependence of the frames. The focus is on a total picture. Switch
to the popular efforts to create a Sandhill Crane Refuge near Kearney, Nebraska on the Platte River. Here are some of the frames a person might select
to develop a model. The economics of visitors coming to see the birds. The
cost of building the refuge. The quality and quantity of food in fields around
the refuge. The water quality in the river. The amount of water available for
cranes as opposed to irrigation needs. Existing agreements or compacts existing along the watershed. Possibility of disease if the cranes become too
concentrated. Enforcement of protection around the refuge. Are there enough
Rangers? How many private or commercial visitor sites will be allowed?

Then consider all the organizations that might be involved. State game
management, The Nature Conservancy, irrigation groups, Chambers of Commerce. It is not a simple, single answer. There are too many relationships involved. There will be pluses and minuses. Then consider that while we are

protecting cranes in Nebraska, there is still a season for hunting them in Texas.

THE MANY QUANDRIES OF GAME MANAGEMENT

We went through a long and sometimes painful series of stages to get to where we are in game management. This all started around 1910 when laws were passed to curb market hunting. We passed laws, but there was little money for law enforcement. That condition still exists in some parts of the country where one game ranger covers hundreds of square miles, where arrests will often go unprosecuted, where violators receive little nor any punishment. Protection is preached but lawlessness is practiced.

But the idea of protection is really Stage I in the process of game management. The second stage is sometimes identified in game circles as the "Regulation" stage. The Endangered Species process uses this stage when a species is declared to be endangered. All of the information about the species may not be known, nor the complete circumstances researched, but available evidence indicates regulation is called for. There may need to be "harvest limits," or, the species is declared endangered. It evolves from "Protection" to "Regulation."

On the east coast, the declining health or condition of ellgrass in 1931 and 1932 was widespread. The biological implications were dramatic. Suddenly, and with no known cause, eelgrass died out and disappeared. This grass was the primary food for the Atlantic Brant, one of the smallest birds in the goose family. It is native to both coasts, being a seagoing bird that breeds in the far reaches of Canada and winters along the coasts. The change in diet changed patterns of migration. The federal government closed the hunting season and treated the bird as a protected species. This was two steps, protection and regulation.

The culprit turned out to be a slime mold fungus that had emerged as a new enemy of the grass. Now we are into stage III, the biology involved. The ellgrass made up 80% of the food supply of the Brant so the specie was especially vulnerable to a depleted food supply. On the good news side, some of the Brant were able to change feeding habits and began using sea lettuce and even ordinary grass. The change in diet changed migratory patterns. Eelgrass began to recover in the 1940s and in the 1950s became almost as plentiful as before the die-off. By the 1960s, a 30 year cycle, the Brant population had recovered to an estimated 265,000 birds. The Brant had evolved from Protection to Regulation to Biology in a Wildlife Management model. It was close to Stage III or a normal sustaining resource.

The Brant story provides a mini illustration of our model at work. It also provides a good story of increasing efficiency in game management. In this case fungus had caused the Reform stage. But it could have been hunting pressure or industrial pollution somewhere along the migration route. If the Brant provides a story of a natural return to a normal condition, the story of the Canada goose, especially the large Canada goose, provides a story of exceptional restoration. Today, some folks would argue there are too many of the honkers around.

THE YELLOWSTONE PARK STORY

The protection stage in game management will usually have an initial stage marked by acquisition and regulation of protected areas, refuges, parks or protected area. Yellowstone Park makes a good illustration. Emerson Hough was a transplanted Iowan who became a working journalist who then became a conservationist and chronicler who produced western fiction and articles for magazines. He commented on a wide variety of western subjects from cowboys to carnivores in the period of the late 1800s.

His writing attracted the attention of Theodore Roosevelt. They developed a friendship and Hough developed into a historian of the western expansion. He traveled widely. Beginning in 1882, he contributed occasional articles to *Forest and Stream*, a New York sporting journal that was a pioneer publication in conservation. In 1889, he became its western representative.

Early in the winter of 1893, Hough and his friend and guide, Billy Hofer, and two soldiers from Fort Yellowstone, had crisscrossed the Yellowstone National Park in sub-zero temperatures to photograph wild game and count the bison. Yellowstone authorities said there were 500 bison in the park. Hough could locate less than 100. Poachers were at work. In fact, on this trip, Hough was a party to the capture of a poacher. When the army imposed a penalty, it only confiscated the poacher's outfit. Hough was outraged by the lenient penalty and wrote a report that highlighted the absurdity of park regulations that did not protect the park bison against the slaughter of the animals by poachers. This report was picked up and published by an eastern newspaper. In 1894, the article resulted in Congressional action and in May of 1894, Congress passed a strong law that made poaching in the park a punishable offense.

Later, Hough was quoted as saying, "I have always thought this was about as useful a thing as I ever was able to do in the somewhat thankless attempt to be of service to the wildlife of America." This article was in the *Saturday Evening Post* in the June 5, 1915 issue. By now, his views of conservation and western life were gaining a wider audience in the *Post* as well as in *Forest and Stream*.

But the point about grassroots apathy is worth noting. It took mainstream media, and their wider circulation, to force legislation that put teeth into the idea of protecting wild animals in a national park. What had been a rather cursory attempt at protection evolved into what came to be regulation. Today, bison are probably in the Biological Stage. There are still reservations about genetics and brucellosis. But the animal has reached a level of sustainability.

SUSTAINABILITY

From an ecological point of view a sustainable condition is a combination of social contract and law that allows a society to satisfy its needs without diminishing the resource for future needs. This ideal is the polar opposite of unlimited growth and unlimited consumption. It is one of the conservation ethics we have identified. In a pure ecological point of view, an ecosystem operates sustainably if the inputs of renewal and the outputs of extraction are balanced and over time, the system does not lose substantial amounts of the elements that provide the energy needed for a high performance ecosystem.

In wildlife management this idea of sustainability sets up the third stage or passage which is often referred to as the "Biological" stage. This stage introduces the scientific aspects of wildlife management and is usually based upon research and careful observation. Usually, this stage is focused primarily on one habitat or one species. In a state game department it would be focused on conditions that result in data that would be used for setting seasons, stocking or planting certain species, and the management of the habitat and species to produce the most good for the most people.

The "Biology" stage then usually evolves into a fourth stage or level and that is the "Ecology" stage. This expands thinking into the whole range or spectrum of the ecosystem and begins to include the game and fish species involved, but also includes non-game species in the given habitat, the mammals, the insects, the reptiles, amphibians and the plants, trees or grasses involved. This fourth stage of wildlife management is more complex and more holistic. If we tracked the history of Ducks Unlimited as a conservation management organization we can see the stages.

THE DUCKS UNLIMITED EVOLUTION

Initially, the focus of this conservation group was on duck population and sustaining that population. Members wanted ducks to hunt. The unspoken goal was the sustainable harvest of ducks. From an economic point of view one could argue that this singleness of purpose accounted for the organization's success in fund raising, especially at the grass roots level. Duck hunters "did it for the ducks" but they also did it for a duck dinner. Then as DU evolved from an organization focused only on ducks and duck habitat to a more scientific and then a holistic agenda, they came to admit that sending money to Canada for restoration of breeding habitat was only part of a very complex ecosystem.

Today, they are involved in restoring wintering grounds in Mexico. They have begun to partner with state and federal agencies in the development or restoration of flyway habitat for ducks. They have come to accept the tough fact that restoration and sustainability of the duck population is much more than duck biology. It involves all the complexities of ecology.

The latest development in the ongoing evolution of the duck scenario is the establishment of working relationships with the private sector. And where law and regulation have been the norm to date in this arena, new social contracts are beginning to emerge in the relationship between a willing private land owner and a conservation organization.

At this time, there are whispers of a fifth level or passage in wildlife management. In wildlife circles, it is dubbed "Sociological." In pure ecology, it is referred to as "deep ecology" and it has many spiritual overtones. Basically, it seeks to learn what the user of the resource wants from the resource. In this emerging stage of wildlife management, the thinking moves beyond the study of wildlife to the study of the users of the natural resource. What do these "users" expect? What are their needs? What are their values? Their attitudes? What do they think is their relationship to the natural resources? In this effort or stage, the wildlife managers have moved from a reactive posture to a proactive posture, trying to anticipate how the public will perceive wildlife problems in advance of the emergence of the problem. This new thinking has led to many of the requests for comments that are being used by wildlife management on both the state and the federal level.

Any organization is a living thing, be it a State Game and Fish Department or a conservation organization. It has determinants or frames. If you are a member of such a group, your organization is someplace in our model. The details of this idea were discussed in Chapter 5.

The left hand frame we label "Value Decisions." It is the statement of purpose. For the North American Grouse Partnership that is . . . " to promote the conservation of grouse and the habitats necessary for survival and reproduction." Included in the "Value Decisions" would be PURPOSE, PROCESS OR PLAN and PAYOFF.

STRUCTURE is the bottom frame in the model. Basically, it is the way the organization is put together. Will it be a President, a Manager, a CEO . . . have

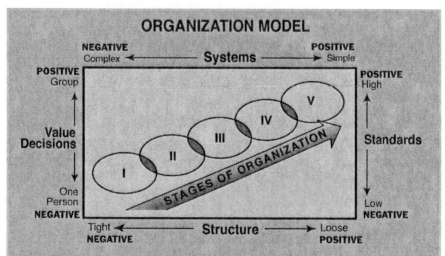

ORGANIZATION MODEL

VALUE DECISIONS. This determinant becomes the mission statement for an organization. Value decisions should state the purpose of the organization, the process by which the purpose is achieved and the desired end result or payoff. As an organization evolves or matures, the Value Decisions may be modified. They serve as the guideposts for management or leadership decisions. Group Value Decisions are more powerful or positive than those made by a single individual.

STRUCTURE. This determinant describes how the organization is put together or assembled. This could be teams, regions, divisions, departments ... all titles commonly used to describe the structure of an organization. The structure will have characteristics. It can be very tight. People do not cross lines. Usually, this is negative. Or, it can be very loose where people can work in more than one area if that becomes necessary to get a job done or accomplished.

STANDARDS. These are nearly the social contracts of the unit in the organization. The continuum can run from very negative and virtually nonexistent to very high, even nearly non-attainable. Sports teams have standards ... or some of them do. Pride is often involved. Standards can be Codes of Conduct. For salespeople, they can be goals. Standards often become the measurements for performance.

SYSTEMS. This determinant would include the economics of the organization and how performance, production, cost and pricing is measured. They could include such areas as benefits, vacation or leave time, compensation plans. They range from very simple systems to very complex systems. Usually, the more complex the systems, the more rigid the organization. Each of these components or determinants can be in different stages. A low stage in one determinant can pull the whole organization down or backward. None of them can be put in a free frame. A high performance organization would have all determinants in Stages 3 or 4 and would be guided by superior value decisions, have a loose structure, high standards and rather simple systems.

Organization Model

a Board of Directors, will they be appointed or chosen? The more layers, more complex the structure, the slower the organization will react or act. Small and loose will beat big and rigid when it comes to getting things done.

The right hand frame is STANDARDS. In many of the State Game and Fish Departments with which I am familiar, this is the weakest part of the organization. You can't have standards if there is no performance review. In too many organizations, there are no standards. Consequently, there is no reward for performance and no consequences for non-performance. As a result, the people who do perform well will usually get additional assignments. The incompetents are sweating out retirement.

The upper frame is SYSTEMS. Will the systems employed be simple or will they be complex? Simple systems enable flexibility. Complex systems are hard on forests . . . they use lots of paper. And tempers flair!

LIKE NATURE

. . . any organization is in a constant state of evolution. Managers try their best to freeze frame an organization. If it gets to a high level of performance, the natural action is to lock everything in place and just oil it now and then. But an organization is like an ecosystem. It is difficult to manage either. The best choice is alignment so the four frames are working in harmony. This is the difference between management and leadership. A manager tinkers with only one frame or another. A leader searches for alignment

Wildlife management will chaff a bit at any indictment of the way their organization is being run. They will tell you they have lots of bosses, not the least of which in some states is the state legislature. Some states have tried to eliminate political interference with constitutional reform that has resulted in more independence. The value to ecology of the well run state wildlife operation can hardly be overstated. They are closest to the grass roots. They are down on the field where the action is.

Wildlife management, in the broadest context, represents the hope for coming generations. Ecology gives us a chance. A chance is nothing more than a limited probation. As space for the human being becomes more valuable, the need for the strongest possible conservation ethic becomes more paramount. And while science is sounding bells in the global spires, we still have a responsibility, a stewardship, for the game in our local ballpark.

In a short book, *Ecology*, Ernest Callenbach provides a summary of the idea.

A. "that human and non-human life and the richness and diversity of life forms have intrinsic value in themselves;

B. "that humans have no right to reduce this diversity except to satisfy vital human needs so current increasing human interference with the non-human world is excessive and calls for basic and far-reaching changes;

C. "that civilization could continue to flourish during substantial decrease in human population needed to reduce our ecological impacts, with an improvement in 'life quality' rather than increasing levels of consumption.

D. "and that people who agree with these points must try to express them in daily and political lives, not just talk about them."

REFERENCES

Beal, Merrill D. *The Story of Man in Yellowstone.* Yellowstone Library and Museum.

Berry, Wendell. *"God and Country in Christian Ecology"* (Holt, Rinehart and Winston)

Botkin, Daniel B. *Our Natural History.* (Oxford University Press) 2004

Callenbach, Ernest. *Ecology.* (University of California Press) 1998

Maass, David / McIntosh, Michael. *Wildfowl* of *North America.* (Brown and Bigelow) 1999

Chapter Twenty

Can You Put It Back?

It was America's busy age. As the country pushed west, the groves of oak, walnut, pine, hickory and buckeye were cleared for building houses, ship masts, barrels, furniture, wagon wheels, barns, fences, fuel, church spires, sidewalks, wooden ties, trestles, trusses for mine shafts. The sawmill industry spread from the clapboards of New England to the redwood missions of the west coast.

American democracy, thought, character and attitudes were shaped by this push westward. It became known as "frontier character." Certainly, Hollywood has played a part in creating this image. In truth, the frontier did shape the way America used its natural resources for extraction and the search for wealth. To extend the ecology involved with burning into the frontier mentality requires that you sew some little discussed patches into Grandma's quilt. In the ecosystem world, small patches of burned areas created the small areas of fresh growth that attracted the birds and animals. These edges were where the biological action was. In the sociological world, when the trail or the railroad pushed westward, towns would spring up along the road. A small community would burn a patch into an undeveloped area. To this area came a variety of people. Most were on a mission; to find wealth. Any conservationists in this group of "patch pioneers" were an endangered species. Concerns for the natural resources simply were not on the agenda.

From 1846 to 1896, both movement and resource use were at their height. The movement ended on the sandy beaches of the Pacific. At the close of this busy period, a few people were beginning to examine the idea there could be an end to how much could be extracted. There were limits to water, forests, grass and the other natural resources. We paid a price for what we took from the earth.

THESE BUDDING CONSERVATION IDEAS

. . . were another "crazy quilt," patches of ideas coming from a wide variety of sources. Colleges began to embrace forestry schools. Yale, Harvard, Cornell on the east coast; Michigan and Wisconsin in the upper Great Lakes and Kansas State and Nebraska in the prairie country all had schools of botany, forestry or horticulture taking scientific approaches to what would become ecology. By 1910, affluent families in the eastern establishment were providing grants and support for a variety of scientific projects dealing with natural resources.

Interestingly enough, our case study state of Nebraska, usually thought of in terms of grass, milo, corn and wheat, plays an important role in the evolution of the national attitude toward our forests. It is 1873. Franklin B. Hough, a physician and historian, presents a paper to the American Association for Advancement of Science. He entitles it, "The Duty of Governments in the Preservation of Forests." On August 15, 1876, there is a rider attached to a free seed clause of the Appropriations Act of 1876. It provides $2,000 for the study of forestry. Hough produces, in 1878, a 650 page report entitled "Report on Forestry."

Until 1881, as a forestry agent, Hough had been the only person in the Department of Agriculture Division of Forestry. He was replaced by Nathaniel H. Egleston in 1883. Both men pushed two ideas. One was that timber land still owned by the government be withdrawn from sale or grant and two . . . that timber should be cut under lease and young growth protected on government land. In 1886 Egleston was replaced by Bernhard E. Fernow. These developments would dovetail into the Nebraska scene.

THE ROLE OF NEBRASKA

Charles Edwin Bessey arrived in Lincoln, Nebraska in 1884. He was to be the Dean of the Industrial College, professor of botany and horticulture and Nebraska state biologist. Bessey was convinced the sand hill country of Nebraska had once been forest land and could again produce trees. Bessey was not the run of the mill botany professor of that time. He had different ideas about grassland too. By 1893, political radicalism was spreading across the Kansas and Nebraska and populism was growing. Bessey had become a bit of an exponent of scientific radicalism, or so it seemed in those times. In 1887, he had been appointed to a joint committee formed to encourage the state legislature to reforest the sand hills of the state. Bessey got in touch with Dr. B.E. Fernow, whom we left just moments ago in the Forestry Division of the Department of Agriculture.

The exact nature of their correspondence is not known, but in the early 1890s, Bessey indirectly got a donation of seedlings which were planted in Nebraska by William L. Hall, a student at Kansas State College. He would soon be with the Forestry Division. Hall put Scotch pine, Austrian pine and Douglas fir in ¼ acre plots near Swan Lake in Holt County Nebraska. His effort produced twenty foot trees in 10 years and the knowledge that trees planted in grassy furrows would grow. They also learned that if the ground was plowed and trees planted, the soil would blow away. The implications of this lesson were a reminder of the temperament of the soil. Back in the spring of 1891, or just about the time they were planting trees in Nebraska, another rider, Section 24, was attached to a bill in Congress designed to revise some land laws. It was related to the Timber-Culture Act of 1873. This is worth quoting because it is still used today by Presidents to set aside and establish forest reserves and wilderness areas.

"The President of the United States may, from time to time, set apart and reserve, in any state or territory having public land bearing forests, in any part of the public lands, wholly or in part covered with timber or undergrowth, whether of commercial value of not, as public reservations; and the President shall, by public proclamation, declare the establishment of such reservations and the limits thereof." This has been referred to as the "Creative Act" or the Forest Reserve Act of 1891. It was used by President Harrison, just after it was passed, to set aside the first forest reserve, the Yellowstone Park Timberland Reserve, now a part of the Shoshone and Bridger-Teton National Forests of Wyoming.

By the end of his term, Harrison had created fifteen forest reserves containing 13 million acres. His successor, President Grover Cleveland used this rider to create two more reserves in Oregon. He then waited four years, while Congress passed legislation for the management of these preserves. Then in 1897, he proclaimed thirteen more new forest reserves. The outcry from the foes of forest reserves rose from the western U.S.A.

Meantime, John Muir had taken several long walks around the country and had finally settled down in California, more specifically, Yosemite. Muir became a strong advocate of preserving public forests, an equally strong opponent of sheep grazing on forest lands. His articles and letters got Yosemite transferred from a state park to a national park. His goal was to get national forests more like national parks. When the forest resources came under the purview of Gifford Pinchot and his ideas about grazing on the forest land, Muir and Pinchot were in disagreement. They did remain friends as Muir concentrated his ideas and efforts about forests on the west coast. He wanted to preserve natural forests. Pinchot preferred to use the forest as a sustaining resource.

Back in Nebraska, C.E. Bessey wanted forests, too, but he wanted to plant them and his efforts would also involve Pinchot. In 1902, President Theodore Roosevelt proclaimed the Dismal River Reserve as a national forest area. It was really three areas, one being in far northwest Nebraska, one in north central Nebraska along the Niobrara river south of Valentine and one along the Middle Loop River in central Nebraska just west of the little town of Halsey. The Halsey National Forest, as it came to be known, would have 92,000 acres when it was surveyed. It would also have a very unique feature, a nursery for growing young trees.

The idea of trees in prairie country was a hybrid opinion. The scientists, largely from college locations, were pushing for research, species substitution and experimentation. The Congress, prodded by the more liberal eastern conservationists, was passing a collection of bills and riders dealing with land use, agriculture and natural resources. The exponents of the open range were fighting a rear guard action to protect their interests. Organized grazing and stockmen groups were emerging. These were known as the "western interests" and became more aligned with the extractionists. Still other forms of social contracts were emerging. Sportsmen's clubs were taking on conservation issues. Women's clubs were becoming involved. In Nebraska, in 1872, J. Sterling Morton had started Arbor Day. Morton had been Secretary of Agriculture and President of the American Forestry Association. He had aroused great interest in the idea of planting trees. Eventually Arbor Day would spread from Nebraska and become a national holiday. The social contract was shifting from the domain of the individual to social group contracts and government intervention.

NEBRASKA SAND HILL ACTIVITY

William C. Hall, now graduated from Kansas State University, reappears on the Nebraska scene in 1900. Now disparate events surrounding forestry would include the sand hills of Nebraska. The trees Hall had planted that survived in the Holt County experiment had grown to 18 to 20 feet in height in ten years. The growth was greater than similar species had attained in eastern Nebraska. The experiment was important because it seemed to prove that certain species of trees could survive in more arid country. The promise of a man-made forest seemed realistic.

Hall is joined by Charles C. Scott who became the first Forest Supervisor of the Nebraska National Forest. The men were members of a crew headed by Royal S. Kellogg. This crew of eight would survey the new forest land. They would head to country where Moses could have lost his sandals. They assembled their gear. A team of mules, covered wagon and harness, saddle horses, two tents, eight cots and a sheet iron cooking stove. They carried crosscut saws and axes to fell different species of trees found in their survey so stem analysis studies could be made. They headed up the Platte River, following a trail that had been marked by the covered wagons of immigrants 50 years before. They took the Platte valley to the Wyoming state line, then went north to Crawford, Nebraska, then east to Rushville and from there south to the intercept with the Burlington Railroad near Lakeside.

They found Ponderosa pine, red cedar and stumps of some of these species. Early settlers had used the trees for houses and barns. Many of the stumps were well over two feet in diameter. They estimated the trees had been sixty feet in height when cut. They found Ponderosa pines with over 300 annual growth rings. This research convinced them trees would grow in the sand hills region and live to attain marketable size. Note again, the mindset of these men was more agricultural than it was conservation. They wanted to grow forests for fence posts, lumber, trusses and building purposes. Any recreational uses were not even in the picture.

NOW THE BESSEY IDEA

. . . of growing trees in sand hill country, supported by Fernow in Washington was slowly moving into fruition. In the spring of 1902, Charles Scott and L.C. Miller were sent to the Black Hills region of South Dakota. Their mission was to locate seed. They found two and three year seedlings in cut-over areas of the Black Hills. They also found an area called the "Nemo Burn" where several thousand acres had been destroyed by fire. They met with Captain Selph Bullock who was head of the Black Hills Forest Service. Bullock asked Scott why he was interested in seeds. Scott confessed he had always wondered where the trees came from because "I had never seen any under four feet in height."

The story has it that Scott was asked how he would handle the burned over area. His reply was that he would sow seed from horseback on freshly fallen snow. Four years later he got the chance to prove his theory when he was called back to the Black Hills. Thirty years later he saw the results of his work on the Nemo Burn. He wrote in his notes that "it was a dense stand of sapling pines twenty feet and over in height." These stories, indicate it was a time of experimentation, some disappointments and a few successes. But note the time spans.

An observer could logically ask, "what were the credentials of the early appointees to the Forest Service?" I asked Herb Gerhardt, who is writing the history of the first 100 years of the Halsey National Forest, and is a Forester, this question. His response was, "They were Mormon, white, male and usually had an agricultural background."

Staffing was another chapter in the muddle of the formation of the Forestry Service. When the Organic Act of 1897 was passed by Congress, the General Land Office had a forestry division. This was in the Department of Interior, a place with a long and tawdry record of graft, fraud and sleazy dealing. State Superintendents were appointed by a combination of political appointment, the GLO in Washington and state legislators. In Arizona, it was not uncommon for superintendents to come from the ranks of the Sheep Growers Association.

TO PROVIDE SOME IDEA OF THE QUALITY

. . . of the appointments, one of the first men to be appointed as an early day Forest Ranger was Frank N. Hammit, from Denver, Colorado. He had been Chief of Cowboys in Colonel William F. Cody's Wild West Show. He had spent the summer of 1898 in Yellowstone National Park. He had absolutely no knowledge of forestry. He died in 1903 when he fell off a cliff in what is now the Shoshone National Forest in Wyoming. On the other side of this point and counterpoint drama, Gifford Pinchot had, in 1898, been appointed as the "Chief" of the Department of Agriculture's Division of Forestry. His close association with President Roosevelt put him in a position to provide forestry with some organization. He was highly qualified for his job, by education, study and practical experience. Finally, in 1905, the management of forest reserves was transferred from the Department of Interior to the Department of Agriculture. In 1907, forest reserves were renamed National Forests. Pinchot set about the job of building a professional cadre and expanding the holdings of the Forest Service.

In 1905, there had been sixty forest reserves covering 56 million acres. By 1910, there were 150 national forests covering 172 million acres. Pinchot was the primary founder of the Society of American Foresters. His philosophy was that a forest was, and could be, a sustainable resource. These are his words. "When the Gay Nineties began, the common word for forests was 'inexhaustible.' To waste timber was a virtue and not a crime. There would always be plenty of timber . . . The lumbermen . . . regarded forest devastation as normal and second growth as a delusion of fools . . . And as for sustained yield, no such idea ever entered their heads. The few friends of the forest were spoken of, when they were spoken of at all, as impractical theorists, fanatics or denudatics more or less touched in the head. What talk there was about forest protection was no more to the average American than the buzzing of a mosquito, and just about as irritating."

THE BROAD PHILOSOPHY IS CHALLENGED

In fact, to this day this idea of sustaining natural resources is still being challenged. It is one of the ten conservation ethics we have offered. On one hand, there is the idea of sustained use for the highest and best purpose. On the other side is the idea of wilderness for the sake of wilderness. These differences in thinking would be stirred into the forest bonfire.

While all these developments were playing out on the national scene, the plans for the first manmade forest were moving along in Nebraska. Scott and the contingent in the Dismal River Reserve had been joined by Wallace I. Hutchinson who had come directly from the Yale School of Forestry. After the survey work had been completed, and after sources for seed and seedlings had been located, the development of the nursery site on the Halsey National Forest was begun. It was the first Forest Service nursery in the United States and became one of the most famous in the country. It was named in the honor of Dr. Bessey.

The seed pine cones and seedlings arrived at Halsey by freight train. All told, Scott had procured sixty bushels of cones and 6,000 seedlings. From the Wyoming area around Guernsey they secured a good supply of red cedar seed. The seedlings were "heeled in" upon arrival and passed the winter in good condition. All were planted in the spring. Anxious watching began. Fewer than 100 of the 6,000 seedlings survived. Then the red cedar seeds from Wyoming refused to sprout. No one could figure out why? The Ponderosa pine seeds came up when they were planted in the nursery beds, then began to wilt and most of these died. Scott appealed for help from the plant pathologists. They couldn't provide answers. Scott would learn he was pretty much on his own in the new undertaking of a forest nursery. It took time, but eventually they found ways to make the idea work.

The preserve records show that in 1963 approximately 3 million trees were shipped under the Clarke-McNary program. These trees went to Kansas, South Dakota, Colorado and Utah as well as Nebraska. In 1964, the nursery had 17 million trees on it with 8 million ready for planting. Another of the major contributions of the Bessey nursery was trees supplied to the Shelterbelt program that began during the depression years of the 1930s. The shelterbelt program extended from northern North Dakota to the northwestern Texas panhandle. The Civilian Conservation Corps was enlisted to do the planting.

In March, 1935, the first tree was planted on a farm near Magnum, Oklahoma. In the ensuing years and until 1939, millions of trees from the Bessey nursery would be planted. In 1938, the workers, many unemployed farmers, would plant 52,000 cottonwood trees in just one blown out area near Neligh, Nebraska.

Today, when you get off the main highways of Kansas, Oklahoma and Nebraska you can find the vestiges of this great conservation effort that attempted to replace eroding soil with mile long strips of trees. In many places, the work of the bulldozer is evident and in much of the country it is back to roadside to roadside farming, aided in no small part by center pivot irrigation. In early spring or late fall, the wind stirs restlessly. Small clouds of dust weave across the level land. Furrows no longer run in straight lines. New plowing techniques now curl across the sandy soil. The result has been some reduction in top soil erosion.

Two other stories merit weaving into the Halsey quilt. One involves John Mandeville, an old-timer who knew the land. There seems to be one such character in every region of prairie country. Charles Scott had enlisted Mandeville to show him the country and help him locate the government corners so he could complete his legal descriptions of the preserve. Mandeville had the reputation of being the only local person who could still find a deer in the almost depleted deer herds of the area. Scott brought up the question of deer hunting and Mandeville was suddenly very silent. Scott writes that he gave Mandeville a lecture and told him if the deer were protected for a few years, they would come back as a result of the forestry efforts.

When Scott paid Mandeville for his services, the old man said to Scott, "Mr. Scott, if you will see that no one else shoots a deer in the reserve, I'll give you my word of honor that I will help you protect them and that I will not shoot another until they are plentiful." So was born another social contract.

NEARLY FIFTY YEARS LATER

. . . in 1945, Nebraska had an open season on deer. Five hundred hunting permits were sold. Management and conservation had brought the population back. "By official count, 368 deer were taken home by hunters. It took a long time for recovery of a natural resource, but patience had paid off. It was another recreational value added to the forest.

Today, white tail deer, mule deer, antelope, turkey, grouse, prairie chicken and quail are all part of the wildlife population of Halsey National Forest. Some 600,000 people annually visit this small piece of the big forest picture. They picnic, shoot a lot of film and, in season, shoot some game. Boy Scouts,

Future Farmers, 4-H Clubs and church groups are part of the mix that enjoys the man-made forest. When I visited the setting in the fall of 2001, several groups of visitors were prowling around the forest. I met two young members of a fire fighting group of foresters stationed there. During the summer, Loren Eaton and Tedd Teron had been off in Montana fighting fires. They were especially proud to tell me about their swimming pool. "Halsey Forest is the only National Forest with a swimming pool."

I asked Herb Gerhardt if the forest had ever been timbered. "No," he said, "not to my knowledge. We have taken some dead wood for use as posts and fire wood. Mostly, today, the forest is used for tourists and recreational purposes."

Then I asked him what, in his opinion, were the pluses and minuses of trying to create this forest, in essence bring back what might have been there many years ago. "On the minus side," he said, "there has been no production of wood. That is something we could have measured. On the plus side, a lot of people use it for recreation. The economic benefit of that is harder to measure. If this land had been used for agriculture, who knows what the economics would have been.

He paused. I made notes. "The nursery . . . that was an unexpected bonus. We have shipped out thousands of seedlings. But yeah, I'd say we had put a forest back. But look at the time it took. Minimum ten years . . . up to thirty years sometimes to see the results. Out of that came a better understanding of what would grow and what wouldn't. You know . . . some things weren't meant to live, or grow . . . exist . . . in some places."

Whether it is a man-made National Forest or a natural forest management of the resource has been a long standing thorny problem. In early times, the Forest Service opposed legislation to establish a national park service. In 1916, Congress passed an act that set up the National Park Service. To counter this, the Forest Service set up an extensive outdoor recreation program. Additional Congressional acts and a number of riders to other bills, set up various agencies that would participate in this effort. Today, it is a maze of bureaucratic turf. A more recent issue is privatization of park employees. Depending upon what wilderness area you choose, what areas around U.S. Corps of Engineers dams or lakes you select for research, you find a contentious condition. Add the U.S Fish and Wildlife Service and their refuge system to the mix and it is often difficult to find out who is doing what, with which and to whom.

THE SHIFT TO RECREATION

As southern areas of this country have shifted to man-made forests for timber production, western forests have become a playground for an estimated 1.5

million "pleasure seekers." A million of these used the forests for day use only but the other half million were campers. Human waste became a problem as people messed up the nest. In some parks and wilderness areas, hikers and packers are requested to weigh human waste when they leave the area. The idea is, you pack it in, you pack it out. Too often, the telltale signs of the human presence are marked by bits of toilet tissue or paper napkins.

More recently, the issue of another sort of waste has surfaced. Technology has invaded the forest and the presence of the four wheel vehicle has put more pressure on roads and trails or has made new roads. Add to this, in the winter months, the roar of the snowmobile. Some claim the exhaust fumes are fouling up the winter air. Others claim they are disrupting the natural routes and paths of the animals. Others claim they are disturbing the beauty of silence by making too much noise. It isn't all peaceful in forestville.

Modern communication has probably made the general public more aware of our forests. What was once a local condition gets enflamed on TV screens and Los Alamos or the Bitterroot fires become a living room event. In 1871, the Peshtigo fire in Michigan burned 1 million acres and took 418 lives. In that same year, fires in Oregon and Washington burned more than 1 million acres. These fires received little national press coverage. In 1910, there was a big blowup in the northern Rockies that burned 3 million acres. There was reaction to this in Washington and Congress passed the Weeks Act which allowed Forest Service personnel to cooperate with states in firefighting and protection.

The Halsey National Forest did not escape. In 1910, a prairie fire started sixty-five miles west of the forest and covered 125 miles in one afternoon. It swept through the young plantations of trees. But a greater damage occurred in 1965 when, on May 5, lightning struck and caused a fire that destroyed 20,000 acres on the forest. The burn included 11,000 acres of man-made forest and 9,000 acres of grassland. 90% of the trees in the burned area were destroyed.

SO WHAT CAN BE SUMMARIZED?

. . . from this brief overview of Halsey National Forest? One conclusion would be that Dr. Bessey was right. Given time, enough money and enough human sweat, man can create a forest. A second conclusion is that the Halsey National Forest, while unique in being a man-made forest, represents many of the trial and error experiences of the National Forests. A third conclusion would be that through the years, the Forest Service has been more closely aligned with agriculture than with conservation. Recreation has been a side benefit.

Still, the history of the Forest Service is worth some study because it was the first natural resource management component to apply science to the management of natural resources in this country. It was the first to hire college trained people, the first to apply the benefits of research and education to natural resources. It has had the benefit of dedicated people, starting with Gifford Pinchot. Another dedicated Chief of Forest was Ferdinand A. Silcox who served from 1933 to 1939, a time of considerable examination of natural resource uses. This is what Silcox would write of the evolving philosophy of the Forest Service.

"Civilizations have waxed and waned with their material resources; dwindling means of livelihood have set rolling great tidal waves of migration and have been a prolific cause of domestic disorder, class uprising and international war; but never have the people of a great country still rich in the foundations of prosperity sought to forestall future disaster by applying a national policy of conservation—-of which planned land use in the central core."

It is important to note that Aldo Leopold came out of the National Forest Service. He would be one of the voices that would speak on behalf of a land ethic, a more balanced view of natural resources. He would be part of the evolution of the Forest Service from a collection of political appointees to a more professional organization. That organization would evolve and eventually come to espouse what would be called ecosystem management, the management philosophy of today's Forest Service. Their effort has been to see the forest as well as the trees. This is what they say about their management philosophy.

1. Multiple Analysis Levels. Use different levels of analysis, from the site-specific location to the broad watershed perspective or even larger.
2. Ecological Boundaries. Define ecosystems by analyzing and managing them across political and administrative boundaries.
3. Ecological Integrity. Protect the natural diversity, ecological patterns and processes. Keep all the pieces.
4. Data Collection and Data Management. Require more research, better data collection methods and up-to-date information.
5. Monitoring. Track results of management actions. Learn from mistakes. Take pride in successes.
6. Adaptive Management. Use adaptive management, a process of taking risks, trying new methods and processes, experimentation, and most of all remaining flexible to changing conditions and results. Encourage better public participation and involvement in planning, decision making, implementation and monitoring.
7. Interagency Cooperation. Work with agencies at the Federal, State and County levels, as well as the private sector, to integrate and cooperate over large land areas to benefit the ecosystem.

8. Organizational Change. Change how the various agencies work internally and with partners to encourage cooperation and understanding with other agencies and the public.
9. Humans Are Part of the Ecosystem. People are a fundamental part of ecosystems, both affecting them and affected by them. I involve people at all stages in the analysis and decision making phases.
10. Human Values. The human attitudes, beliefs and values that people hold are significant in determining the future of ecosystems as well as the global environment. Seek balance and harmony between people and the land with equity across regions and through generations by maintaining options for the future.

Jack Ward Thomas, the 13th Chief of the Forestry Service summarized the direction of the service. "We don't just manage land," he said, "we're supposed to be leaders. . . . What does ecosystem management mean? It means thinking on a larger scale than we're used to. It means sustaining the forest resources over very long periods of time . . ."

A small book, Ecology of a Cracker Childhood, written by Janisse Ray, captures the spirit of another feeling. "If you clear a forest, you'd better pray continuously. While you're pushing a road through and rigging the cables and moving between the trees on a bulldozer you'd better be talking to God. While you are cruising timber and marking trees with a blue slash, be praying; and pray while you're peddling chips and logs and writing Friday's checks and paying the diesel bill—even if it's under your breath, a rustling of lips. If you're manning the saw head, or the scissors, snipping the trees off at the ground, going from one to another, approaching them brusquely and laying them down. I'd say, pray extra hard when you are hauling them away.

"God doesn't like a clear-cut. It makes his heart turn cold, makes him wince and wonder what went wrong with his creation and sets him to thinking about what spoils the child."

Our technology has produced a lot of big boy's toys. As we confessed in the chapter on hunting, it is a blood sport. The gun can be lethal. So is the chain saw lethal. There is nothing deader than a fallen tree in a forest. Trees are hard to put back. Both the gun and chain saw need to be used very wisely. This thought should be a part of a national conservation ethic.

REFERENCES

Hunt, John Clark. *The Forest That Men Made.* (University of Nebraska Press) 1981
Interview with Herb Gerhard. October 2001.

Ottoson, Howard W. *Land Use Policy and Problems in the United States.* (Beard Books) 2001

Strong, Douglas H. *Dreamers and Defenders.* (University of Nebraska Press) 1988

Ray, Janisse *Ecology of a Cracker Childhood.* (Milkweed Editions) 2000

The Bulletin—Rocky Mountain Region. Vol. 18—Number 10   1925

Timber Resources for America's Future. U. S. Forest Service Report 14, 1958.

Worrell, Albert C. *Principles of Forest Policy.* (McGraw Hill)

Chapter Twenty-One

Tom's Dream

In deference to our founding fathers, most folks always refer to George Washington, James Madison, John Adams, Thomas Jefferson, Alexander Hamilton and use their full name. When writing, those men were rather formal. Certainly, their written arguments in which they espoused their causes were formal. Jefferson's letter to Merriweather Lewis was very formal, to Lafayette it was "Dear Friend." To Adams it was "Dear Sir." Mostly when they wrote to each other it was "Dear Sir."

Evidently, in those times you just didn't head a letter, "Dear Tom, I was wondering how your dream was coming? You must be up to something. You send Monroe to France to try and buy Louisiana and all that land to the west. You claim you want to find a waterway to the West Coast and now you are sending this Lewis chap up the Missouri River. I have read your instructions to him and they are beautifully explicit. You want to know about the soil and the face of the country, its growth and vegetable productions, the animals of the country, the mineral productions, and you want to know the climate as characterized by the thermometer, by the proportion of the rainy, cloudy and clear days, by lightening, hail, snow, ice, by the access and recess of frost, by the winds, prevailing at different seasons . . . Tom, you want to know an awful lot about the country. If you ask me, you are up to something. Sincerely, James."

The signature of Thomas Jefferson rests heavily on the legacy of the prairie country. Not just his mark of the squares into which the land was arranged, but in the idea that a willing worker could make a passable living and therefore become a backbone type of citizen if they owned land. He passed this dream on to a lot of people. The legacy of freedom stuck on the prairie, embraced by stoic, fiercely independent people. As to making a living on 160 acres of prairie land . . . that idea may not have been realistic.

THE REAL ASPIRATION OF JEFFERSON

Lost in the expanse of grass was the real aspiration of Thomas Jefferson for that land. He may have wanted to find a way to the west coast to open up trade with the Far East. That may have been the professed reason for the Lewis and Clark trip. It is more like Jefferson to have been thinking about land and the extension of his idea of the yeoman farmer from Virginia to the western reaches. His idea of an agrarian society that would extend from coast to coast seems to have guided his thinking. That was 250 years ago.

It might have worked out if there had been adequate water. In the 100 years following the date when Jefferson sent Lewis and Clark up the river, it was proven that it was nearly impossible to make a living farming 160 acres west of the Missouri River unless there was water. True, you could do it when the weather cooperated. But the weather was never normal. It was farming with cycles. Dry spells were a constant possibility.

It was a constant process of recruiting people to live on prairie land. Abandonment of homesteads was a constant process. The boosters would produce another wave of people. Keeping a population on the prairie required increasing federal subsidy in one form or another. By 1940 there was survival for a farmer or rancher if you learned how to play the subsidy game. But again, your timing had to be good because that cycled too. Now there was a two wheeled cycle. . . . weather and federal support. Not everybody could find the pedals on that cycle.

The focus was on production, not conservation. The search was for some technology, some wizard that would make the land produce. Improved seeds were tried . . . wheat, corn, milo, soy beans. Some crops were susceptible to salt and leaching from irrigation. Invasions of bugs and pests seemed to make spraying a necessity. The ecology of the prairie was really never understood by those who worked the land. Even the large corporation with skilled tax lawyers is often in a prone position. Farm coops aren't in much better shape. The nation's largest farm coop is Farmland Industries, Inc. In 2003, it was busy selling off assets in an effort to emerge from bankruptcy. Meantime, the quality of the land is slowly reduced. Sometimes it is erosion. Blowing and silt continue to take a heavy toll. The land requires more fertilizer and more spraying. We go slowly backward. For any new conservation effort introduced, economics applies a back hold.

The shelterbelt program was an effort to curb blowing. Those were depression times and the economics could be justified because of the jobs that were created. Many farmers saved their farms by working at paid jobs in shelterbelt programs. The economics of such a program today would be highly questionable. The Soil Conservation Service also built thousands of small, upstream dams to curb silting and store water in likely watersheds. After 50 years, those dams have largely silted up and need to be repaired and updated. There are 2000 of these dams in the state of Oklahoma alone.

WATER IS THE ACHILLES HEEL

. . . that keeps the farmer's feet from reaching the pedals of the cycles. Efforts to eliminate the water cycles were the stated goals of the Army Corps of Engineers and the Bureau of Reclamation. Initially, they focused more on irrigation and hydroelectric power. Later, barge traffic on larger rivers entered the equation. The upstream, small dams fell to the lot of the Soil Conservation people. Water would be manipulated to stabilize agriculture, would create jobs, jobs would cement the society, the society would protect the infrastructure and maintain the tax base. In essence, technology would make Tom's dream work. Today, most of the jobs being created in the agribusiness world in prairie country are minimum wage or going to a wave of Hispanic immigrants. These are basically low pay people. This is the demographic of the feedlot operation, the packing plant operation, the vegetable, fruit and nut farms. At the other end of the spectrum you have the big operator, big technology, big fertilizer, big spraying and big tax support. It has been an endless cycle of trial and error.

The ecotourism industry sometimes is cross bred to agriculture. Some water recreation has developed along rivers. Large reservoirs have produced the bi-product of recreation. But in most instances, when water is needed for power, navigation transportation or irrigation, the recreational or natural resource use of water is curtailed. Water use in the prairie country, like the arid west, is focused on power, irrigation or transportation. Conservation, ecological sustainability and recreational use of natural resources are near the bottom of the priority list. The purpose of water in prairie country is production, consumption, and economics. Sustainable conservation is a small wheel on nature's bicycle.

DOGMATIC STATEMENTS

. . . such as these are not popular in prairie country. But what is the reality? This news from the Associated Press. "But more and more, farmers are reaching out for help, often finding it through farm crises hot lines that connect them with legal aid, financial assistance and mental-health counseling. It's a new way of thinking for farmers, who are usually a fiercely independent lot."

But, Nebraska's state economy lost $600 million on crops, hay, pastures and ranges in 2002. What does all this do to the mindset of the farmers who choose to tough it out? Dave Mussman farms about 1,000 acres in Nuckolls County In Nebraska. This is his third year of losing money. He says, "I say by holding this in, it's a poison and it screws up your mind." He sought counseling and mediation. He got help in understanding a complicated new farm bill, low commodity prices and increased competition from large corporations, some of which own big land units, the local feed lot and maybe even the packing plant.

None of this quite squares with the Jeffersonian dream as drought puts more pressure on water supplies, suffering habitat and any wildlife. In areas that once regarded the pheasant season as the second cash crop, that source of revenue has dropped. Using the land to produce a natural resource crop seldom enjoys a subsidy like land in production. As a farmer, faced with tough choices, where would you place your bet? Do you turn some of the land into "game farms" where deer or elk can be hunted for some type of fee. That might work out, given enough time. Then Chronic Wasting Disease shows up in the state, and wildlife authorities ask that you shoot all your penned up deer or elk.

THE PARALLELS TO VERMONT

Farming is tough in Vermont too. Farming is just tough anywhere, even if you get regular rain. In Vermont, poor soil quality is still a problem; big yields hard to come by. Much of the farming community works in a nearby town and farms as an avocation. Some of the big farm operations have been turned into tourist attractions. Some larger farms are being operated with minimum return and are usually owned by more affluent people.

The words of George Perkins Marsh still hang over the land, not just in Vermont, but wherever in this country there is land producing something. More and more, the thoughts of this ecology pioneer seem prophetic. This from his classic thesis, *Man and Nature.* "The changes for evil and for good have not been caused by great revolutions of the globe, nor any inaction of the peoples, or, in all cases, even of the races that now inhabit their respective regions. They are the products of a complication of conflicting and coincident forces, acting through a long series of generations; hence, improvidence, wastefulness, and wanton violence; there, foresight and wisely preserving industry. So far as they are purely the calculated and desired results of those simple and familiar operations of agriculture and of social life which are as universal as civilization—the removal of the forests which covered the soil required for edible fruits, the drying here and there of a few acres too moist for profitable husbandry, by draining off the surface waters, the substitution of domesticated and nutritious for wild and unprofitable vegetable growths, the construction of roads and canals and artificial harbors—*they belong to the sphere of rural, commercial, and political economy more properly than that to geography . . . (my emphasis)* and hence are but incidentally embraced within the range of our present inquiries, which concern physical, not financial balances." He was talking about Vermont but the application to the Midwest is eerie.

Marsh was concerned about waste. His study of the effects of erosion due to timbering (resonance with grass) . . . the subsequent run off of soil into Vermont streams may have been one of the first connections in ecology to stitch together the relationships between poor timbering practices, soil run-off and erosion, water quality and reduced fish populations. His book could have been the Old Testament and Marsh was a prophet!

Vermonters will tell you they worry about their two great challenges, water quality and development. Water quality is important because it will determine to some extent the quality of life of the Vermont resident. Development because it threatens the pastoral charm of Vermont, the very reason for the influx of retired people coming to the state and the reason so many tourists enjoy the state. Meantime, Vermont has found a way to generate enough of a tax base to support a good infrastructure. There is a premium on education and roads are generally well maintained.

Vermont would seem to have fewer challenges than prairie states. It usually has a steady supply of rainfall. Lack of consistent rain will always be a prairie problem. So government subsidy for the grasslands is a continuing reality if our society wants people to live there. This will be the case as long as we share Tom's dream of doing something with land the land doesn't want done to it. The land wants to grow grass! It grows best in years of enough rain, then exists in dry times.

Stand in western Nebraska prairie and you are struck by the long sunsets. It takes longer, it seems, for the sun to give up. Agrarian social contracts die in the same lingering way. How much longer can the Jefferson dream be prolonged? Could the billions of dollars being spent to prop up this country be spent in some more productive way? The moments of reflection fade into the west. You recall John Wesley Powell's unappreciated efforts to get folks to rethink the 160 acre squares. He may have been another prophet.

DAM BUILDING FOLKS

Lord, we built a lot of dams! Mostly, for power. Electric power and political power. What a strange culture we have evolved...we have very little concern for water quality but we have a fixation for dams. Some person has estimated there have been a quarter million dams built in this country, some for the purpose of irrigation, some to raise fish. There are 50,000 dams in this land of ours that would qualify as "major works," Major being tall, some 500 feet or more high. That's two city blocks stood on end. I once visited a major dam that had created a man made lake in which striped bass prospered. The Corps of Engineers hadn't figured on the bass. The fish prospered when the Wildlife Department stocked them as an experiment. This might be called "casual conservation." Deep in the bowels of the dam was an elaborate conference room. I asked what sort of meetings they held in this unique location. The room had never been used.

Perhaps one day, an anthropologist will double check his notes on the western United States and say to an associate, "Franklin, you remember that story of the fellow who got the sea to part and folks walked across....well, here is a civilization where people walked west and built dams....five dozen on the Missouri River, 14 on a little river in California they called the Stanislaus, and here is a river that starts out in Teton country in Wyoming and ends up as the Snake which flows into the Columbia. It must have more than 25 dams built along the way . . . and some of them got built but fell apart before they ever filled. Now this one . . . The Teton Dam . . . Idaho . . . it doesn't make any sense at all. They were dam building folks."

IT'S A DAM GOOD STORY!

The Book of Dams may get inserted someday just after Revelations. The verses dealing with the Teton Dam would go this way. And it would come to pass that the Mormons who had sprung from Vermont, had walked across the land to a place now called Salt Lake City. Crickets struck their first crops and great swarms of birds descended and ate the crickets and lo, the land prospered. Determined to go forth and multiple, and aided in that cause by several wives, the Mormon men again walked across the land, this time north, into a land called Idaho. They were attracted to the water along a river the natives called the Snake. Some, walking on north on land, found more land increasingly hospitable in terms of climate and rainfall, but without the agricultural possibilities of the land along the Snake. So it was, the Saints would spread out along the valleys of the Snake River, establishing towns, potato and barley patches, their commerce, and lo, their political positions.

Satire to one side, the Mormons built a few of their own dams along the streams running into the Snake. Then in the early 1900s, the Bureau of Reclamation arrived and there was a burst of dam building activity. They built Jackson Lake Dam but not before the Mormon faithful extracted an agreement from Wyoming that in the event of dry weather, Idaho had the rights to the top 50 feet of water in Jackson Lake. They built American Falls Dam and the Bureau was on a roll. In fact, sometime in the middle 1920s they ran out of decent places to build dams. Some of the dams they had built in this grainy, porous rock had been suspect. Inspect the forty mile corridor that runs along the Snake River, an irrigation area that has created more millionaires per acre than any other in the U.S. The soil is so porous it requires annual watering in excess of 10 feet. In Idaho, they call this irrigation corridor "the bench." It is not lowland as such, but gently rolling hills that sit back from the river itself. Wells produce the water.

NORTH OF THE SNAKE

. . . which coils to the south before heading west and north again, is the Teton. There had been talk of building a dam somewhere on the Teton River since the 1920s. Any study had been shuffled into a backwater because the land was fractionated and fissured and had more than its share of faults, not all of the faults being in the ground. The immediate farmland was not of high quality. It required a lot of water. In the 1970s, Mother Nature got cranky and dealt some dry cards. The Mormons in the Fremont-Madison Irrigation District talked about potatoes and sugar beets that "could be finished out if they only had a little more late season water." Of course, Idaho has more ground water than most states in the west, but pumping is expensive. Why spend your own money when you can jawbone the feds into building a dam. By its own figures, the BLM figured the finished project added only five inches of water to the 132 inches of irrigated water already being used annually on 111,000 acres of land.

You walk up to the dam site from a parking lot to the south of the river. It is fenced off with chain link fence and weeds. There is a small observation platform, also fenced, maybe 6 feet by 6 feet, fairly close to the bank where you can look down into the river. Swallows swope down a ravine and for a moment you imagine they might be $100 bills. The project had swallowed $85 million of the taxpayer's money. Parts of the dam still abut the river bank. You wonder about the bulldozers caught in deep holes when the dam finally went. It did. On the 5th of June, 1976, at about 11:30 am the crest of the dam fell into the reservoir, and almost without any noise, the 300 foot high Tabernacle of concrete, rip rap and steel became a Tabernacle of Waste. It was the second largest flood this country has experienced.

The leading wave that swept down the valley was 20 feet high. It roared into the little town of Wilford east of Rexburg. The next published Atlas showed Idaho but not Wilford. The lower half of Rexburg would be a total loss. The area, by nightfall would be a standing pool of water, 11 people dead, 350 businesses lost, with damages estimated at $2 billion. In 2001, all this information was enshrined in the small museum in Rexburg. None of the statistics were quite as awesome as pictures in the museum showing land stripped of soil, as though a huge bulldozer had come along and scrapped the land down to rock. I queried an older gentleman attending the museum. He said he understood more land had been permanently destroyed by the flood than would have been put into production by the water from the irrigation project.

Blame for what had happened was tossed around but nobody wanted to catch it. There were no apologies. Claims the water had been absolutely vital became an attitude water really wasn't that important. The landowners on the bench made do very well with what they already had and they continued to prosper. The landowners and business people in the valley set about putting their lives back together.

IRRIGATION IS NOT A WIZARD

In prairie country irrigation prolongs life. In arid country it is a simple matter of life and death. Contention is part of daily life. I have found dams in southern Idaho that never filled because of porous soil condition, but fill the air with strong opinions and invective. On the Idaho/Nevada border, near a dam site, a Casino now provides the economic prop for the area and may offer better odds for success than farming. In Colorado, proposed damming of the South Platte River is in litigation. In Wyoming, the Midvale story is one of a $50 million trail of legal tears. Oregon has the Klamath Valley. It is a damming legacy.

Irrigation in prairie country is used chiefly to grow crops. Sometimes these are surplus crops. When you get to arid country, irrigation is used mostly to grow alfalfa. Even in Idaho, where growing potatoes makes most of the money, most of the water is used to grow alfalfa. In Colorado, tourism is an industry topping $5 billion a year. The alfalfa crop, which uses most of the water, now amounts to less than $100 million. In arid country, there is a growing awareness of the value of trout and salmon and the tourists that chase both with flies. There is more plowing being done for rafters and hikers. Folks may be catching on to the exhaustibility of what was once considered inexhaustible, catching on to the idea the cheap water spread on irrigated land is really not so cheap. There is a feeling water left in rivers has the potential to create more jobs, more revenue, more economic benefit than water taken out of rivers to grow cows.

Rivers can produce a higher quality of life. Our country has tended to look at rivers for what they produced economically. There may be no such thing as a perfect balance. Bear in mind, things fell apart even in the Garden of Eden. One rule does apply in all places. Biological populations are limited not by annual resources, but rather they are limited by the minimum food supply at the scarcest time of the year. Water fits in that condition too.

TO BE FAIR

. . . if it had rained a little more, we might have pulled off Tom's dream. Take out the drought years, the blizzard of 1888, and what American agriculture did was actually pretty amazing. When you grow more food than you need that is not all bad. What American technology did in the arid country...in the west for that matter, was not all shabby. We did create the greatest desert civilization in the world's history. We created more wealth, more great cities, more cattle herds, grew more potatoes than any people on earth had ever done before. We could do that because the Army Corps of Engineers, the Bureau of Reclamation and Uncle Sam combined to masquerade as the Wizard of Oz and got the western United States more water than any race of people ever had under controlled conditions. But one of nature's laws kicked in. When things are abundant, like we learned with land, they cease to have a value. And because water value is nebulous, we created a mirage. We sold it cheap or gave it away.

We didn't need to do a lot of things we did. We didn't need to build dams on rivers that carried vast loads of silt. We didn't need to pour water on Wyoming land until the soil turns white with salt. We would build the world's greatest dams. We could create lakes of incredible size. The water lobby could raise incredible amounts of money and political support. Our engineers could exceed what the imagination could conspire. Now, as we step back and reevaluate some of our water projects, new equations and new values seem to be emerging. We may have entered the stage of REFORM in our dam mentality.

THINKING ABOUT TOM'S DREAM

What can you locate down on the grass roots, down where the hunters sit in their trucks and look at their maps or the river trout fisherman checks a fly box? People who think about nature are like a big parenthesis. The left side people are almost a "pastoral" group. They see nature as in a state of rest, at peace, in balance, there is harmony. They say, "Don't worry, this too shall pass." On the other side, like an exclamation mark, are the folks who suffer from "Apocalypticism." Nature is painted as an awesome end. We will out breed the ability of land to support us. If something isn't done, the world's forests will be destroyed. If we don't clean up our air, we will not be able to breathe. These folks sing the heavy songs. They like the last verse, the end.

Between these two groups there is a small number of people who have contributed to environmental thinking. We have tried to identify those people and provided brief biographies and set them up in a sort of "Big Footprints Hall of Fame." These people did not all think alike. George Perkins Marsh who was so farsighted in those spectacles he was almost over on the right side with his warnings. There were folks like Muir who obsessed on trees and not everybody saw forests as he did. Not all naturalists are sensitive to the countless species of birds to better paint them as Audubon did.

So why would we expect the grass roots folks to think the same when so many of the naturalists think about nature differently? And how could we expect a prairie farmer or a dry land rancher, intent on getting water to their cows or crop, to think like a naturalist? Their focus is on paying off the bank loan.

Could it be there are three natures? One is the external nature. It can be seen, heard, observed, recorded and tabulated. It is outside. Another is an internal nature. This is the result of internalization, a spiritual feeling that captures the external. There may be a third nature; an image or vision of what the individual person hopes nature would or could be. This would be their idea of the ideal, an improved condition, the utopian circumstances. The conclusion would then have to be that none of us looks at nature in exactly the same way. There is no consensus.

ADD ANOTHER MYSTERIOUS ELEMENT

. . . to the mix. That is "the sense of place." Many people experience a sense of place in nature. You hear the urban dweller explain they "want to escape to nature." You hear the hunter explain, "I like getting nose to nose with nature." Trout fisherman have "special places" on a stream. It's their secret place. They find solace there. Some people will tell you they feel "more aware when they are in their special place." Nature seems to enable a sense of place.

E.O. Wilson has an essay, *The Right Place*. He believes that whenever people are given a choice in their external nature, they will move to a tree studded land overlooking water. He implies that the human sense of place, though it may not rank on the human needs chain as high as survival, is so imbedded, so strong, so instinctive no person may ever quite explain it.

Native American people, usually camped along a stream or river. When the early Euro American people settled New England and Vermont, they settled along the water, river, stream or lake. And when the immigrants moved into prairie country and the arid west, when they couldn't find water, by God they dug for it or they dammed up a river to get it. Then most of them planted some trees.

Water gave them a sense of place. It was the liquid of life. They just didn't quite know when to stop digging and damming. And Tom's Dream? He got the land divided into squares and people would embrace his idea of the yeoman farmer. He just didn't quite get a handle on the water dipper.

REFERENCES

Attfield, Robin. *The Ethics of Environmental Concern*. (University of Georgia Press) 1991

Buell, Lawrence. *The Environmental Imagination: Thoreau, Nature Writing, and the Formation of American Culture*. (Belknap Press) 1996

Ellis, Joseph J. *Founding Brothers: The Revolutionary Generation*. (Vintage Books) 2002

Marsh, George Perkins. *Man and Nature*. Harvard (University of Washington Press) 2003

Reisner, Marc. *Cadillac Desert: The American West and Its Disappearing Water*. (Penguin Books) 1993

Schrnitt, Peter J. *Back To Nature: The Arcadian Myth In Urban American*. (Johns Hopkins University Press) 1990

Stegner, Wallace. *Heaven Is Under Our Feet*. Henry Holt And Company)

Wolfe, Linnie Marsh. *Son of the Wilderness: The Life of John Muir*. (University of Wisconsin Press) 2003

Notes from the Rexburg, Idaho Museum.

Personal visits to the Teton Dam site, Rexburg area and Walden Pond and the Concord area.

Various news clippings from Associated Press.

Chapter Twenty-Two

Dry Land

A real Wyoming cowboy walks into a town tavern in Wind River country. His boots are scuffed so whether they were alligator, lizard or Wal-Mart is not discernible. Heels are run down to the nubbins. Only the dudes have shiny boots. The real cowboy pushes back his sweat rimmed hat which is less than 3x and probably is looking for parenthood. He produces a couple of crumpled bills from faded jeans. The fade is more from wear than from washing. He nods to the bar maid. Male bartenders have become part of folklore. "I'm dry," says the cowboy.

Now he could be describing his present physical condition or announcing that he had run out of water for irrigation. He might also be talking about 80% of his land, if it is his, or land he happens to be grazing or leasing. Dry in Wyoming, as it is in most Western areas, is a relative condition. It ranges from dry to very dry to damned dry. That's the reality of the arid west. Of course, reality is relative in dry country.

Any study of ecology in the western part of the United States is akin to listening to the Handel Oratorio, the Messiah. Listen to the songs in the wind and you can hear the soprano parts, the high whistles that carry a message of hope, expectation and possibility. The low passages, the bass parts, rumbling or raging and sonorous seem to almost be a requiem. It is a double song of rejoicing and retribution, of optimism and warning. Wallace Stegner, in his essay, *Living on Principal*, calls it "love and lamentation." Like the dollar bills lying on the bar, the songs have two sides.

WHEN WE TALK ABOUT "DRY"

. . . we need to understand the dry "dry" of western lands. This land mass contains a lot of public lands. It is Wyoming, Colorado, New Mexico, Arizona,

Utah, Nevada, Idaho and Montana. Wyoming serves as our case study and is a good prototype of the region. Within this area, there is a certain commonality. Usually, great space separates any urban presence. The country is predominately brown or tan except when sage or rabbit-brush is blooming or where there is water, be that reservoirs, streams or irrigation. There are brown and white signs that proclaim the landlord is either BLM, Park Service, Forest Service or State Game and Fish. Occasionally there are dams that interrupt the flow of the Green River, the Columbia, Colorado, Platte or Missouri rivers. And there is an attitude, spoken or unspoken, that Bernard Devoto once captured in a sentence. "Get out and give us more money."

The second commonality is lack of rainfall, 14 inches or less. Winter snow provides the runoff for streams and creeks. Snowpack in the high country does not count in the annual rainfall figures. That is an extra perk that provides water for creeks and rivers until it is exhausted in the early fall or just runs low. Water is the commodity of exchange. It is also a crazy quilt.

Land is spikes of rocks, granite, red sandstone, chert . . . some of the land was once ocean bottom . . . most of the land is a geologist's idea of a picnic in the rocks . . . it is long curves of horizons that disappear when you get to them . . . it is windy tablelands, endless plateaus. What fertile land there is rests in the valleys spawned by creek and river beds. The forests are a variety of pines. Landmarks are more often rocks . . . Slippery Rock, Chimney Rock, Castle Rock or "a big, black rock on the right," when some person is giving you instructions. There is another commonality in the rural areas. Most of the humans who provided the settlement of this land were either told to be there by the Federal Government or were invited to be there by the government and promised free land. Or they sneaked in. All share a feeling the Federal Government cannot be trusted to keep its promises.

THE CONTRIBUTIONS OF POWELL

One date and one life span changed the course of history for this vast, dry land. In 1869, Major John Wesley Powell would emerge from a long journey down a canyon he would name "The Grand." He had started in the Green River country of Wyoming with a grub stake of $11,000, some rations from the Federal government, the blessing of President Grant, and with nine associates, mostly hardened mountain men. He would wear a red beard, have only one arm, the other having been left at the battle of Shiloh. He had some scientific equipment borrowed from the Smithsonian. Three of his associates would desert him on the trip so he came out of the canyon with only six.

The trip did two things. It gave Powell something of a grubstake as a hero and adventurer. He would later use this aura as he lectured, sought exposure in the press and conducted a public relations campaign for his concepts of what to do with western land. More than that, he brought the authority of science to his study of the land, or the water of the dry land. This contribution would not change until the years from 1880 forward. From that date, science would be a part of the debate surrounding the use of land and water in the dry west. Technology had been the hand maiden that had helped subdue plains country. The Winchester, Colt, railroad, windmill, barbed wire and plow had contributed to some small foothold on the prairie. Today, we argue the value of wind power. The windmill was the settler's power-plant for water. The western dry land promised more new freedom. Cattle and sheep provided new opportunity for individual enrichment. Each of these promises would turn into myths. The lack of water would prove to be the wild card.

Powell would labor twenty years of his life and use science and rhetoric to dispel the common wisdom. Another reward of a $10,000 grant from Congress would enable a second trip down the canyon. It would help fund a United States Geographical Survey of the Rocky Mountain region. He had a long journey of twenty years of government supported science. Our sixth conservation ethic emerged . . . that nature is a property, or condition that can be quantified and defined and therefore can be managed as part of a scientific process. Powell became the advocate for orderly settlement of western land and the scientific conservation of the dry lands of the west. He told people what they needed to hear, not what they wanted to hear.

THE CONCLUSIONS OF POWELL

In 1874, Powell released his first conclusions. It was his opinion that "all the region of the country west of the 100th or 99th meridian, except in California, Oregon and Washington is arid . . . no part of it can be redeemed for agriculture except by irrigation." On April 1, 1878, Powell presented newly appointed Secretary of Interior, Carl Schurz, his *Report on the Lands of the Arid Regions of the West.* The book came to be regarded as one of the most revolutionary books on land use ever written by an American. It attacked and refuted most of the philosophy and the existing body of law being used to that time. Many of his recommendations had to do with water law.

Eastern water law had been adapted from English common law. It had been placed band aid fashion on water law as the country moved westward. It was essentially riparian. Water could be used by the owner of land along a stream for milling, to produce power or transportation. The unspoken idea was water would be returned to the stream. Dry land water use consumed water. It more nearly fit the idea of water used for human consumption, golf courses, irrigation or big reservoirs where evaporation uses water.

Water law in the west emerged in any codified fashion in the Gold Rush times in California when the miners diverted streams for use in their rockers when placer mining. What evolved from this came to be known as the Colorado Doctrine; first appropriation or first come, first served. In arid country it worked this way. If you were the first to build a ditch that would divert the water to some dry land, even miles away, you had the senior water right. You were upstream and any water use below you was subject to your "water rights."

This law was clearly understood by the early settlers in Greeley, Colorado which sits in the foothills of the Rockies north of Denver. The Poudre River flows down Thompson Canyon and spills into the farming country around Greeley. What was dry land farming of wheat, would evolve into sugar beets and then into irrigated land. This came about because the early settlers of Greeley staked a claim to the water coming down the Poudre. 125 years later, after at least two court cases, those early claims still stand.

An interesting sidebar here on the implications of water and the resultant economics. Greeley had the water and would become the agricultural and cattle center of Colorado north of Denver. Fort Collins, west of Greeley, turned to education, recreation and tourism. Today, Fort Collins is flourishing. Greeley has become about 40% Hispanic and is languishing.

PRICELESS

Of all the resources in the dry west, water has become the most precious. As people have come to understand this better, some of the early settlement practices that were ruthlessly exercised have been curbed. Still, some curious pacts remain in place. Developers of the Jackson Lake area northeast of Jackson, Wyoming and south of Yellowstone Park have learned water lessons. In a preceding chapter, we mentioned how Idaho wanted water from the Snake basin to irrigate potatoes. Before they would agree to any "upstream" use of the water, they secured agreement that says when it gets damn dry in Idaho, they have the right to the top 50 feet of water in Jackson Lake. In the summer of 2001, Idaho made a call on that water. Owners of large boats moored on Jackson Lake found themselves scrambling for other locations. In the equation of economics where agriculture and recreation were in direct conflict, agriculture prevailed.

When people moved into dry land country, they quickly learned that 160 acres of land, deeded, proved up, or in some cases appropriated, was like playing with a hand of two deuces. They were in the game, but they weren't going to win very often. Water was the aces.

POWELL'S RECEPTION

Ironically, 160 acres of irrigated land in those times was more than one family could handle. Powell, in his report, had recommended irrigated homesteads of 80 acres and grazing homesteads of 2,560 acres. He recommended each allotment have access to water. Every water right would be tied to the title of the land. He further recommended and would argue for, laying out political boundaries along the drainage divides. This was the only way, he argued, to manage foothill grazing lands, the surrounding headwaters and timber lands extending down to the irrigated lands of the valley.

When he advanced these ideas to the Irrigation Congress that met in 1893 in Los Angeles, he was met with boos. His warning of a heritage of litigation went unheard. To this day, his prophecy goes largely unheeded. Even now, the boosterism that surrounds the land grabs of the dry West for grazing, farming and mineral purposes, still does not fully understand the delicate ecology of the dry land, the dryer land and the damned dry land. What was consumption and production push in times past has been replaced by a recreational consumption today. Whether it is the part-time resident moving into the arid land around Phoenix or Las Vegas, or the corporate condominium builders redesigning the timbered slopes of Jackson Hole, Aspen or Big Sky, the new shovers and pushers do not take into account the fact there is only so much water in the dry west. The consequences of aridity have not been modified.

The western states that share this condition complain the states are not masters of their own house, and from time to time, their politicians introduce legislation on the federal level to move federal land to the states. States, in turn could sell it to private interests. As it presently stands, the states do own some land within their state boundaries. Then there is federal land or what is owned by Native Americans. And while laws have been passed through the years, the Homestead Act, The Desert Land Act, the Timber and Stone Act and even as late as 1934, the Taylor Grazing Act, states have not been able to shake the heritage of federal land ownership in the dry west. For all the failed efforts of various stockmen associations to get a piece of dry land, their actions may have been only a small part of a much bigger effort to extract the natural resource riches of the west.

NATURAL RESOURCE EXTRACTION

Early efforts in Wyoming to extract oil, uranium, phosphates, coal and natural gas have ebbed and receded as the demand for these resources ebbed and receded. Wyoming has at least one ghost town, South Pass City, where the supply of gold ran out. The state does have its share of coal ghost towns and at least one uranium ghost town. But, coal production does continue and today long trains of coal cars on the Union Pacific Railroad wend their way eastward. Extraction is still an important part of the Wyoming economy. Resource extraction too often is only a temporary benefit. Extraction has a life certain. That may be true today around Pinedale, Wyoming, the site of the tussle between natural gas and the grouse, extraction and preservation.

Mark Twain, who once extracted the stories of the west, tells of a Christian sitting calmly in the card game and holding four aces. Water sits calmly in the card game of the dry west and holds the four aces and maybe even a joker to make it a heckava hand.

The joker is that while federal agencies are major landowners in the arid west, the states retain the water rights. And while there is some federal ownership of large reservoirs in the eight states, the watersheds and the water in the stream has, by law and tradition belonged to the states unless that ownership was challenged by some Native American tribe. Over time, we have set aside large areas of the most spectacular parts of the west as parks or protected areas, or have declared them wilderness areas. We have taken steps to protect the land. The issue of water has remained a legacy of pain and legal action. Whether this was more by intention or oversight, and you can get legal arguments on either side, the ownership of the water in the dry west is far more valuable that any land ownership.

In Wyoming, the Wind River flows south and east through the Shoshone Indian Reservation, then into Boysen Reservoir. It leaves the reservoir as the Bighorn River and heads north. You have another example of why square miles and sections are so difficult to apply to drainage systems. The water rights, and who has them, are a continuing source of legal action along this area. Irrigation and water rights are regularly bought and sold. Naturally, they add to the value of the land being bought or sold, if in reality the water rights can be proven.

STRUGGLES OF THE MIDVALE PROJECT

Around the edges of the Shoshone Reservation is land that is deeded to people who also believe "this land is my home." These folks arrived after World War II. Many were veterans. They were invited by the Federal government to buy 160 acres of land. They were promised water for this dry land. Of course, the Shoshone people had also been promised water for their land.

The newcomers became a version of the "hard working, independent, stubborn" farmer. "We worked like hell," one gray haired, stooped shouldered man told me one afternoon. "We were promised water for irrigation." When the Wind River Degree was finally delivered by the State Supreme Court after 13 years of litigation, this farmer, like many others, felt like the Federal Government had not kept their promise. He also felt a little guilty, he said. The Riverton Project, which was to have delivered him water, had by 1980, absorbed over $70 million of Federal subsidy. In 1988, the unpaid debt owned by farmers on the project amounted to $73.40 per acre per year. It was the biggest water project ever in Wyoming.

Today, there is a strong feeling in the area that became known as "The Riverton Project" and would later include "The Midvale Project," that the Bureau of Reclamation did not deliver on their promises. They never completed the project. The gray haired farmer first asked not to be identified and then said, "The government did not keep its promises." Of course, the Shoshone people had said this many times in years past. Again, water controlled the ecology of a whole area in the west.

THE NEW MOBILITY OF THE WEST

The ATV, the four wheeler, the snowmobile have brought a new kind of struggle to the dry west. It is another form of mobility. And to understand the dry west, or even the west for that matter, you must consider the role of mobility. Sometimes it seems only rocks are at rest. The animals of the country are moving. The cowboy was always on the move, either with the cattle or on his way to the bar. The mountain men were constantly moving, following the beaver. The Native American moved around. Dust bowlers moved to California. Today Californians are moving to Colorado. Mobility is the earmark of the west.

So, we have a new source of pressure on the ecological balance of the dry land in the form of the Polaris, the Honda and the Harley. Why wouldn't this be the case? Americans have always been wanderers. It is a national habit, whether it is wandering down the Oregon Trail or wandering around on four wheelers in the Teton National Forest. And this new "wandering," this new search for freedom and independence puts our third conservation ethic on display? "Happiness depends upon material satisfaction and nature can be used in any way that accomplishes this end." This latest invasion by technology is in direct conflict with our seventh ethic which would probably contend that when a person gets on that Polaris, they should accept the moral ethic of personal stewardship. Exactly what the impact of this four wheel activity is on the ecosystem is a question still being debated. Jaw exercise is another earmark of the culture of the arid west.

FINGER POINTING

A popular pastime in contemporary "green environmental circles" is to point long fingers of blame at large companies and corporations for their lack of care for the conservation of natural resources. But the so called "common man" has not been without blame, and the arid west did not escape his presence. Great expectations were placed on his character. Sometimes this popular image got smirched by individual instances of disregard for the earth. We have tended to sweep the dirt of these grass root misdeeds under the rug of self-righteousness. Stewardship was not exactly a consensus condition in the west. It is well and good to point fingers at the manufacturers of smoke belching ATVs or four wheelers. It is usually the "common man" who saddles these mechanical horses and proceeds to cut a new trail.

There were common people among the Puritans who moved rocks in New England to "fence off his land," only to see the soil wash into once clean rivers. Meantime, he was searching for witches. The prairie settler turned virgin grassland into cultivation, only to see his soil become vengeful clouds of airborne dust, all the while blaming the railroads and the weather. The western cattleman saw hoof ground soil become less and less productive as the dry land cooked in the sun. He blamed the government. The common man did not create The American Dream. It was sold to him, or her, for purposes of the well being of the uncommon man. We professed a dream, even in dry, arid country.

The stories of common people are hidden in a variety of places in the arid west. The graves in the old church yard in Sutter's Mill. The church yard is grown over with weeds. It is hard to find a tombstone listing a name where the age is over 30. There are other sad memorials where human life became cold numbers instead of names. Exactly who died at Sand Creek or Washita or Fort Robinson is hard to sort out. Mostly, they were treated as numbers. They were searching in arid land for a piece of earth that would offer security. They were looking for a place.

I have found small stories of people on the Oregon Trail who arrived at Fort Laramie. They would often arrive with wagons overloaded with family heirlooms, huge metal kettles, or in one case, "ten tons of bacon." They had hung on to such things on the prairie part of the trip, hoping to sell their excess baggage at Fort Laramie. The traders weren't buying, only selling. "Sugar was $1.50 a cup." The immigrants dumped their excess. They had been introduced to the economics of The Oregon Trail. They were about to learn the value of water as they moved on west.

I found a reference to these prices in a book by Frances Parkman. He had witnessed one such trading incident where he estimated the profit exceeded eighteen hundred percent. He would write a book called *The Oregon Trail.* It contributed to the boosterism of the west, even when one critic called it "the genteel New Englander's conception of progress." Parkman never finished the entire trail through arid country. He turned back east, seeing only the first third of the journey, missing the last two thirds, the most arduous part of the trip. It was a trip eventually, if you can believe the signs in South Pass, Wyoming, taken by 600,000 common people. Most of these stories are not on the tablets erected by BLM or the Forest Service.

THE FORCES OF NATURE IN ARID COUNTRY

When you kick around in the dust, you find these stories of the common man. These are hidden stories of what the playwright Arthur Miller has called "small minded self-righteousness." The big stories are better recorded. Mother Nature chose 1886 to join the game. The years from 1884 to 1886 were notable for at least one major event linked to conservation. Nature dealt the great cattle and ranching kingdoms a devastating hand. Overgrazing had depleted or destroyed native grass that had protected loose soil. Blizzards and sub-zero temperatures invaded the broad spaces. A twenty year cattle picnic came to a deadly demise. Anywhere grazing or limited agriculture was most intense, the damage was the greatest. The eastern or European ideas of land use had been blended with raw greed and imposed on arid land. The dry land didn't like the game.

In this period, John Wesley Powell was pushing his ideas for government sponsored science that would set the standards for arid land development. Small cattlemen didn't like his ideas for larger land grants and were fearful the government would close the public domain while Powell's methodical, but slow classification of land continued. The boosters, speculators and those who already held water rights didn't like his ideas for obvious reasons. Powell began to lose his political support. He also believed local people should, working through their local communities, build, finance, regulate and control their own irrigation projects with the federal government establishing fair division of the water. He argued that people who directly used a watershed could best protect it because their livelihood was at stake. His ideas were explosive issues. Powell was branded as "un-American" or a "dreamer blind to reality" and any political support he once enjoyed turned against him.

Powell, probably better than many conservationists, understood arid country. He was just ahead of his time. His focus on water as the controlling card in the arid land card game has proven to be totally justified. His ideas on watersheds as the control nerve for government or economic boundaries were valid. This idea was largely ignored as surveyors continued to drive square spikes into the heart of the dry land country. His ideas could have resolved many of the issues existing today.

DRY WEST LAND USE EVOLVES

The arid land demanded adaptation and accommodation. Finally, in 1934, the Taylor Grazing Act belatedly closed the public domain to homesteading and established some local grazing districts. It gave the Forest Service and BLM new responsibilities. The legacy of Powell and his view of science had emerged as models. But it is a land of elusive ecological solutions.

It is a lonely land. For all its stark beauty, it can be a grumpy land. It is a space where dawn takes its time to happen; where sunsets can command the whole sky, compelling and brilliant. Huge urban centers have emerged. They drain their own land of water and then covet the water of their distant neighbors. Between these urban sprawls and splotches of asphalt and cement are the small towns or the ghosts of small towns. Usually these are along a road. The town's presence is announced by highway signs that proclaim a motel or a trailer park, both symbols of the mobility in the dry country. This is the road into town. The road out of town has no signs except that the next town is seventy some miles distant. The town is done with the traveler or the tourist. Affirmation of the legacy of Bernard DeVoto lives on. "Give us your money and move on."

Usually, these towns have a huge four legged sentinel of steel standing guard over the town that huddles along the road. It holds the town's most precious commodity, water. It also provides a space for the enterprising senior high school class to pronounce their presence in large spray paint letters, "Seniors 2000." It may be their last creative expression before they take the road out of town.

A plastic sack, a refugee from a passing travel trailer or the back-end of a pick up truck is impaled on a sagebrush, fluttering and snapping in the wind. The sack looks used, lost and scruffy, somehow lonely and vulnerable. The sage is self sufficient. Evolution has provided adaptability. The plastic sack was made to serve a temporary purpose in a throwaway society. It did not have the native adaptability and will chatter and scold in the wind until it is extinct, snapping blame at the weather or the government for its condition. Something like the rancher or cowboy in this dry country, who, like the sack, cling to the land.

REFERENCES

Cronon, William. *Under an Open Sky: Rethinking America's Western Past.* (W. W. Norton & Company) 1994

O'Gara, Geoffrey. *What You See in Clear Water: Indians, Whites, and a Battle Over Water in the American West.* (Vintage) 2002

Parkman, Francis. *The Oregon Trail / The Conspiracy of Pontiac.* (Library of America) 1991

Stegner, Wallace. *Living On Principle.* (Henry Holt and Company)

Strong, Douglas H. *Dreamers and Defenders.* (University of Nebraska Press) 1988

Wilson, Edward 0. *The Future of Life.* (Alfred A. Knopf) 2002

Conversations with the Shoshone Cultural Council. Reba Term—Sonny Shojo.

Personal Visits to Greeley—Diversion Dam.

Chapter Twenty-Three

Beginning with the Beaver

It was close to being high country. Still. Hardly any breeze at all. Unusual for this stretch of stream. A quiet place. Cottonwood trees could find pockets of soil that here and there mingled with gravel and rock. A fallen tree offered evidence the beaver was about. Castor Canadensis. A British chap named Kuhl had labeled the little critter when science finally took note of her presence in 1820. Her fur had adorned hats as early as the 1600's. It just took some time for her to make the switch from fashion to science.

Castor carries the heavy burden of having become a symbol . . . for industrious behavior. She has survived because the orderly, inoffensive creature had the innate ability to adapt. Her fossil ancestors have been found in the bone beds of the Niobrara River bottom country in Nebraska. They rest among the remains of the mastodons. There is some conjecture that at one time there was a very large beaver, perhaps about the size of a big dog. Probably through time, the beaver became smaller. Maybe because she had to dodge a lot of traps.

BEWARE OF MAN

Man was really the only enemy the beaver had. Some of the predator animals, the wolf, bobcat, lynx, bear or fox might from time to time kill a beaver. The beaver's eyes, though small, allow the animal to detect motion quickly. They have a sharp nose and a very keen hearing. They were built for survival. The tail has been romanticized. It really isn't used to haul mud, to slide on, or to trowel mud onto dams. The beaver does not sleep with her tail in the water to warn her of a falling water level. If habitat conditions are stable, Castor stays rather close to home. She uses her tail to swim.

When you place the beaver in our model as the natural resource, it is also possible to put the other three primary frames in place. First might be economics. Beaver pelts were the first great export of the North American continent. The "Beaver Rush" began in the early 1600's and lasted until roughly 1845. In some of those years, French, Spanish and English trappers and fur companies were all exporting furs to Europe. One company alone might export as many as 25,000 pelts. One trapping operation in one year might represent 100,000 pounds English changing hands. When you consider the total scope of the beaver trade, it was a big commercial enterprise, even by today's standards.

Certainly, the role of government played a part. There were the land disputes between English, French and Spanish. The interests of the United States were emerging. As to social contracts, the beaver trade created a culture of its own. The trappers had only a loose allegiance to any government. Add the presence of the Indian tribes, their alliances, their participation in trading and trapping and you had another "government" element in the mix. Trappers took Indian women as wives. So government and social contracts often intermingled.

The Indian seems to have been a prudent steward of the beaver. Long before the white man's tempting trade goods and his rum and brandy arrived on the scene, the Indians lived in some accommodation with their neighbors, the beavers. While the settlers might keep swine or cattle for their own use in times of food shortages, the Indian would use beaver from nearby ponds for meat or skins. While Indian trapping contributed to the extraction of the beaver, spiritually, they did not believe in the extermination of the animal. Whether by the consensus of the tribe or the individual social contract, the Indian regarded the beaver as a rather excellent sustaining resource.

EARLY HISTORY

There is some evidence of Spanish beaver trade activity as early as 1598 in the New Mexico area of the United States. French activity began around 1605 and moved westward across the Northeastern part of the continent. The French began what we call the "cross cultural condition" with their pattern of intermarriage with women of the tribe where they were trapping. Often they would join the Indians as trappers. This "cross cultural lineage" extended into the time of Lewis and Clark and the Mountain Man.

Lewis and Clark became a sort of time line in the world of the beaver. To that point, water had been a controlling natural resource. The canoe or the perouge carried the explorers or the trappers along various water routes, whether river or lake. Most of the Indian trade and traffic was along the rivers. Trapping was generally along rivers or creeks. Forts or trading posts, signs of both government presence and economic influence, were usually along a river.

Water travel and the beaver trade shared 140 years. Horse and/or wheel travel and the beaver shared only 20 years. What had been essentially an Indian-trapper rendezvous involving local trade goods would evolve into trapper gatherings and supply wagons arriving from St Louis. The beaver pelt became an item of reckless pursuit. For the men who worked the traps, it was an endeavor of little financial return. The trappers planted their own seeds along the rivers, extracted the pelts and moved on.

THE FOLKLORE PART

. . . of the beaver story began on March 20, 1822 when Jedediah Strong Smith saw an ad in a St. Louis paper. It advertised for enterprising, young men to join a party to ascend the Missouri River. That party started what became the big rush for beaver in the upper Missouri country. It would marquee such names as Smith, Sublette, Fitzpatrick, Bridger, Manual Lisa, Osborne Russell and John Coulter. Coulter would be the first of his kind to view the bubbling sulphur pots of Yellowstone country. Smith, with his uncanny sense of distance, would create maps from the South Pass of Wyoming, to California and back to the Green River country of Wyoming.

Most of the "enterprising, young men" didn't last long in beaver country. Folklore says that of every 20 young men who aspired to the solitude of the mountains, only 3 made it for any period of time. The last rush for beaver lasted only about 25 years. Then fashion changed and the Army or Navy, the Wellington, D'Orsay or Regent, all hat styles, went out of fashion. So did the rush for beaver pelts. The little critter went back to being busy.

THE GOLD RUSH

. . . that spilled out in California would rush into Nevada, Montana, and Colorado. This rush eventually included silver, copper and lead and later, gold in Alaska. Mining demanded a type of extreme individualism that modeled the mindset of that time. It reinforced individual attitudes toward extraction. Whole ecosystems would be affected. Gold, discovered at Sutter's Mill in 1849, started the extraction. Sutter was a land baron. In 1846, he had sent a group of his men from his fort on the American River to look for timber. They found some excellent sugar maple between two creeks. They named one Sutter's Creek. Sutter really had little nor any part of mining. In 1850 . . . another one of those strange historic connections: he hired a young San Francisco attorney, Frederick Dillingn to do some legal work, Billings, just arrived from Vermont, was admonished by Sutter that he worked too cheap. Evidently he adjusted his fees. He became the Billings of the Northern Pacific Railroad, the town of Billings, Montana and a farm in Vermont.

MORE EXTRACTION FASTER

Anthony Chabot, a common miner, devised a canvas hose with a large nozzle in 1852. He could squirt water against a hillside along a stream and wash gravel and sand into lower levels where it could be examined or screened for gold. In short order, this new way of washing gold caught on and by 1870, the first basic hose evolved into a huge hose that would ravage a whole hillside. Hydraulic mining had joined the rush. The stream became a caldron of debris, soil and rocks. Along the Yuba in California, the river town of Marysville had to build higher dikes to contend with the runoff. In the year 1875, Mother Nature cooked up a big storm. The Yuba went over its banks, the town filled with silt.

In this period of time, there seems to have been no restitution. You could tear it up, pillage, destroy and walk away. I saw the results of this practice outside Dawson, Alaska where huge dredges had left mountains of rock and gravel along the stream. Gold miners can be natural resource predators too.

CUT AND RUN WAS THE RULE

. . . in the timber industry. It wasn't until the early 1800's that finished lumber was shipped to distant places. In early times it was largely a local use. The wood harvested in Vermont was used mostly in Vermont. The colonists, however, had brought the ability to organize with them. By the middle 1800's, the entrepreneurs got to work with the steam mill and circular saw. The steam mills were usually close to water. The first trees cut were along the water. This started erosion. Cheap labor, plenty of wood, the technology to produce lumber fast and the timber rush was on. The sawmill had a life expectancy of about twenty years. That was long enough to strip most of the nearby land. The mill moved on. The denuded land went back into public ownership, many times because of unpaid taxes.

MORE SYMBOLS

Of course, Grandma would need a symbol patch or two for this part of her quilt. The writers-promoters created Paul Bunyun and the other Nordic figures that symbolized manhood, strength and a great war whoop . . . TIMM-MMMBBBBBAAARRR . . . and in the midst of these heroics, the trees danced to the chainsaws. Today, some ardent environmentalist would chain themself to a tree, the TV crew would show up, it would make the evening news and several lawyers would have risen from the sawdust.

Our model is working in all these extreme cases of extraction. The left frame is natural resources. You have millions of acres of land. They are in a stage of high performance. The right frame is government which is based on tolerance and little regulation. At the bottom there are the social contracts of that time. These are guided by high consumption and are totally negative toward conservation. The frame of economics is based on consumption. In just a few years, in actuality or in the model, the forest went back to a stage of Form. Economics, social contracts and government all impacted negatively on natural resources.

The bison rush is a shaggy, sad story. The animal eventually becomes a symbol. I have often wondered, in a slightly cynical way, what would have happened if the USFWS had been setting seasons for bison as they now do for ducks. In the spring, they would have gone to the usual pastures on their list where they had found bison. They would have advanced computer programs that magically produced the bison count. Based on those figures, every bison migratory route would have had a season. There might even have been a limit for every hunter. Like ducks, how would they have known how many there were? When it comes to counting bison, some folks like the "prairie windage" method . . . "millions of them." I have seen numbers ranging from 10 million to 100 million as the original number.

We do know that one day in 1874, in Nebraska, a train stopped because the tracks were so slick from grasshoppers the wheels couldn't get traction. Finally, they got the train moving, went west only to be stopped by a bison herd headed north. They waited 24 hours for the animals to get across the tracks. However many, that's a lot of critters! The bison rush would last only about 15 years. Railroads wanted shipping revenue from hides and bones. Hunters wanted money for tongues and hides. The army wanted them killed because they were food for Indians. The "sportsmen" just wanted some great shooting. Most sources, and there is some agreement on this, say that fewer than 300 eventually survived.

I once visited The Conrad Mansion in Kalispell, Montana. It had been built in 1895 as the home of C.E. Conrad, a Montana pioneer who had made his fortune as a trader, freighter and founder of Kalispell. On the wall was a brief article that said he once owned a few head of bison and made a genuine effort to protect them. The article didn't say what had happened to them. It was another patch for Grandma's quilt. The final act in this sad and melancholy tale of the bison rush is that in 1875 Congress passed a "buffalo protection act." It was the first piece of legislation ever passed by Congress to protect a wildlife species. President Grant, the most mortal enemy of natural resources ever in the White House, vetoed the act.

CATTLEMEN WEPT FEW TEARS

. . . for the demise of bison. The animals were competition for grazing and range. They had their own rush with Longhorn cattle. To the time of the Civil War, most farmers kept a cow or two. That idea fit into the myth of agrarian self sufficiency. The cattle rush brought together Texas Longhorn cattle that wandered during the Civil War, the coming of the railroad to the central plains and the emergence of the cow town along the tracks. The economic wheels of connection were now in place. For fifteen years, there was a steady stream of cattle driven from Texas to the central prairie railroads and in some instances, on up to Montana and Wyoming.

Mother Nature and ecology put an end to this rush. She got a lot of the Longhorns in the blizzard of 1886, as we have mentioned. An encore act in 1895 hit a more southwesterly area. Drought got so bad cattle were eating bark off trees until both trees and cattle succumbed. Leasing of Federal land for grazing provided probation, but not a solution. In fact, to this time, leasing remains one of the thorny quandaries of land policy.

As the smaller rancher has sold land or found economic benefit in a sale to a conservation organization, private land in the west has become a case of an absentee landlord. The old goal of at least 2500 acres for a viable cattle operation has proven to being a minimum holding. You also need water. When weather conditions come unhinged, no rain, no snow, no moisture . . . the stage is set for grass depletion, land suffers, livestock suffers, and wildlife suffers. Even stockmen suffer. Plenty of suffering to go around.

NOT ALL THE GREAT RUSHES

. . . were on land. The U.S. got Russia to sell Alaska in 1866. There are some islands in the Bering Sea called the Pribilof Islands, named for the Russian navigator who discovered them. When Mr. Seward got money from the Treasury, $7,200,000 to buy his Alaska "folly," the Russians had harvested about half the estimated 5 million seals originally on the islands. The Alaska Commercial Company wrangled a franchise from the Federal government to hunt the seals. The great "seal rush" was on.

American hunters moved in and began shooting the seals during migration. Where the Russians had slaughtered them in rookeries and so spared the pregnant cows, the Americans did not discriminate. Strangely, the United States government showed little concern. I located this story in the Anchorage Museum. It alleges the company took enough seal skins in the first twenty years of operation to pay for Alaska. The debacle ended in 1911, when Canada, Japan, Russia and the U.S. all became parties to the Fur Seal Treaty. At that time, less than 5% of the seal population remained.

The stories of the "rushes" and what the white man did to this country are not necessarily new. They are a replay of what men, of various colors, did to land in Asia, Europe or the Middle East. In the old regions of the world, the process was hand made and a slow one. It extended over so many years complete records are often hard to come by. By comparison, the transition on the North American continent was so swift, the big events so compressed, it is difficult to sort out the over-abundance of quilt patches.

THE SOCIOLOGICAL THREADS

If each generation leaves a legacy of community, a crazy quilt of social contracts, then three things seem to determine that legacy. First, there is the prowess of the human animal to exterminate other species, including the human. In the human equation there is war, genocide, suicide for racial or religious rituals. Terrorism falls in the last grouping. Killing animals, bison, beaver or seals would be included in this prowess. In each case, the human predator has the ability to chose, reason and determine the outcome of their ability to extend or control this prowess. The human choice, whether it is extermination or extension of life, will ultimately alter the ecosystem. In this sense, man controls the destiny of the ecosystem.

The second element in establishing a legacy of community is the ability of the human animal to manipulate the biotic community. The human has the ability to sow and harvest. They can, with study and reasoning, enhance the process with further manipulation. New England newcomers learned that a fish added to a hill of corn produced more corn. Three hundred years later, a catfish farmer finds that feeding the fish corn makes catfish fatter. Whether it was fish for more grain or grain for more fish, the human animal manipulated both trade-offs in the biotic community. The ecosystem had been manipulated.

This leaves a third element in the assessment of a legacy. This is the compassion of the human animal. Here we encounter the mysterious, the spiritual, religious, ethical, moral. The customs, traditions that in reality produce social contracts. The elements that make the human animal think and feel about their community, their place and their relationship to the community of living things. The concept of social responsibility and stewardship is necessarily a part of conservation compassion.

When the human animal gets in a hurry and rushes into extermination or manipulation, the lesson seems to be he loses this conscience. Lacking this third element, conscience, man soon finds a way to mess up the equation. This is the ugly lesson of the "rushes." The lessons are not all pretty.

As a nation, we seem to be fumbling into the future and losing a lot of environmental yards. We have turned the ball over and even lost it more than a team can do and still win the game. There is one thing to be said about our past. When it came to the predation of natural resources, man has surpassed the busy beaver.

Stephen Tyng Mather

Was born in San Francisco in 1867. He attended the University of California, worked five years as a reporter for The New York Sun, then quit to join his father in the borax business. He succeeded in developing the famous 20 Mule Team Borax brand, became wealthy within a decade, and dedicated his interests to conservation.

Mather, a member of the Sierra Club, made a trip in 1914 to Sequoia and Yosemite National Parks. Upon his return home, he wrote his college friend, Secretary of Interior Franklin Lane a highly critical letter. Lane responded that if Mather didn't like the way the parks were being run, to come to Washington and run them himself. Mather went to Washington and agreed to take the job of Under Secretary for one year.

He had exceptional success in enlisting PR support and money, from railroads, influential people and his life long friendships. For a time, he personally paid part of the salaries of staff in the start-up Park Service. He would resist resource exploitation in the parks and so became a John Muir preservationist. He succeeded in establishing a professional ranger/manager core that was free of political patronage and at the same time cultivated a favorable public opinion of the parks.

His epilogue, on display in many national parks says, "There will never come an end to the good that he has done." He has had a highway in the Cascades, a forest in New York state and an Alaskan peak named in his honor.

Mather died in Boston, Massachusetts on January 22, 1930.

Stephen Tyng Mather. National Park Service Photo

Emerson Hough

Was born in 1857 in Newton, Iowa. He graduated from the University of Iowa with a law degree. He moved to White Oaks, New Mexico where, after a brief law career he began to write articles about his camping trips.

Hough was selected for our All-American team because he is a good example of a writer who became a conservationist and in a peculiar circumstance started a string of events that made a mark. He was visiting Yellowstone National Park where he learned of the poaching of the last of the bison. He wrote about this in his "Out-Of-Doors" column in the Saturday Evening Post. This led to a law being passed by the U. S. Congress that protected the bison, the first time a particular species had been singled out for Congressional protection.

Hough was especially tough on "sportsmen." He said, "At the hands of the ignorant, the unscrupulous, and the unsparing, our game has steadily disappeared until it is almost gone. We have handled it in a wholly greedy, unscrupulous and

*Emerson Hough. Photo courtesy of the Library of
Congress, LC-DIG-ggbain-04852.*

selfish fashion. . . . If there is to be any success for any plan to remedy this, it
must come from a few large-minded men, able to think and plan, and able to do
more than that. . . . to follow their plans with deeds."

Hough is a good example of the power of a writer to influence conservation thinking. He did this by getting the word out to the grass roots level and
by his political activism. He could, by no small stretch, be credited with saving a species on the brink of extinction.

Hough passed away in 1923.

REFERENCES

Clyman, James. *James Clyman, Frontiersman: The adventures of a trapper and covered-
wagon emigrant as told in his own reminiscences and diaries.* (Champoeg Press) 1960

Cronon, William. *Under an Open Sky: Rethinking America's Western Past.* (W. W. Norton & Company) 1994

Curtis, Jane; Jennison, Peter; Lieberman, Frank. *Frederick Billings: Vermonter, Pioneer Lawyer, Business Man, Conservationist: An Illustrated Biography.* The Woodstock Foundation. 1986

Klotz, John W. *Genes, Genesis and Evolution.* Concordia. 1955

Sandoz, Mari. *The Beaver Men: Spearheads of Empire.* (University of Nebraska Press) 1978

Stegner, Wallace. *Marking the Sparrow's Fall: The Making of the American West.* (Owl Books) 1999

Beaver Habits, Beaver Control and Possibilities of Beaver Farming. U. S. Dept. Of Agriculture #1078.

Notes made in the Anchorage, Alaska Museum 1995.

Chapter Twenty-Four

The Trail of the Wapiti

No two wildlife species provide a better synthesis of the conflicts of natural resource thinking than do the elk and the wolf. Wapiti was the Shoshone name for elk. The animal has all the themes found in previous chapters . . . patterns of extraction, disregard of natural resources, lack of national conservation conscience and the wildfires of litigation.

An estimated ten million elk once occupied every continental state in the United States except Florida and Alaska. The Euro Americans and their market hunter cousins decimated the herds. By the late 1800s, the best estimate of the remaining elk was fifty thousand, found mostly in the Yellowstone Park ecosystem. Some predicted the animal was headed for extinction.

Early in the 1900s, policies of protection were begun as more enlightened ideas prodded federal and state action. Hunting was curtailed. Poachers were hounded into diminished activity. Gradually, where the animal still existed, habitat improvement helped the survivors. As they prospered, elk were often relocated in other areas of the country. In fact, the story of elk restoration is a proud chapter for conservation.

THE SITUATION TODAY

Wildlife managers tell us America is home to about one million elk. Most of these belong to the Rocky Mountain subspecies. There are, however, populations of tulle elk in California, Manitoba elk in Canada and Roosevelt elk, the largest of the species, in Pacific Coast forests from California to British Columbia.

North American elk are in the same generic family as the red deer in Asia and Europe. They are second in size to the moose. An adult male may stand five feet tall at the shoulder, weigh as much as one thousand pounds, and support a very large rack of antlers. Evidence of elk mystique and their massive racks adorn many walls. The cow elk are shorter, usually weigh two hundred pounds or less than the bull. The animal is mostly tan or brown. They sport a distinctive white rump.

The elk is a herd animal. The early method of hunting them was to locate a herd. Knowledge of their wanderings was usually based on seasonal movement. Unlike the deer, which seems to live in almost a matriarchal society controlled by the does, the elk is a socializer. They often form sizable herds. In the fall, bulls compete for breeding rights, with each would-be patriarch forming his own harem. The time of rut is fairly predictable. It is marked by some spectacular antler clashing and the bull's frequent bugles, a blend of whistling and screams.

In terms of our model, the three other determinants, economic, government and social contracts all apply. Each fall, in Jackson, Wyoming all three frames are on display. Nearing Jackson on the highway from Moran Junction, it is not uncommon to see crowds of hundreds of people. There may be an interpreter from the Park Services. The elk, in time of rut, are on display. Sex has become a spectator sport!

The binocular manufacturers love these moments. The Fuji and Eastman Kodak people see the sales of film mount. The merchants of Jackson are busy swiping credit cards. The Jackson Chamber of Commerce estimates the fall elk display is worth 3 million dollars in tourist generated revenue. Elk contribute revenue in eastern elk states. According to Pennsylvania state sources, elk are responsible for added revenue of some $1.7 million. Economically, elk are a very big business, almost a commodity.

Social contracts have supplied a fair share of conservation efforts. One of the main movers has been the Montana based Rocky Mountain Elk Foundation. It is a private group, non-profit, that has spawned grass roots chapters in most states. To date, it has provided 1.5 million dollars for relocation projects in areas like Kentucky, which state hopes eventually to have a population of ten thousand animals. Arkansas began relocation in the 1980s, and Wisconsin started its herd in 1995. From the standpoint of restoration of a species, elk restoration is one of conservation's prouder achievements. Partnerships between the public and private sectors have worked well.

THE ADAPTABILITY OF ELK

One reason for their comeback is their adaptability. Wyoming is a good study because the animals range from virtual red desert floors to rugged mountain areas. Although primarily grass grazers, they will eat other plants, and even trees in the winter. They are opportunists and will move to find food. Game managers often use food as "bait" to encourage them to change feeding areas.

The cow elk will imprint the calves. The cow is attentive and protects calves as best she can from predators. The wolf is a primary predator. Only rarely do you find evidence of road kill. Elk are tough animals, but they are also wary, with excellent eyesight. Now we come to an area of contention. The Wyoming Wildlife Federation has studies claiming elk are migratory. The Wyoming Game and Fish Department says they are using old data, that elk stay in one general area. They will move to avoid deep snow.

Pete Petera is the former head of the Wyoming Game and Fish Department. He is a highly regarded elk authority, once a Game Warden just south of Jackson Hole, prime elk country. He has "mothered" orphan calves until they could be returned to the herd, or to the Department research station. He has transplanted elk to areas of the state that could handle the animals. His position is that elk migrate seasonally. They need about 450 acres for a herd to remain intact. If conditions remain fairly normal, they will adjust elevation to find food, but basically the herd is indigenous to a general area.

"The department," says Petera, "has used ear tags and neck bands for nearly 40 years to secure an accurate check on where the animals seem to be. We have had a few instances of elk wandering over into Idaho, but very few. The Montana reports are even fewer. The idea of elk moving great distances over migratory routes would run contrary to what we have learned about them."

SO HERE IS THE FIRST CONFLICT

. . . the opinion of the Wildlife Federation versus the information from the State Game and Fish Department. Checking other sources close to the scene, there seems to be a feeling the Wyoming Wildlife Federation reports may be open to professional questioning. Point one is that radio collars and helicopter sighting have improved the ability of any group to do a study. Point two, when a group, or an individual studies life spans in animal husbandry, research may take several years. Our populace likes speed and single solution fixes. Nature moves very slowly. There has been public impatience with elk management in Wyoming.

Petera then cited another instance that involved controversy about an elk herd between Yellowstone Park and Teton National Park. One group claimed a belt of habitat was under such heavy grazing by elk the area was in poor condition. "They said the elk should be moved," said Petera. "We inspected the area and found it was ground squirrels that were causing the problem." I later verified this story with Jeff Hogenson of the Forest Service. "Ground squirrels can eat the root systems of both grass and trees and in time it will look like an area has been over grazed," he verified.

One of the major accomplishments of the Wyoming Game and Fish Department under the tenure of Petera was the successful acquisition of the Spence-Moriarity property in northwestern Wyoming. A valley ran north and west along the East Fork River watershed and had become a rather important wintering area for elk. The state acquired some 20,000 acres of land which can never be subdivided so it is fairly immune to development. This provided good habitat for elk and remains open for hunting. Nearly fifteen years after acquisition, the work on improving irrigation and roads continues. Not that anybody is dragging their heels. Restoration takes time.

DENALI NATIONAL PARK

Several years ago, my wife and I visited Denali National Park. A side benefit was learning about the management policy of that park. When it was declared a national park, the Park Service commissioned three different sources to make a study. Each study had the same objective, how to manage the park. Each study lasted 10 years. Each came back with the same answer. Leave it alone and let natural cycles of wolves and caribou find their own levels.

This was a far cry from early thinking about elk management in the Yellowstone area. Depending upon who was superintendent, Yellowstone has seen a wide variety of management views and styles. The area may rank as one of the higher spires bureaucracy has erected in the Tabernacle of Arrogance. Of course, the greatest deception men suffer is from their own opinions. As a self professed "landscape naturalist" who sometimes slips into the jacket of arrogance, my view of an ecosystem is that it is "space." An ecosystem is both a macrocosm and a microcosm. It is multidimensional. To my mind, it is not a geographical area. It is a space more like a cube. An ecosystem can have acres, boundaries, perimeters, sections; all geographic terms. But it also has space. Space is more complex. It could include wind, or the lack of it, sunlight or lack of it, contaminated air or pure air. Trees, grass or none of the above. All would impact the space of the ecosystem. Roads can alter the way animals use the space. Space becomes an issue anytime you pack too many animals into a concentrated area.

"Roads are another big issue," says Petera. "Generally, if the road is gravel or dirt, it doesn't seem to bother elk. It is the noise, traffic or human intrusion that becomes the issue. If you are talking about asphalt highways, that is a different issue. You just have to be careful what kind of roads you are talking about." Roads can change or modify space quality. Lack of quality space results in crowding. Crowding encourages disease.

MORE RECENT RESEARCH

Employees in a research station in Colorado noticed some captive mule deer were drinking and urinating frequently, had lost weight and were drooling and stumbling. Soon, the animals died. This was in the early 1960s. Researchers found a new disease. They labeled it Chronic Wasting Disease (CWD). As best they could determine there was no treatment and no survival. In 1981, CWD was found in elk. In 1985, it had spread into wild deer herds.

The disease may have come from another animal. It may have existed in the wild and nobody knew about it. The only thing for sure at this time is that it scares the hell out of researchers and wildlife managers. Recently developed technology allows wildlife managers to test dead animals for disease and has speeded up that process. The cause remains elusive. Chronic Wasting Disease has already been identified in deer in Wyoming. CWD has appeared in captive elk in Montana, South Dakota, Oklahoma, Nebraska and Saskatchewan. Researchers are convinced it's only a matter of time before it jumps from a game farm into the wild herds. How it travels is unknown. Beth Williams, Professor of Veterinary Medicine at the University Wyoming and a leading CWD researcher says, "However it travels, CWD is an extremely persistent agent. The specific vehicle could be saliva, urine, feces, contaminated soil, even simple nose to nose contact through a fence."

Concern about the disease has become widespread. This report from a short article in the South Dakota Wildlife Federation *Out Of Doors* magazine. "An outbreak of CWD in six game farms in Saskatchewan last year caused the Canadian Food Inspection Agency to order the slaughter of seventeen hundred elk between June and December. The disease was confirmed in a dead elk on one of the farms in June . . . It is unknown how many of the 1700 elk were infected with the disease." Missouri officials recently voted to postpone a planned introduction of elk in Missouri to some future date. To date, elk in the Yellowstone Park region have shown no signs of infection. But another problem has emerged. Now the spokesman for the U.S. Geological Survey and an ecologist, is quoted as saying. "Yellowstone is grossly overgrazed by elk and the negative impact on a variety of wildlife species is extraordinary." Richard Keigley, just quoted, goes on to say that the northern range for elk in the park area is in deplorable condition. He cites the juniper, spruce, aspen and other woody plants and shows you photos from decades ago and then photos of the conditions today. And, of course, this condition affects habitat for birds, it impacts on erosion and soil condition.

Critics of present management systems that have resulted in current conditions say the problem can be traced back to the early 1900s when two things happened. Federal intervention stopped any hunting. Trappers, hired by the

Federal Government eradicated the wolves. But as long as elk can get to the diner on a regular basis, their population seems to thrive. So the focus was on elk food conditions and habitat. The National Park Service tried to control the population by killing the elk or shipping the animals elsewhere. Public opposition to this (social contracts at work and some economics, too) forced the NPS to suspend that policy and rely instead on "natural regulation," a concept that elk will limit their own numbers by reduced fertility and survival when the population gets too high.

POINT AND COUNTERPOINT

This policy did not meet with a standing ovation. Utah State University wildlife ecologist Charles Kay is quoted as saying, "Natural regulation is a scientific fraud and it may take ten thousand years to repair some of the damage." Not so, shoots back John Varley, Director of the Yellowstone Center for Resources. "If we killed every elk that ate a lower limb of a tree, we would have no elk and a lot of bushy trees. We have more faith in the processes nature employs than in our arbitrary judgments about how things ought to be."

Comes another study from Oregon State University and lo, the hand of the researcher is placed upon the wolf. According to this scripture, aspen began to disappear from the park in the 1920s, the same time the last wolf was carted off. The study theorizes that wolves benefited aspen by keeping browsing elk away from the groves. The elk were more vulnerable to wolf attack when the wolves were better hidden in the aspen.

This debate illustrates the many conflicting views. A man who may be the most surrounded by both elk and views, Barry Reiswig, manager of the Jackson National Wildlife Refuge, says the feeding program of elk and bison on the refuge is "a giant house of cards." A million dollar study of the problem with the feeding program on the refuge is headed into stormy waters according to Reiswig. He points to three serious flaws in the present program to feed elk and bison at Jackson Hole. The first flaw he believes is that feeding hay or pellets instead of using natural range for the animals concentrates the animals in higher numbers on limited range. This, in turn, sets the stage for his second point, the possibility of disease.

His third point is that if elk migrate each winter, the animals face a firing line of hunters on the east side of Grand Teton National Park. Reiswig claims forcing the animals through what amounts to a firing range teaches the hunters poor habits and ethics. This view introduces the many opinions of using hunting as a method of elk population control. If hunting is used on any refuge as a means of population control, the wildlife managers run the risk of public disapproval and a possible law suit from some environmental group. The reality is that without hunter predation, elk will starve. The question is . . . who will be the predator, the wolf or man. These issues surface when disparate groups gather to write any sort of partnership plan, such as the proposed National Elk Refuge Plan. In this case, the parties can't even agree on who should be involved in writing the plan, the Park Service, U.S. Fish and Wildlife, staff from the Elk Refuge, state officials from Wyoming? To quote Reiswig, "I don't think we have come to agreement on anything."

IN A THOUGHTFUL INVITED PAPER

. . . in the *Journal of Wildlife Management* 65 (2): 173–190, Bruce Smith from the National Wildlife Refuge in Jackson, Wyoming tries to objectively sort out some of the many issues involved. His points from his abstract include the cost of feeding which averaged $1.6 million annually from 1995 to 1999, to feed 31 thousand elk in five states.

Smith lists four rationales for feeding elk. His points are these. (1) Feeding elk can maintain a larger number of elk which enhances hunting possibilities. (2) Feeding can make elk more available for public viewing and commercial benefits can result. (3) Feeding may reduce winter mortality of elk and assuage public concerns about animal welfare. (4) Feeding alters winter distribution of elk, helping to keep elk off private lands where damage to crops, orchards and fences occurs, and off roadways where motorist safety may be of concern.

He then gets into the economic issues. He points out that feeding enhances economic opportunities for guiding and outfitting and related businesses that benefit from guiding and outfitting. In Wyoming, state law does not permit nonresident hunters to bag big game without a guide in any of the fifteen national forest wilderness areas within the state. In 1998, Wyoming generated $7,770,000 in revenue from the sale of elk licenses. If you used the accepted "four factor" . . . license fees times four produces an estimated total economic benefit . . . then Wyoming generated some thirty million dollars from Wapiti hunting alone. This would not account for tourist revenue from viewing. Elk, in Wyoming, are not only wildlife, they are serious economics.

For the Wyoming Game and Fish Department, the rewards are not as significant. If revenue from licenses was $7,770,000, the cost for the department to feed elk was $8,820,000. Again, I asked Pete Petera, the man who had spent a lifetime in wildlife management, how he would evaluate the present condition of the elk, an animal that seems to have the ability to walk around in a constant snow storm of controversy "Well," he said, "you've got a lot of points of view involved. The state, in Wyoming, has stayed pretty close to the individual. Money is many times the controlling issue. But the state is subject to political pressure. Politicians have a vantage point and sometimes there is no recourse to this kind of pressure. The Federal people are too far from the scene. Many times they are too slow to react. They are not always proactive."

He inspected some ice in his glass. "Fish and Wildlife has the authority in the Endangered Species Act. In the case of wolves, the USFW Service implemented it. The state pays for damages, even in cases involving the endangered species. In the case of Teton National Park, and that gets involved in elk, it operates under a deed from the State of Wyoming. The state still retains some authority under the agreement. The Park Service can close a trail anytime. The state must hold a public hearing to close a road. Utilities can get easements over state and federal land to deliver power. Usually a road for maintenance goes with this easement . . . so, effectively, utilities open a natural area. You could say that cheap utilities still have a higher priority than natural resources. Over the years we have just built a lot of conflict into our management systems. You could argue it is a balance, not unlike the legislative . . . judicial . . . executive balance. Not always efficient. Sometimes painful. Not always beneficial to natural resources."

The elk have been through every stage of our wildlife management evolution; protection, regulation, biological, ecological and even the emerging ideas about sociological. They provide an excellent example of our model at work. At some point in time, they have involved every one of our ten conservation ethics. Even the Bambi ethic is on display now and then.

They also offer hope. For all the court cases, all the contention, all the pontification and wrangling, there is an elk population today, however imperiled it may be in Wyoming by a growing wolf population. Wapiti represents challenges and opportunities. If some positions on arrogant maps left us wondering where we were, there are plenty of stories where people really cared about the species and honestly tried to do the right thing. We have not only preserved the species, we found ways to help it multiply. And we didn't need an ark to load them up on. In that fact alone, there is hope for Wapiti.

REFERENCES

Beal, Merrill. A *Story of Man in Yellowstone.* (Yellowstone Library and Museum) 1956

Gearino, Jeff. *"Group disputes federal elk study."* Casper Star Tribune 9-1-2002.

Martin, Claire. "A *Fall Treat.* "Denver Post 9-4-2000.

Osborne, Russell. *Journal of* a *Trapper.* (Syms—York Company) 1965

Smith, Bruce C. *Journal of Wildlife Management.* 65(2): 173–190.
South Dakota Wildlife Magazine. Fall of 1995.
Wyoming Wildlife Federation Studies. 2000 and 2001.
Interviews with various professionals associated with elk in Wyoming.
Numerous visits with Pete Petera, former Head of Wyoming Fish and Game.
Visit to Denali National Park. Fall of 1995.

Chapter Twenty-Five

Wolves

The wind blows down the river. It is May and light hints of snow ride the moaning of tree limbs still naked of leaves. You feel the presence of power, the sting on your cheeks.

The wind can be a cleanser, a sort of gusty Windex that allows you to swipe away some of the turmoil of wolves. It can help you sort out the absolute people who think they know precisely what is right or wrong. Some wise person has said, "Our differences define us, but our common humanity can redeem us." And there are differences in how people think about wolves and how they feel about them! Redemption may not fit the equation.

The wind is big today. As it passes down the river, it blows into big country, places you find heroes of times pasts. Some stories are "tall tales," sort of true. Once an old settler told a man he was a liar. "Pardner," said the other, "are you doubting my veracity?" "Nope," said the first fella , . ." I just wondered if you brought it with you today."

For the settler, the wolf rated as the incarnation of evil and depravity. For the modern biologist, they rate as a symbol of struggle and survival. For some people, they are respected as intelligent, canny and faithful; the progenitor of man's best friend, the domestic dog. Kids may relate them to Three Little Pigs.

As a symbol of struggle in ecology, the wolf raises the most profound feelings regarding procreation and survival. In a deeper sense, to understand the wolf, the observer must drop all pretense of social veneer. The wolf forces the intransigent acceptance of death and life as absolute in an ecosystem.

EARLY WOLF TIMES

Russell Beck was one of the first ranchers along this river, maybe a model Owen Wister used for the "tall, rangy, hard working, dependable" rancher. He was born in 1870, died in 1955. There are stories about early day ranchers rounded up by Esther Mockler that quote Beck. "The most feared predator was the wolf. Before cattlemen came to the area, wolves preyed on elk, deer and antelope. But after the tame, docile cattle came, the wolves found them easier prey. The wolves began to multiply and soon became a menace."

"We had the habit in the early days to wean the calves in the fall and turn them out in January. One time we took about 150 head up Mosely Draw to work their way to the range. Three days afterward, I rode up there and found five dead calves. They were laying so close together you could almost throw a stone to any one of them. Wolves had eaten off what they wanted, which didn't amount to much. They had killed them more for sport than for food. Coyotes usually followed the wolves and would finish up what the wolves did not eat."

You turn around and walk with the wind. Enough struggle. You remember more of the story. How in 1899, Beck had joined John Landis and Captain Jay Torrey, formed the Wind River Livestock Company, using land 12 miles east of what is now Dubois, Wyoming as their base for the Circle-M Ranch. More from the Mockler narratives. "John Landis was an old bachelor and the superintendent of our company. He did a lot of riding, mostly looking for wolves. One day he came into our house and when he walked through the door I could see he was greatly agitated. I asked, 'what's the matter?' He said, 'I saw a sight today that has made me all in. When I was riding I followed the track of five wolves. Pretty soon I came across some cattle that were in a close circle. Sure enough, the wolves had a yearling down and were eating on it. It wasn't even dead yet! They were eating the insides. It's hard to believe what I saw today. There can't be a just God after what I saw. I feel sick all over!'"

The ranchers, as the story goes, hired a French Canadian to poison the wolves. One day he found he had poisoned 12 pups in one den. A bounty was placed on wolves. First it was $10. Then it went up to $100 for mature animals and $50 for pups. A final chapter in the saga was one last, large wolf hunt that eliminated the animals. For the next 60 years, the wind along the river would not ruffle the fur of wolves. It was not an easy life in those times. Some folks placed a sock in a can of water and stuck it someplace in their one room cabin. They next morning they judged the cold by how hard the sock had frozen. When you lost five calves, worth about $10 each, that was a hard loss. Each time this happened, you tended to become a little more brutal in your opinions about wolves. This was the heritage that faced wolf introduction!

THE HERITAGE BEGINS TO SHIFT

Sometime in the early 1970s, our social contract with the wolf began to change. Wildlife painting began to romanticize the animal more. Until that time, painters like Charles Russell, Bob Kuhn and John Clymer had them as less than desirable pets. This new shift in visual art and social contracts was echoed by environmental writers and mainstream media.

It was decreed the U.S. Fish and Wildlife Service would get the program for reintroduction. The focus of the effort would be Yellowstone Park and areas adjacent to it. The book, *The Wolves of Yellowstone*, says 14 wolves were "translocated from Alberta (Canada) to Yellowstone Park in 1995. In January of 1996, another 17 wolves were "translocated from British Columbia to Yellowstone National Park." The wolf release in 1995 had included 6 wolves named the Crystal Creek Pack. These were released in the Lamar Valley in the park. Press releases claimed the pack had been very visible to park visitors. "It is estimated over 4,000 people saw these wolves in 1995. This pack has allowed visitors to witness first hand the amazing life and death struggles in what has been referred to as the Serengeti of North America."

True? Half true? Whatever? In their effort to build public approval, the PR people churned out a lot of "stuff." Meantime, the remnants of the old settlers south of Yellowstone Park oiled up their Winchesters. They knew from experience, because they had been sighting an occasional wolf, that not many people see them in the summer time. Park Rangers do fewer exit interviews in the summer. So much for 4000 people who had seen wolves.

THE WOLVES PROVIDE ANOTHER EXAMPLE

. . . of a long, complex and expensive journey of an environmental issue through the legal process. It probably began in 1973 with the passage by Congress of the Endangered Species Act. The Rocky Mountain Wolf, the original wolf of the Yellowstone ecosystem, was listed as "endangered." In 1975, the Federal government established a Rocky Mountain Wolf Recovery Team to do an environmental assessment on wolf recovery. In 1980, a formal plan for recovery was completed by the USFWS. It would focus on gray wolves. This is one of the biological disputes. Mike Jamenez, Project Leader for Wolf Recovery in Wyoming with USFWS told me the Gray Wolf and the original

Rocky Mountain Wolf are essentially the same animal. Sources in the Wyoming Game and Fish Department told me the replacement Gray Wolves were 25% larger and much tougher predators.

If you aren't careful when you ask questions about wolves, you can get your house blown down. My friends in game management told me there are some truths about wolves and if I wanted to know more, go read a book by Dr. L. David Mech, a noted authority on the animal. I read it, and several other books dealing with wolves. This would be a consensus from those sources.

Wolves usually operate and survive in packs. Now and then, there is a loner, often looking for a mate. They have a pack pecking order. There is an Alpha female who rules the roost. There is also an Alpha male. In Wyoming, 6 to 12 pups are born in dens, usually in April. These mature and usually leave the pack to find new mates and start new packs. If there is available food, they will remain in a general area. If the food supply gives out, or moves, or weather conditions force a wider hunting range, they will roam. Seven years after the original Yellowstone release, there have been unconfirmed reports of wolves in southern Wyoming. Recently, there was a confirmed report of a Yellowstone wolf that had roamed to Buford, Utah.

As of January 1999, according to the National Park Service and USFWS, "115 to 120 wolves" comprised of about eleven packs, inhabited the greater Yellowstone area. That estimate was down from a summer count of 160 individual wolves. This is due to natural mortality and control measures. It is estimated about 8 packs have established territory within Yellowstone and Grand Teton National Park ecosystems. The wolf may have heard the admonition "go forth and multiply." They did. Meantime, there had been local reports of wolf sightings before any releases.

REPORTS OF EVERYTHING EXCEPT RED RIDING HOOD

The "control measures" casually mentioned in press releases and news stores involved several wolves. The USFWS removed 6 wolves from the Sheep Mountain Pack and 2 wolves from the Chief Joseph pack. But the bottom line of this numbers game is that from 1995 and 1996, until 1999, radio and numerical monitoring indicated that from the original release of 31 wolves the number had grown to 115 to 120 known animals. Then still later information to verify the early growth numbers. There are now 3 packs in the upper Dunoir valley west of Dubois, Wyoming. There has been a pack of 17, sighted and counted, in the Green River area south of Jackson Hole, Wyoming. Wolf tracks have been found in the Lander area. The wolves are spreading, but I could locate no credible source who either could or would, give me any kind of count of what the present wolf population is in northwestern Wyoming and eastern Idaho. One USFWS person did tell me they "felt there were 8 packs in the park.

Back when the many disparate wolf views were being incubated, I sat in a room in Jackson Lake Lodge in Teton National Park, a part of the legacy of the Rockefellers. Our host for the meeting had arranged for a young man he introduced as Doug Smith to speak. This day, Smith had slides with him that told the wolf story from the Park Service side. He was exceptionally bright and articulate. He was also zealous about his cause and spoke with all the passion and excitement of a disciple.

On May 20, two years later, I again talked with Doug. I asked him what the status of the wolves was. "In Yellowstone and the surrounding ecosystems, which include Targee, Shoshone, Bridger-Teton, Gallatin and Custer, we estimate we now have between 210 and 222 wolves." This meant the incremental growth had slowed so I asked him about that. "Well, three reasons," he said, "the natural law of dispersion, human intervention and some disease. So yes, the pack growth has slowed down." (Natural dispersion is biology speak for "they spread out beyond a given area.")

I asked about rancher reaction. "They still don't like the idea. They have come to respect us because we have done what we said we would do in terms of animal removal and reparations for lost livestock." In visiting with ranchers, I found this was not their view. They spoke of slow response from Federal authorities and less than adequate restitution. And a final question to Smith. "In retrospect, what have you accomplished?" There was no pause. "We replaced a keystone specie in an ecosystem. We repaired what had been neglected for a long time. The Yellowstone ecosystem is now about as close to original condition as we can make it at this time." We chatted about the things they had learned. As is often the case with biologists who are on the cutting edge of a new project, they will tell you they learn as they go. Smith's focus was entirely on biology.

LEGAL INTERPRETATION

Wolf relocation carried an "experimental/nonessential" designation. The law does not like ambiguity mixed with "whereas." But ambiguity sneaks in. So, the Endangered Species Act, while the intent was to be specific, is often subject to interpretation. Examples of this would be one part of the rule that says "wolves will be completely protected in national parks and national wildlife refuges." On the other hand, it also says, "land owners and livestock producers may harass wolves on private property *or the vicinity of livestock*. My emphasis on the last phrase points out the question of what to do about livestock grazing on federal land by virtue of a livestock grazing permit?

Sections in the act deal with states and tribes having the option to move wolves to reduce predation on local deer and elk herds if "the action does not hinder wolf recovery." When the private landowner takes individual action if cattle have been killed, they need hard evidence to prove their actions were not arbitrary. By the time a Federal officer shows up to examine the evidence, which may be 3 or 4 days, the evidence may be gone.

There is no shortage of biological disputes since they teach biologists to master the Hokey Pokey in college. One of the root issues revolves around what government entity will manage wildlife resources. The issue is even more complex in the mountain west where so much land is federally owned and managed. The wind really blows things around in Wyoming. Remember, that state deals with the wolf, the grizzly, the coyote and several predator species of mountain cats. Now upset the wildlife equation by either federal or state law replacing the human predator with the animal predator, or vice versa, and there will be strong opinions, squabbling and court action.

THE VALUE OF ELK

In the preceding chapter, we mentioned the economic value of the elk to the state of Wyoming. The year this book was finished, Wyoming sold 165,223 resident hunting licenses. These generated $3.9 million for the state. They sold 75,846 nonresident licenses. These generated $15 million, 144 thousand. Antelope, deer, moose and elk were important parts of the economy of the state. Put a normal economic multiplier on license income and Wyoming sources say you have an estimated $90 million contribution to the state economy. Then add guide and outfitter income.

The deeper worry in the minds of the Wyoming outfitter is not just the presence of wolves, it is the numbers that seem are emerging on elk/cow/calf ratios. Bud Betz is a Dubois, Wyoming outfitter. In 2003 he served as the chief lobbyist for the Wyoming Outfitter and Guides Association. This is his take on the situation. "In the Spring Mountain Herd (elk) on the south side of the Dunoir River road, where there are no wolves, the calf ratio was 39.1 to 100 cows. On the north side, where there are wolves, the ratio was 13.4 calves to 100 cows. In 1994, before the introduction of wolves, the calf ratio was 54 to 100. Betz, who is also a rancher, runs no cattle so he had no cattle loss. He has, however, lost three dogs. Defenders of Wildlife compensated him for one of the dogs. He says his elk hunting business has dropped 50% in the past two years. He puts much of the blame on the Canadian Imported Gray Wolf. He says that "was not clean ecology."

Dubois is a proud, little western town hanging onto life by its board sidewalks. Hunting, snowmobiling, dude ranching, fishing and tourism help it scratch by. Darrell Graft runs the Wind River Meats in the town. He says, "If something isn't done about the wolves, we're going out of business." He has numbers to show that in the year 2002, the plant was 100 elk short in processing numbers against 2001. "I employ," he says, "9 people from mid-September to mid-December. If our animal count drops much more, there is no work for these people."

If the biological and economic trail is part of the huffing and puffing about the wolves, it is equally so on the government frame of our model. It started in 1987 when Utah Congressman Wayne Owens filed his bill to initiate wolf restoration. In 1989, Congress asked for an Environmental Impact Study. This was completed in 1993. It contained an amazing 160,284 comments. In November of 1994, the final rules for restoring wolves were published in the Federal Register. Three days later, November 25, the Mountain States Legal Foundation and the American Farm Bureau filed suit in the U.S. District Court in Wyoming asking for a preliminary injunction to stop government agencies from introducing wolves into western states. On November 29, government attorneys filed a stipulation agreeing not to import wolves into the U.S. before January of 1995.

In January of 1995, U.S. District Judge William Downes in Cheyenne, Wyoming denied the preliminary injunction filed by the Farm Bureau. Work was underway on wolf transfer and on January 11, they were on their way to Wyoming. The Farm Bureau filed another last minute appeal to prevent release. This, too, was denied. Wolves were moved to acculturation pens in Yellowstone to prepare them for ultimate release. Some folks involved in the release had been working on the project for 20 years.

During this period, interesting stories had been appearing in various press releases and on wire stories. Friends who knew of my interest, would send them to me. They made an old radio reporter wonder about the validity of the media. This was coverage from Denver, credited to Reuters, dated January 13, 2000. "A Federal Appeals Court panel on Thursday ordered a reprieve for the thriving gray wolves of Yellowstone Park, handing a victory to environmentalists and a defeat to ranchers. In a 3-0 vote, the Circuit Court of Appeals in Denver reversed a Wyoming lower court ruling that the wolves, regarded by ranchers as livestock predators, be removed."

"'The wolves are here to stay. It's a great day for wolves' said Mark Van Putten, president of the National Wildlife Federation, talking by phone.

"'Ranchers represented by the Farm Bureau launched their legal battle against the wolves when they were reintroduced in 1995 after not being seen in the area for 60 years. The original pack of 30 wolves was captured in Canada. Fifteen were sent to Yellowstone in Wyoming and the rest to the wild lands in neighboring Idaho.' Van Putten said there are now more than 300 wolves in Yellowstone and Wyoming."

It is difficult to sort out press stories when you are walking into the wind. The figures flutter around. So, back in the warmth, you poke into old files to be sure your numbers are in the same ballpark as your memory. Yep, the US-FWS says there are 160 wolves, less what have been put down or have disappeared and that really the number is closer to 115 to 120. Van Putten says there are 300. The old rancher outside Yellowstone says it another way, "Too damn many and gettin' moreso."

MORE WOLF NEWS

The wind whistles around the corner of the cabin and I locate a clipping from *Field and Stream*, a story about a recent wolf event. Well, it's in the wind today! You read . . . "A controversial new project hopes to produce kinder, gentler gray wolves in Montana. The plan involves four wolves captured from the notorious Sheep Mountain Pack . . . a band that routinely preys on cattle in the Montana's Paradise Valley. Presently, the wolves are being held at Ted Turner's Flying D Ranch near Bozeman, where they are being trained to avoid livestock through the use of electric shock collars . . . the same kind used to train dogs. When the wolves have completed their "rehabilitation" they will be released back into their home range with the hope they will remain averse to cattle. If it is successful, similar projects will be planned for the future."

You recall a story written by Matt Kelly for the Associated Press, datelined Washington. Well, all the wolf lies aren't in Montana and Wyoming. "Captive wolves in Montana are being trained to shun cattle in favor of natural prey such as buffalo in an experiment in which the wolves are zapped with an electric shock when they approach livestock. The federal agencies and private groups involved say they are trying to avoid killing the wolves, which belong to an endangered species that has been reintroduced to the wild from Canada. Critics call the experiment cruel and say that western ranchers, not the wolves, are the ones that need behavior modification.

"'We think it is absolutely ridiculous that we should be trying to alter the natural behavior of wild animals, particularly to benefit a private industry that used public lands,' said Andrea Lococo of the Fund for Animals. "'Government funded livestock protection programs all but wiped out gray wolves from the lower 48 states by the 1960s. After the wolves were put on the endangered species list, the U.S. Fish and Wildlife Service began breeding them and reintroduced then into the wild in 1995.'" (The breeding part would be news to USFWS!)

"Now, more than 250 released wolves and their offspring live in the northern Rocky Mountains and more than 20 are in the Southwest. The Fish and Wildlife Service has proposed taking the species off the endangered list." The story then provides some more interesting information, especially interesting to the Park Service and USFWS people who are seeing some numbers and hearing details for the first time. The article concludes with this. "Critics say ranchers should learn to live with a few animals being picked off by wolves and say it's particularly cruel to shock wolves to avoid cattle. 'If we want wild animals in the wild areas we can't be turning them into Pavlovian dogs, because they are no longer wild animals,'" said Lococo.

I shared the foregoing with a local Wyoming acquaintance. His talk is liberally sprinkled with obscene words, some knowing no origin. He claims he was trained by a one eyed carpenter. But, he has a sharp mind that often cuts through the wind of conflicting stories. His first reaction was that somebody couldn't add or was "just flat lying." And if you check the numbers in the various stores you find more than gentle disparities. His second observation was more earthy. "What I wooda damn sure done to solve all that ruckus was to release them wolves in some sumbitchin' park in Washington and then that loco lady would be able to see what a real Pavlovian dog was like."

GRINDING GOVERNMENT WHEELS

It is easy for government wheels to grind in Wyoming because the soil is so sandy. Three pieces of legislation were introduced and passed in recent sessions, pushed primarily by Thermopolis, Wyoming legislator Mike Baker. S.F. 97 was a strongly worded challenge to the Endangered Species Act. They toned the bill down and it became essentially a bill to transfer wolf management outside federal parks to the State Game and Fish Department. This Wyoming action would move the state into line with Idaho and Montana.

If most of this legal wrangling is done on higher levels, what is it like to deal with wolves on the grass roots level? The Diamond G ranch is north and west of Dubois. Dave Moody, of the Lander office of Wyoming Game and Fish is a Regional Manager who deals with wolf issues and will probably write the Wyoming Wolf Plan if delisting comes about. He says the wolf problems at the Diamond G are more an exception than a general rule. "It has been a hot spot, the most consistent, long term trouble area. I know federal authorities have issued kill permits," But, he told me his office "had no specific data on depredation . . . nothing specifically tied to wolves," He also told me he had no knowledge of any reduction in elk tag hunting permits. That last statement does not square with press releases.

"Actually," he told me, "we are trying to reduce elk herds, getting closer to management objectives. Our first concern is cow/calf ratios. We're not hiding our heads in the sand . . . it's just hard to deal with drought and predation." So, concern on the federal level is the wolf restoration. On the state level, concern is managing the elk population. On the Diamond G Ranch the concern is damage control. Start with Deb Robinet, wife of ranch manager Jon Robinet. She had two Great Pyrenees dogs the couple kept because this breed would bark at visiting grizzly bears. "The older dog," she told me, "got attacked 3 times in a week by wolves. The younger dog . . . I was taking him to the barn . . . he got out ahead of me . . . wolves came out of nowhere . . . I ran to the house, got the jeep, drove back, scared them off . . . we found him in the morning . . . I saw his eyes flicker . . . we fed him by hand but the injuries finally got him."

"My husband, Jon," she went on, "has a video of 17 wolves. We know we have 21 this year. One went to Yellowstone and got a female, so now we probably have 3 packs. We lost a saddle horse, found him in the morning with a leg broken where it had been chewed. We had to put him down. We also lost a very valuable colt. We had to supply all the papers. We wanted the colt replaced but the man from Defender of Wildlife said that would cost $7,500 so he paid us $2,000. Then we had to pay income tax on that."

Moose are also a concern. In 1995, with no Gray Wolves, the count was 4900 moose in northwest Wyoming. There was good reproduction. Nearly every cow moose sighted had a calf. In 2003, they issued only 145 moose tags. This compared to 811 tags issued in 1991. Back with Mike Jamenez, USFWS in Lander, who does much of their work with wolves. "We have very little data on moose." He told me a pack of wolves might kill 20 elk per month. "In the winter," he said, "that might go a little higher."

UP IN MONTANA

Joe Fontaine, a federal wolf biologist, wants to "chase down and kill wolves that killed four sheep in the Ninemile Valley." He said he hoped killing two wolves from the six wolf pack would stop attacks on area livestock. The pack is "suspected of killing 11 sheep, 7 llamas and a horse in the past 1½ years." We don't know what makes them flip-flop from natural to domestic prey. But once they do, it gets into the pack and just continues."

Then one more thought from Mike Rinehart, who represents a version of a Utah conservation organization in Wyoming. "Our organization believes the environmentalists are using the Endangered Species Act to stop hunting. If they can reduce the moose and elk population, that discourages hunting. The environmentalist," he says, "is using the wolf to stop hunting."

This chapter is a choppy collection of opinions about wolf restoration from a variety of sources. From the high levels to the grassroots levels, there is little consensus. In their zeal for restoration of a wild specie, biologists may have focused too much on biology with too little consideration of grass roots economics and social contracts. The search for a simple, single solution hides in the aspens. The good intentions of the biologists may succumb to the reality of economics and pressure of local interests. And some ultimate balance, some eventual accommodation will be a long, tedious, painful and litigious journey.

Ding Darling

Was born in Norwood, Michigan in 1876. His real name was Jay Norwood. He joined the Des Moines Register as a cartoonist in 1906 and began signing his work as "Ding," and so he was thereafter.

Jay N. Ding Darling. Courtesy of the "Ding" Darling Wildlife Society

He was an avid hunter and fisherman who became concerned at the possible loss of species and wildlife habitat. He became a vocal wildlife conservationist and often worked this theme into his cartoons. His convictions drew the attention of President Franklin D. Roosevelt who appointed him Director of the U. S. Biological Survey which became the U. S. Fish and Wildlife Service.

Darling would initiate the Federal Duck Stamp program and personally drew the first stamp. He also increased the size of the National Wildlife Refuge System with the duck stamp revenue providing needed funding. In 1918, the Migratory Bird Hunting Act required all waterfowl hunters to purchase a stamp. He also designed the Blue Goose logo, the national symbol of the refuge system.

Perhaps his greatest conservation contribution was the partnerships he formed with state land grant universities that trained biologists in this emerging field. In 1936, he founded a society of wildlife clubs which, in time, became the National Wildlife Federation. These were dust bowl times and these conditions drew more attention that did the plight of ducks whose population, Darling argued, had reached an all-time low.

Darling died in 1962.

John Wesley Powell

Was born in 1834 in Illinois. Very early in life he began taking exploring trips and collecting natural objects. In 1856, just 22 years old, he descended the Mississippi River alone from St. Anthony's Falls in Illinois to the mouth of the river and in 1857 he rowed the length of the Ohio River from Pittsburgh to its mouth.

In the Civil War he was mustered into the 20th Illinois volunteers and lost his right arm in the Battle of Shiloh. Later, he was made a major. He accepted the position, after the war, as professor of geology at Illinois Wesleyan University, where he had earlier received degrees of A. B. and A. M. He also had a Ph. D. from Heidelberg in Germany and a LL. D. From Harvard.

Powell's claim to fame came when he scratched together enough money and supplies to make the first of his two trips down the river that he named The Grand Canyon. It was a 217 mile trip. Three of the eight men in his expedition left before the trip was completed. Powell returned a hero, hit the lecture tour, and raised money for a second trip in 1891.

Powell then became a prophet. He was convinced no family could live on a 160 acres of land in the arid west. His report to Charles Schurz, then Secretary of Interior, "Report on the Lands of the Arid Regions of the West" was a prophetic treatise. It refuted much of the philosophy of the western lands as well as existing laws. In Powell's opinion, irrigation was the only recourse and his plan regulated that to prevent a monopoly on land owned by a few in-

John Wesley Powell. Photo courtesy of the Library of Congress, LC-USZ62-3862.

dividuals. Powell met with opposition and even defeat of his ideas. He helped establish the Geological Survey and the Bureau of Ethnology and his ideas influenced the Taylor Grazing Act of 1934. He helped establish Washington as one of the influential scientific centers of the world.

John Wesley Powell died at his summer home in Haven, Maine in 1902.

REFERENCES

Leopold, Aldo. *Journal of Forestry*. (Falcon Publishing)
Mockler, Ester. *Wind River History*. (University of Wyoming Press)

Phillips, Michael K. and Smith, Douglas W. *The Wolves of Yellowstone*. (Voyager
    Press) 1998
Ridgeway, James. *The Politics of Ecology*. (E. P. Dutton and Company) 1970
Letters to the Editor, Casper Star and Tribune 2002,
Personal interviews 1999–2003 with:
    Doug Smith National Park Service Biologist for Wolf Recovery.
    Bud Betz Wyoming Outfitter and Guide.
    Mike Jamenez Project Leader for Wolf Recovery, Wyoming.
    Darrell, Graft Owner, Wind River Meats.
    Deb Robinet, Wife of Diamond G Ranch Foreman, Jon.
    Dave Moody, Wyoming Fish and Game Ranger, Lander, Wyoming.
    Gene Linn, Outfitter, Wilson, Wyoming.
    Susan Stone, Defenders of Wildlife, Bozeman, Montana.

Chapter Twenty-Six

Judgment Day for the Cause

"The greatest unspoiled wilderness is the search for truth, and the one who seeks this wilderness will find the trail untraveled." A grump biologist who walked a trail with the conservation organization, Delta Waterfowl, said that. He spent a lifetime searching for answers in the tangled web of duck ecology. H. Albert Hochbaum had a problem. He told people what they needed to hear rather than what they wanted to hear. John Adams once wrote to a friend that Thomas Jefferson's view of history would be used by succeeding generations because Jefferson told people what they wanted to hear. Adams felt he too often told people what they needed to hear.

Judgments about the value of conservation efforts may not be what people want to hear. Most people who belong to conservation organizations do so because it makes them feel good or better. It's a little like a Sunday service under a spire. Folks are there for a variety of reasons. Many drop a contribution in the plate when it comes by. They feel better. Grassroots people support conservation organizations in pretty much the same way. "Putting something back" or "doing the right thing" is part of the mix. They are less concerned with what happens to the money in terms of return on investment.

CONSERVATION AND RELIGION BALANCE SHEETS

The two share one essential commonality. What goes into the cause is more easily measured than what results are achieved. Not many communicants of either take time to read balance sheets. Very seldom does the average member vote on an issue. This is usually handled by committee. Annual meetings and newsletters are the most common means of communication. Seldom is the bottom line mentality of the business world applied to either by the average member.

There is another shift in the mentality of the member. This idea comes from the sociologist, Irving Babbitt, who suggests there is a human crusader. "The United States is rapidly becoming a nation of humanitarian crusaders who believe that one may dispense with awe and reverence . . . and spiritual standards . . . provided one be eager to do something for humanity. The most palpable outcome of this trend will be a drift toward license and a reign of legalism.

"Under the pretext of social utility, these crusaders are ready to deprive the individual of every last scrap and vestige of freedom and finally to subject him to despotic, outer control. No one, as Americans of the present day are only too well aware, is more reckless in his attack on personal liberty than is the 'apostle of service.' So today, traditional moral restraints are being replaced by regulatory and legal codes."

What Babbit suggested in the early 1920s was a rather chilling statement. But consider what he predicted in the context of the increasing regulation being passed to bring industrial pollution under control. I can not believe it is accidental that today many high ranking officials of corporations under legal duress for environmental malfeasance have been encouraged to serve on boards of conservation causes so they can serve, at the least, as watchdogs for their companies.

The conservation cause in this country began as a peculiar collection of sports/recreation, bird watching, various forms of eco-tourism and a few well intentioned people who believed natural resources had a place in the quality of life. It should serve the altruistic needs of common man. This is a tenuous umbical at the best. To fund this belief and possible intercession by their group, they began to pass the plate. Common people gave because they felt it was right to support the cause. This is the reason most conservation organizations have local chapters. They needed grass roots support. Large foundations eventually gave because it was to their financial benefit to do so. It involves tax benefits. Sometimes, government will support the private sector if the politics are right. But money raised and political leverage are the measurements in conservation organizations more than tangible, specific results that are part of a business mentality.

BUT CONSERVATION ORGANIZATIONS

. . . change. They evolve. And they fit our model and help explain some of the transitions. Delta Waterfowl was FORMED as a scientific, non-profit organization. The mission or purpose was research. It was aligned rather closely to the scientific community. It endured a long period of STORM when it struggled with a singleness of purpose and an image of being something of an elitist organization. Through the years, it gained an international reputation for science and research. But, for the average duck hunter, how do you measure what Delta Waterfall is doing for you? If you put in $50, what did you get for your money? Science is hard to measure.

A thoughtful article in a recent Delta Waterfowl Report brought this idea into focus. John L. Devney, Group Manager of Communications, quoted a successful businessman who was on the Delta Board. "Think about it," he said, "all we hear about is how many acres of wetlands or grasslands have been secured, or how many dollars have been spent on habitat projects. Those are inputs, like raw goods into a manufacturing plant.

"We never hear about outputs . . . how many ducks are actually being produced? How long do you think a business would survive if it measured success in raw goods rather than finished products?" This is a business person speaking. He is accustomed to a monthly or quarterly financial statement that shows what was put into the operation as expense, what was sold and the resultant cash flow, if any.

Thumb through nearly any conservation publication. You can locate stories about input. What member gave land or money for a specific project . . . what chapter got a grant to improve habitat for trout, what turkey banquet raised the most money. There might be figures about the number of members, what they raised in money for the year, how much money went into projects. The end results of all this activity are elusive. Output or specific results of projects are hard to calculate. The search for value is elusive.

Delta Waterfowl went through the growing stages, the STORM stage. They slowly reached a stage of NORM and if you counted the number of students they graduated and placed in important conservation positions all over the world, you would say they are reasonably HIGH PERFORMANCE. But, even at this level, there are shifts. Where once the focus was on habitat, and it is still important, nest depredation has become a serious part of their research. They are not satisfied with a 10% nest success ratio for mallards. This is an input and output mentality. What are our supporters getting for their money?

Now we are into Hochbaum's "unspoiled wilderness of truth" because what I have just said is easy to generalize. After some forty years of observing conservation organizations, writing for them and about them, working with and for them, I have found they are different. Most begin with a worthwhile purpose. They differ as they evolve through the passages. But in most all cases, as they have grown and gotten bigger, more successful if the measurement is money raised, it has become more difficult to locate what the average member is getting for their money.

THE NATURE CONSERVANCY

. . . was spawned in 1951. It was essentially an east coast collection of groups with common interests in protecting species. From an initial focus on endangered species, they have evolved into protection and acquisitions of ecosystems that provide habitat for species TNC scientists believe are endangered. From a group interested in local protection, they have evolved into a world wide protector of conservation "hot spots," areas they feel deserve attention and protection. Today TNC manages 7 million preserved acres, they own 2 million acres outright. They run the world's largest private sanctuary system. They have become bankers of benefactor annuity plans. In 2002, they had an annual income of $972 million. As they evolved from a conservation group into a business, their practices drew fire in a series of articles in the Washington Post newspaper and Range Magazine.

In the 1980s, TNC had adopted a "non-confrontational land-habitat acquisition approach." This successful approach blended conservation with business interests and produced a decade of growth. Revenue went from a reported $58 million to $222 million. The Conservancy rode the wind of the excesses of the 1990s. They were one of the top ten money raising non-profits in the U.S. They had 3,200 employees in 528 offices, headquartered in an eight story building in Arlington, Virginia that cost $28 million. They had evolved from a modest conservation organization into an operation that was more like a Fortune 500 company.

In fact, the Board of Directors had evolved from a group of concerned conservationists into a collection of officers of the nation's largest companies, some of which had rather splotchy records of dealing with environmental issues. Boards of Directors also go through passages as the organization matures. As is sometimes sadly the case, TNC began to stray from its original purpose, protection of species. From Martha's Vineyard, to Virginia's eastern shore, to Gulf Coast Texas, their "conservation" practices were challenged. In the early 1990s, Mobil Oil made one of their largest corporation donations. They gave TNC a patch of land near Houston, Texas because, they said, "this land was the last best hope" for saving the Attwater's prairie chicken, a cousin to the prairie chicken on The Tallgrass Prairie Preserve in Oklahoma. TNC took the position they wanted this sensitive area to be a model for "compatible development." They drilled a well for natural gas under the last breeding ground of an endangered species. To carry the drilling scenario to one other issue, TNC then sold gas owned by another party. In 2002, they settled this case for $10 million. The value of a species had been replaced by the value of settling a law suit. This was not what many of the members had given for.

Conservancy President, Steven J. McCormick is quoted as saying, "There are tradeoffs in conservation. We make a judgment that less than 100% is acceptable. Another TNC official, Michael Horak, is quoted as saying, "Some of our brethren say we are dealing with the devil, but I say quite to the contrary. Some of the deals we're making are quite extraordinary." Some of the TNC scientists would agree but in an entirely different context. They feel the value of environmental science and conservation has now taken a back seat to accumulating money. David Morine is quoted in the Washington Post expose of TNC as saying, "It was the wrong decision to get so close to industry. Business got in under the tent and we are the ones who let them in. These corporate executives are carnivorous. You bring them in and they just take over."

There is a strange chemistry that exists within an organization that is in a state of HIGH PERFORMANCE in terms of financial growth. Greed replaces the original mission statement about value. The TNC mission has come to be guided by what I would call "a significant disconnect." That premise is . . ." if you protect the habitat, you increase the populations of the species using that habitat." It isn't that simple on the grassroots level. There the question is more likely to be, "don't give me the amount of money you raised this year as compared to last year . . . tell me why the prairie chicken got lost in the gas down in Texas?"

LOSING SIGHT OF THE CAUSE

The biologist dedicated to science will tell you research is the gospel. In fact, this premise has become part of the mission of Ducks Unlimited. Bruce Batt has been a scientist with Ducks Unlimited for many years. Our paths crossed in 1985 when I video taped an interview with him that was later used on OETA, PBS in Oklahoma. In 2003, we talked again. He said, "Habitat is the most important part of the duck population equation. If weather conditions are right, ducks have the ability to produce large populations quickly. If conditions are dry, there is drought, they either won't nest or they become more susceptible to predation. But predators will cycle in population because of weather conditions too. We believe if the habitat is there, when the weather is favorable, the duck population will rebound. What limited money is available for conservation ought to be used for habitat."

"It is not simple," he adds. "There is no single, simple answer and we have too many people running around looking for simple answers." I then asked him why, in the same time period, 1995 to 2002, the goose population had grown while the duck population had declined? I had figures that in 2002, mallard ducks had declined from an estimated population of 10.8 million to 7.5 million, Gadwalls were down 42%, Green Wing Teal down 26% and Pintails down 48%. "Because," he responded, "there is a biological difference. They nest differently. The goose mates for life, the male helps protect the nest, they are bigger birds . . . can protect the nest against raccoons, fox and crows. They lay fewer eggs, usually 4 and can cover them better. The mallard drake leaves the hen while she is on the nest. She can't deal with a predator. You have a very different set of circumstances." *(Note that his response is based on nest success.)*

You get a different view from John Devney, at Delta Waterfowl. He says, "you have measured conservation of ducks by the dollars invested for the acres conserved. We build wetland preserves, then walk away and provide no maintenance. The problem is basic . . . getting baby ducks out of eggs. The measurement of shoreline acres preserved is not the issue. Reproduction of ducks is the issue. Predator control is an important part of the equation." Then he added an interesting comment in our discussion. "Are we chasing the money or are we addressing limiting factors. I'm afraid, too often, we have not asked for accountability."

DIFFERENT VIEWS OF RESULTS

So one highly respected conservation organization takes the position that predator control is an important part of duck population management. Another equally respected organization chooses to put the emphasis on habitat protection. And the hunter in the blind chooses to judge the equation more by the number of ducks they see on a given day. And where is the truth? How does the average conservation organization member measure the value of the organization they support? Are not they entitled to some sort of grass roots accountability?

I have asked many sportsmen, game people, game wardens about this question of accountability. From some I get a bench press answer . . . "Hell, I shootem . . . (catchem) . . . why shouldn't I put something back?" From another, the fair weather reply. "When the forecast is good, I buy a license or the stamp. When the forecast is bad, I do something else." Said a few who had some understanding for the situation, "I understand the weather cycles . . . I don't need some space cadet biologist telling me about dry weather. I support what he does because it is his job to take the cycles out of hunting . . . to manage habitat so there is a regular supply." In the recreational view, there is limited tolerance for wild rides provided by Mother Nature.

Like the business man or woman who likes the profitable times all the time, and may even enjoy taking credit for that condition, the sports/recreation folks use the good times as the measure of success. Their judgment is based on personal game or fish counts. However, in fairness, there are many outputs to be found in conservation achievements. There is an accounting in the increased populations of deer, elk and turkey. Counting ducks gets more nebulous and certainly counting trout is not an absolute checkout stand. Duck hunters and trout fishermen just need more faith!

When I ask game managers from a variety of sources, which organization do they think is showing the most accountability, giving the hunter or conservationist the best return on investment, the response generally gets around to the National Wild Turkey Federation. They are a relatively young organization. They don't have 60 years of baggage behind them. They haven't been around long enough to accumulate detractors or become defensive. They have also been helped because members see the results of their contributions turn into more turkeys and better hunting.

THE RESTORATION OF THE WILD TURKEY

. . . population in this in this country has been one of the grand conservation stories of recent time. The common statement is there are more turkeys in the United States today than when the Pilgrims landed. When you compare this story to the Passenger Pigeon or the bison, this is a remarkable story of restoration and renewal. It has been achieved by enlightened state wildlife departments and NWTF, often in cooperation with each other. If the true purpose of biodiversity is to provide an annual crop of sustainable wild game, animals or birds for whatever purpose, then the Federation has stuck closely to that mission. They are a rather lean organization. Their reports say about 87% of the money they raise goes into projects.

These might be actual stocking projects or something like their Guzzlers for Gobblers project in the arid western states which installs large watering tanks for not just turkeys but all wildlife. "National project, industry and agency cooperators have provided almost $620,000 for water projects in 14 western states." The tanks cost about $5,000 each and hold from hundreds to thousands of gallons of water which is delivered by helicopter. The projects can include research in the Napa Valley which proved turkeys were not eating grapes, but the insects under the vines. This put to rest untrue allegations.

Their 450,000 members carry on trapping and relocation in nearly every state. Some New England states now have a hunting season. Protection of the hunting heritage is a strong part of the NWTF action plan.

THE FISH CONSERVATION EFFORTS

Call the office of Trout Unlimited in Arlington, Virginia. A voice answers the phone and says what you have reached. Then, the first thing on the recorded message is . . . "if you wish to make a donation" and you are aware of the opportunity to drop something in the plate when it passes by. Conservation is an everlasting business of raising money.

And if you were in the business of trout conservation, how would you measure accountability? Now, the "output" gets very misty. A poor pun would say it "gets a little fishy." The growth of TU in the 1990s was robust. It quadrupled the budget and added 80,000 new members. The mission of Trout Unlimited has changed in recent years, much as the philosophy and direction of fishery managers has changed. The emphasis in many states had been on fish stocking. The emphasis of fish management has shifted more to water quality, riparian restoration and streamside improvement. Of course, this delays the evaluation of what the money invested has produced in returns.

Trout fishing has always had a grassroots image of an elitist sport. Many folks had the presumption you needed to know Latin to look in your fly box. Henry David Thoreau said of fishing . . ." Many go fishing all their lives without knowing it is the fish they are after." In a broad conservation sense, even an ecological sense, we all fish. We just fish for different things. Even authors fish for readers.

Stocking fish was a common practice even before the Civil War. In fact, in the late 1700s, many streams in higher country in the eastern part of the United States had reasonable trout populations. Some were in unsuspected places like the tiny Poncantico River in Winchester County, New York. Washington Irving, better known as the author of *The Legend of Sleepy Hollow*, wrote about brook trout fishing in one of his essays.

Trout fishing should be part of any tracing of ecology in this country because it involves the depletion of the fish in many areas, often from over harvest or pollution. It involves species substitution. Trout have been through all the stages of protection, regulation, biology, ecology and today sociology is emerging. Once the emphasis was on stocking fish, even in poor habitat. Now, the focus is more on improving habitat as a means of providing better value for fish and angler.

The story of the Esopus River in New York's Catskill Mountains could serve as an example of the turmoil in river restoration. The river has long been a trout stream. They estimate it harbors some 100,000 rainbow which were stocked in the late 19th century, some wild brown trout and some 30,000 stocked brown trout, nearly all stocked annually and fished out each year. A few brook trout survive in the headwaters. The river is part of the New York City water supply system. An 18 mile culvert carries water from the Schoharie Reservoir into the stream, introducing some murky water. In 1996, torrential rains and snow melt ripped away banks. Wide spread bulldozing to "repair" stream damage destabilized the stream even more. A developer now wants to build a resort along one of the creeks running into the river. It is a classic case of watershed contributors from many sources putting demands on river water quality. Just stocking more fish won't provide an output that will justify the input.

GOOD NEWS ON THE MADISON RIVER

Out west, when you say "Madison River" you tip your fly rod and genuflect. Near Ennis, Montana, there was a crude, handwritten sign. It said: "Attention Fishermen. This is private property. A parking fee of $3.00 is charged. Please deposit the fee with your auto license No # on the envelopes provided in the collection box. Overnight camping $5.00. Non payers will be band. We thank you for your cooperation. The management." The word "band" was on the edge of the sign. Maybe they meant "banned' and just ran out of space.

Through many years, the sign has stood on the entrance to the area, as it did the day I visited. The fishermen paid the $3.00 . . . some 3,000 of them every year . . . for a chance to fish this legendary stretch of water. The site became know as The Three Dollar Bridge. The owners decided to sell. They were asking $2 million dollars for the 800 acre portion of the Candlestick Ranch where the "bridge" is located. Enter Trout Unlimited, the Montana Department of Fish, Wildlife and Parks, a private conservation buyer . . . support from Orvis in Vermont, contributions from the private sector and TU members. John Wilson is the Conservation Director for the TU Montana Council. He is quoted as saying, "From a fisheries perspective, the decision to move ahead with the project was a no-brainer. There was no way we could stand by and watch this property get subdivided and locked up." Value was measured in terms of fishing opportunity.

So this spot in the hub of four great mountain ranges, a part of the Yellowstone ecosystem and boasting 3 miles of Madison River frontage is still open

to any fisherman who wants to take a cast at the trout. They had a lot of
patches to sew into the crazy quilt of protecting just 3 miles of a great fishing
stream. The judgment of "was it worth it?" would be hard to come by. For the
person standing in the river on an August evening and watching Caddis flies
come off, the dimples of feeding trout rimming the water, there would be no
doubt the cost of conservation was worth it. There are just some natural resources that are hard to reduce to a bottom line economic value.

THE VALUES PLACED ON WATER

It is popular in conservation circles to equate wetland with waterfowl. But the
ecologists will tell you these areas are enormous contributors, and valuable
and irreplaceable resources, to the health of the land. The swamps, marshes,
tidal flats, bogs, bottomlands and river oxbows, as well as potholes not only
serve wildlife, they also serve the human condition. They act as natural flood
control barriers, they filter pollution from water, act as holding basins for water supplies and otherwise contribute to a complex and larger ecosystem.
Leave the birds and the critters out of this emotional issue. Focus only on the
impact on the human condition. As the old Hindu proverb puts it, "Even the
frog does not drink all the water in the pond."

Consider that each year over 400,000 acres of this nation's wetlands are
lost to filling, plowing, dredging and development. It is estimated that when
Thoreau was tromping around in New England, this country had 250 million
acres of wetland. Today, fewer than 100 million acres remain. We have
drained over half the water from our national bathtub! And we have left some
dirty rings around the tub. Wetland reclamation can't keep up.

If measuring the efficiency of a conservation organization is a formidable
challenge, consider how we have measured the value of water. We measure
water in terms of acre feet, gallons, water depth, acres of surface covered. The
Corps of Engineers measures the value of water created by dams by a complicated formula involving power produced, flood damage controlled, economic impact, and if they really want to sell the project, by the recreational
opportunity the effort will produce. In the case of cost of water for irrigation,
we really have no standard. There is no cost equation for salination of the land
or the subsequent loss of land production. Any compass for the best use of
water is hard to come by. Of all our natural resources, water more often falls
victim to the power of consumption.

Add to this conundrum, our national propensity to fix it and then hurry to
the next area to fix it. Natural resources take constant care, constant alignment. A cause, an organization, a restored watershed, a renewed trout stream

takes maintenance. This care is the hidden cost of conservation. What looks like a done deal can quickly become a misdeal.

Accountability should force all of us to consider the broad questions we have raised in this book. We have tried to report, not from high up in the stands, but on the playing field where the action is. Like the Gordy Yeager WMA near Rochester, Minnesota where the restored flock of giant Canada geese have gown to number 20,000 birds. The original management area has shrunk from 705 acres to 340 acres. Olmstead County, where the WMA is located, has asked the Minnesota Department of Natural Resources to sell part of the remaining WMA to develop an energy park.

Then consider a small part of the larger equation of whether a wildlife specie is managed only for the sake of preservation? Should a specie be exposed to possible disease and denied adequate space? At what cost should a specie be transplanted to restore a more balanced, natural ecosystem? Is water better used for alfalfa or stream quality? And what standards do you impose on a conservation organization involved with natural resources to insure that stakeholders realize a reasonable return on investment?

Environmentalism in the United States is not unlike the Wizard of Oz story. We have some scarecrows, environmentalists searching for a brain and drawing upon exaggeration or hyperbole to make their case. We have the tin men, scientists and engineers with technology who take a dim view of the scarecrow and stoutly pledge to economic development and the lure of more jobs. We have the Wicked Witch of the West stirring her pot of pollution and extraction. Then we have the monkeys, filing law suits so frequently they have been able to clog the judicial drains. As one fellow who practices environmental law said to me recently, "I don't care. I make money either way."

IS THE ROAD TO THE EMERALD CITY

. . . of sustainable quality of life going to be cast entirely into the courts? Are we, by tradition, heritage and comfort really a legally minded people? Have we replaced the social contracts of respect for the earth, nature and wildlife with legal pronouncements? Have we shuffled off individual responsibility for stewardship to Babbit's "humanitarian crusaders?"

The Native American did not share the idea of dominance of natural resources, whether by the individual or the organization. They had a very different system of accountability. This was Chief Seattle speaking. "The white man does not understand America. He is too far removed from its formative processes. The roots of the tree of his life have not grasped the rock and the soil. The white man is still troubled by primitive fears; he has in his consciousness the perils of their frontier continent, some of it not yet having yielded to his questing footsteps and inquiring eyes.

". . . The man from Europe is still a foreigner and an alien. And he still hates the man who questioned his path across the continent. But in the Indian the spirit of the land is still vested; it will be a long time before other men are able to divine and meet its rhythm. Men must be born and reborn to belong."

John D. Rockefeller, Jr.

Was born in Cleveland, Ohio in 1874. He graduated in 1897 from Brown University with a B. A. Degree and was later elected to Phi Beta Kappa. Rockefeller would devote his life to civic activities and philanthropy.

He was deeply interested in conservation and personally traveled to study natural areas. One of his largest gifts was $17.5 million to the Jackson Hole Preserve. Rockefeller became involved in the Jackson Hole Plan after visits in 1924 and 1926 to the area. Private property in the area from Moran Junc-

John D. Rockefeller, Jr. Photo courtesy of the Library of Congress, LC-USZ62-3810.

tion to Menors Ferry had become unsightly. With the knowledge of Yellowstone Park Superintendent Horace Albright, he decided to purchase 35,000 acres of land, using the Snake River Land Company as a shield for his intentions.

Intense feelings developed among ranchers and residents of Jackson Hole. Several bills were introduced in the U. S. Congress, hearings were held to investigate the Snake River Land Company and the badly battered dream of a national park continued into the 1940s. Finally, after holding the land 15 years, Rockefeller wrote President Roosevelt and said if the issue couldn't be resolved he would make "some other disposition of it." FDR responded by making an entire area the Jackson Hole National Monument, setting aside 221,000 acres.

More controversy followed. There were bills in Congress. Even some Forest Service Administrators spoke out against it. Finally opposition dwindled and in 1943 President Harry Truman signed a bill that merged the 1929 park with the 1943 monument to form a 310,000 acre park. It was, according to some sources, "perhaps the most notable conservation victory of the twentieth century"

John D. Rockefeller, Jr. died in Tuscon, Arizona May 11, 1960.

Harold Ickes

Was born in Frankstown, Pennsylvania March 15th, 1874. He attended the University of Chicago and graduated in 1897. Ickes worked for Theodore Roosevelt in the 1912 presidential election and for Franklin D. Roosevelt in 1932. In 1933, he was appointed by FDR to the post of Secretary of Interior.

As a Chicago lawyer, Ickes was interested in national parks, wildlife and Indians. He was a believer in the conservation ideas of Theodore Roosevelt. Over the next six years, Ickes controlled over $5 billion in government spending, a sizable portion of which went to conservation projects such as the TVA and the CCC. While these projects created jobs they also accomplished some environmental remedies.

These actions were a shift from the previous Hoover administration as government intervention replaced the Hoover "rugged individualism." Ickes very much believed that natural resources should be used for "the greatest good of the greatest number." But his cantankerous character rubbed many people the wrong way.

He urged President Roosevelt to transfer national monuments and historic battlefields to the National Park Service, he negotiated approval for Olympic National Park in Washington and Kings Canyon National Park in California,

Harold Ickes. National Park Service Photo

and got Jackson Hole in Wyoming status as a national monument. Ickes got
the Bureau of Fisheries and the Biological Survey moved to Interior in 1939.
Ickes left his mark on conservation, perhaps the best friend of the environment to hold office of Secretary of the Interior.

   Harold Ickes died in Washington, February 3, 1952.

REFERENCES

Cronon, Miles and Gitlin. *Under an Open Sky: Rethinking America's Western Past.*
   (WW Norton & Co Inc) 1992
Ellis, Joseph J. *Founding Brothers: The Revolutionary Generation.* (Knopf) 2000
Jacobs, Lynn. *Waste of the West: Public Lands Ranching.* Self published. 1992
Deloria, Jr. Vine. *Native American Testimony: Chronicle Indian White Relations from
   Prophecy Present 1942–2000.* (Penguin Books) 1999

For much of this chapter, I used facts or figures from *Delta Waterfowl*, *Wild Turkey Federation, Ducks Unlimited* and *Trout Unlimited*.

In addition, I used personal interviews with John Devney of Delta Marsh, Bruce Batt of *Ducks Unlimited* and Christine Arena of *Trout Unlimited*. Many of the opinions in this chapter were expressed by members of those organizations since I am a member of all three.

Chapter Twenty-Seven

Naturalist Thinking

Just a little over 170 years ago, a young man who was the son of a bankrupt storekeeper turned pencil maker managed to graduate from Harvard. He then turned to doing odd jobs like doing survey work for local developers. Then he shifted to spinning written opinions in a web of nature, morality and metaphor. Folks in Concord, Massachusetts thought this odd.

Henry David Thoreau had begun his examination of spiritual fulfillment. He extended the idea of nature. He examined the dichotomy between nature as a source of morality and nature as a commercial enterprise. Thoreau was to say, "I went into the woods because I wished to live deliberately, to confront only the essential facts of life, and to see if I could not learn what it had to teach, and not, when I came to die, discover I had not lived."

In any discussion of ecology, it seems the subject of Thoreau comes up. He wrote, what in many folk's opinion, is the New Testament for naturalists. Like the original Aramaic text, the Book of Henry is subject to questions. On one hand Henry preaches "free roaming" in the woods. Meantime, he practices marking private land ownership. He fusses about this in a journal entry. "I fear this particular dry knowledge may affect my imagination and fancy, that it will not be so easy to see so much wildness and native vigor there as before."

In fact, Henry may have sparked much of the conflict found in ecological thinking today. I've often wondered if he ever had a deed to what became Walden's Pond. The place was misnamed, of course, because it is really more a lake and not a pond. Henry just sort of squatted there. He didn't view that as dishonest. In fact, he was always honest with nature. He may not have been totally honest with economic reality.

Naturalists like to venerate Thoreau. Of course, some folks like to venerate hunters and fishermen for their support of conservation although, in reality, they practice a blood sport. We romanticize wildlife although there is little real pretty in the predator-prey relationships. When a wolf mauls the fetus of an elk cow, the result is not pretty. When a bobcat gets in a pen of goslings and kills for pure sport, it is a sad thing to clean up. For all the power on display, an eagle landing on a field mouse is not all beautiful and noble. It is not a good day for the mouse. The reality in nature is death. It is always a possibility. Pruning is the reality of nature. Nature can be messy.

Without having interviewed Henry, I would say he was every bit as dissonant as our national conservation ethics. In his life span, especially in the eastern United States, good minds were beginning to consider the value of nature. In some rare instances, morality was being considered. The passages of this philosophy would go through what such considerations always go through. They would begin as a personal idea, then transition into regional ideas, then become a sort of collective ecology or what we call a social contract. Ideas got patched together.

EARLY CONTRIBUTORS

So what began as the personal thinking of Thoreau on the pond would be discussed later with Ralph Waldo Emerson in a Concord garden. Some of the notions about nature and morality would spread from Concord to Boston. There, a minister, Theodore Parker, Congregationalist by sect, at the Boston Congregation Society by location and preaching by virtue of his rather exceptional intellect and ability to make connections, was lecturing the brethren of Boston.

"The eyes of the North are full of cotton; they see nothing else, for a web is before them; their ears are full of cotton and they hear nothing but the buzz of their mills; their mouth is full of cotton and they can speak audibly but two words, Tariff—Tariff—Dividend." He speaks obliquely of the divine right of Americans to export our beliefs about government, equality and patriotism, then cautions us to get our own house in order before we do too much exporting. But his theme of divine right, the rightness of the course of action was blessed. Human nature can be messy at times.

He says, "for though we are descended from the Puritans, we have one article in our creed and that is to make money, honestly if we can, and if not, as we can." Regardless of his effort to clean up the Puritan image a little, the idea of the God given right to make money was certainly a part of the social contract of that time. Stewardship of natural resources was not a social contract. In fact, just the exact opposite was more nearly true. The natural resources, even from the pulpit, were viewed as being in great supply and had been placed, by God, on this earth for the total purpose of allowing man to make money. That idea was questioned by Thoreau. Was nature about money or did it have another value?

And while Henry hoed his beans and watched the wild pigeons fly by or watched the war between the black ants and the red ants under the tumbler he had placed in his cabin, he labored mentally with these social contracts. In spite of the Fourth of July celebration in downtown Concord faintly reaching his ears, he heard other sounds in his woods. His friend, Emerson would note in his journal, "Cotton thread holds the Union together; unites John C. Calhoun and Abbott Lawrence. Patriotism for holidays and summer evenings, with music and rockets, but thread is the Union." Emerson, it seems, was reconciled to consumption.

Thoreau raised new questions about nature having a value to human kind. But his words drew scant notice in a human tide beginning the extraction of the west. Indeed, notes in Fort Laramie indicate that by 1850, the Platte River had been cleared of wood and pasturage for two miles on either side of the river. Disease along the trail had produced areas where drinking water from the river would produce only more disease. One of the favorable features of Fort Laramie was the flow of the relatively unpolluted Laramie River into the Platte. On their way to the Promised Land, the immigrant had polluted their trail. Mortality was the concern, not morality.

DIFFERENCES IN NATURIST THINKING

It has been common wisdom in tracing naturalist thinking, to lump people involved into a long sequence when they were not the same in their thinking. This sequence style has added to the confusion about their contributions. I would group them more by their type and start by putting their thinking into at least three areas. See table below.

Naturalists	Scientists/Conservators	Social Conservators
Thoreau	Marsh	Powell
Emerson	Olmsted	Pinchot
Muir	Carson	Commoner
Mather	Leopold	T. Roosevelt
D. Brower	H. Ickes	

Not all these names fall neatly into a single grouping. At various places in this book, we have provided a reader with brief biographies of some twenty five major naturalists or conservationists. It is commentary on our conservation thinking that even simple lists can result in contradictions.

If Thoreau and Emerson were naturalists who added the "nature for nature's sake" type of thinking to the mix, then George Perkins Marsh was the first person who extolled the scientific approach. Some say Frederick Law

Olmstead, with his thoughts about planning, architectural design and urban parks should deserve credit for incubating at least some ideas about environmental science and social responsibility.

Stephen Mather, the first Director of the Park Service, articulated his ideas for preservation of park areas. He had three guidelines. First, to protect the parks from commercialization, then to eliminate private in holdings within the parks and then to add new parks to the existing Park Service holdings. He stressed the need for eastern parks. He fought dams and forest interests. He caught rocks thrown by absolute nature lovers who thought his policies compromised parks. He wanted public visits to parks educational and introduced trained interpreters. Many of his battles with political intrusion in the parks were picked up in more modern times by David Brower. Brower would continue the defense of parks and their preservation.

The line we are crossing between these three schools of thought involves three differences. The first difference is the search for a pure sustaining natural condition. The second thinking seeks preservation by applying science. Scientific study and discovery become the avenues for more reasonable use of resources. This thinking would emerge in the more eastern states after the Civil War. Universities like Yale introduced forestry courses. Congress would establish a Commission on Fish and Fisheries and would pass the Timber Culture Act. These actions were designed more for consumption purposes and less for conservation.

But the third type of naturalist thinking we are trying to identify is not the same as the other two types. The first type is focused on preservation, the second uses science to find a reasonable balance between use and abuse. The third group of names adds more social elements to the naturalist school of thinking. John Wesley Powell would use science to study the arid west. He advocated wise use of land and water for consumptive purposes. He also proposed the regulation of these resources, either federal intervention or local control. He also used natural resources to provide social benefits. Another example of this shift in political activism would be Theodore Roosevelt. He would make naturalist thinking a little more mainstream. Citizen action had led to the formation of Yellowstone National Park. Intertwined with this was the thinking of Pinchot, Powell and Muir. But making the concern for nature an important part of the American mindset was still not a national priority. And even Pinchot, Powell and Muir were not in complete agreement about what should be done with parks.

GOVERNMENT BEGINS TO CONTRIBUTE

President Theodore Roosevelt came to conservation by way of the hunter/harvester path. There is no question Gifford Pinchot influenced the shift in his interest. Pinchot's contribution was the credibility he brought to the Division of Forestry. He was aided by the law passed by Congress in 1891 which gave the president the right or authority to set aside forest lands as "public reservations." The government thinking was edging toward protection and conservation of natural resources. Pinchot and Roosevelt would form a very strong alliance. Both came from wealthy backgrounds. Pinchot had traveled to Europe to talk with experts in forestry where naturalist thinking was moving closer to science. He met Sir Dietrich Brandis, the world renown German forester. Brandis persuaded Pinchot to attend the French Forest School. When the National Academy of Sciences established a forestry commission to study forests on public lands, Pinchot became the secretary of that group. The President was Dr. Charles S. Sargent of Harvard University. The two men would differ in their approach to forest management. Sergent, a botanist, wanted forests preserved, by the military if necessary. Pinchot wanted the forests to be managed by trained foresters for both conservation and consumption. Again, you have two early advocates of forest management but two different schools of naturalist thinking.

To this time, preservation and conservation thinking had been almost synonymous. To Pinchot, preservation was not conservation. He felt there could be consumption and conservation if harvest techniques were handled correctly. In spite of these differences of opinion, the National Conservation Commission formed by Roosevelt, united to encourage President Grover Cleveland to more than double the size of the forest reserves. Naturally, western interests rose up in alarm over what they saw as an attack on their domain. Another Congressional action in 1897 gave the Secretary of the Interior broad authority to regulate the occupancy of reserves. This set the scene for professional management of the publicly held forests. It opened the door for economic use of those resources. Naturalist thinking was evolving. It just wasn't going into one neat box.

THE LACK OF COHESIVE THINKING

Very simply, the American mindset was on extraction until early 1900. The egalitarian prose might have enshrined Thoreau, but such words were written on paper milled from trees extracted from clear cut forests. May 13, 1908 would mark an early high point for the fledgling conservation movement. To that point, Congress had been a barrier in efforts to make conservation thinking an important value in the national mind set. Western interests and big business had stymied or blocked any of Pinchot's ideas on resource management. It was decided that a conference of Governors might create a coalition that could generate enough pressure on Congress to make them more responsive. When Roosevelt, flanked by Pinchot, addressed the assembled group in the east room of the White House, the governors were joined by members of the cabinet, the Supreme Court and groups of eminent private citizens. It was an auspicious gathering. At the close of his address to the group, Roosevelt said, "In the past we have admitted the right of the individual to injure the future of the republic for his own present profit. The time has come for a change. As a people we have the right and the duty . . . to protect ourselves and our children against the wasteful development of our natural resources."

World War I would put a brake on conservation thinking. War time is generally accompanied by both people and government putting great pressure on natural resources. The conservation movement would languish until the depression years. The '20s may have been roaring, but they included very little for conservation. In 1935, while the dust was blowing in the prairies, a little noticed event took place in Wisconsin. Aldo Leopold would purchase 80 acres of land in the sand country of Wisconsin. The seeds from that event eventually resulted in more naturalist thinking finally published in 1948 in the *Sand County Almanac*. Leopold spoke of a "land ethic." Meantime, The Civilian Conservation Corps was engaged in the greatest tree planting project ever conceived.

IN 1964, PRESIDENT LYNDON JOHNSON

. . . signed the Wilderness Act. The full authority of the Federal Government was now behind the radical idea that land, and the riches of that land, whether visual or material, had a value even if it is left undisturbed. In had been a long and twisting trail for naturalist thinking and conservation from 1903 and President Roosevelt to 1964 and President Johnson.

If the Wilderness Act was a major piece of natural resource legislation, and was "big picture," then the second piece of major legislation would be more species specific. The Endangered Species Act was passed in 1975 by the House of Representatives by a vote of 390 to 12. The Senate passed the bill by a vote of 92 to 0. You would assume it was nearly a consensus vote. Hardly. The bill has been under severe criticism almost from the day President Nixon signed it. There have been serious attempts to weaken it. Often referred to as "a biodiversity Bill of Rights," it is a clear argument for the use of resources for social causes and the protection of the small parts of a larger ecosystem. To this date, the bill has been neither a roaring success nor a total failure. Late information indicates about 10% of the listed species are showing improved status, 40% are still declining and 50% are either of unknown status or are thought to be stable.

THE NATURALIST VERSUS THE SCIENTIST

If you use George Perkins Marsh as the beginning of the idea that science offers the best hope for bringing the naturalist community into some sort of proactive consensus you would be forced to conclude science had been embraced mostly in the academic community and the professional fish and game ranks. More recently, it has made some inroads in the business world. Conservation still has a negative tinge in our political thinking. The general public still suffers from a bad case of apathy about conservation.

In some few instances, scientific thinking has managed to dent the armor of the political battleground. Rachael Carson would take on the chemical industry, Aldo Leopold and his land ethic cause would move from the University of Wisconsin campus to a wider national screen. Barry Commoner, a professor of biology, would raise questions about nuclear pollution and social responsibility. He was appointed to the Senate Military Affairs Committee and would write a book, *The Closing Circle* in which he explored the economy and what many were coming to believe was an environmental crisis. The book raised two key questions:

1. To what extent are the fundamental properties of the private enterprise system incompatible with the maintenance of ecological stability which is essential to the success of any productive system?
2. To what extent is the private enterprise system, at least in its present form, inherently incapable of the massive undertakings required to "pay the debt to nature" already incurred by the environmental crises—a debt which must soon be repaid if ecological collapse is to be avoided.

The views of Carson and Commoner were attacked by the proponents of economic growth. Commoner, especially, was labeled as "absolutist and visionary." His views on energy actually led him to form a third political party. He became a candidate for President and got less than 250,000 votes. In retrospect, his idea that the environmental cause needed a national, political platform was probably valid. Commoner saw naturalist thinking as a handmaiden of our system of production. He may have been the first ecologist to create a public awareness of the political conflict between our system of production or economics and our use of natural resources. He was one of those who added sociology to naturalist thinking.

THINKING ABOUT NATURAL RESOURCES

. . . is just sort of messy. There is that old saying about stones and the first person to throw some? There are few among us who are totally without compromise in our environmentalist values. I have a good friend who does great work for The Nature Conservancy. He drives a gas guzzling SUV. I have a neighbor in Wyoming who preaches regular environmental sermons in the paper. His horse corral is immediately adjacent to a river which carries away urine and manure soaked sand. A leading Oklahoma member of Ducks Unlimited has developed some of Oklahoma's prime wetlands.

David K. Leff is a Deputy Commissioner with the Connecticut Department of Environmental Protection. I was attracted to his thought because he used the word "messy" which I used when this was originally written. "The Thoreau we need today is not the saint created by admirers or the media. It is the man who observed in Walden, 'Be it life or death, we crave only reality.' This reality is messy and rife with hazards, embarrassments and accommodation. Following Thoreau's example does not require lives of ecological perfection. Instead, it requires that we regularly question our consumptive habits and consider how even our most mundane and routine actions affect the natural systems on which life's diversity depends. Using such inquiries to continually perfect our relationship with nature—not completely flawless action—-is the yardstick for measuring success. By doing so, we will better appreciate Thoreau and foster a more sustainable planet."

Neatly said. Such a condition might produce some neat thoughts about considering how our most mundane and routine actions affect our natural resources. And, of course, the ducks and frogs might appreciate our new habits.

Horace Marden Albright

Was born in Bishop, California, January 6, 1890. He graduated from the University of California in 1912. Later he attended Georgetown University to obtain a law degree and was admitted to the bar in the District of Columbia and California.

Albright would begin a career of public service under the wing of Stephen Mather, serving as secretary to Mather. When Mather was hospitalized in 1916, Albright as acting Director, organized the newly formed Park Service, set new policies and wrote what became the "creed" for the National Park Service.

He was appointed Superintendent of Yellowstone Park and was Assistant Field Director. In winter months he would supervise all national park areas west of the Mississippi River.

Horace Marden Albright. National Park Service Photo

In 1929, Albright became Director. He expanded the national park areas east of the Mississippi and introduced historic preservation into the National Park Service. Four years later in 1933, he asked President Franklin Roosevelt to transfer national monuments from the Department of Agriculture and military parks from the War Department. The national parks were now fairly well consolidated in the National Park Service.

Albright resigned from public service and became first the Vice President and then President of the United States Potash Company. He retired in 1956 and lived until he was 97. A Conservation Lectureship has been established in his name at the University of California.

Horace Marden Albright died in Van Nuys, California March 28, 1987.

REFERENCES

Fox, Stephen. The *American Conservation Movement: John Muir and His Legacy.* (University of Wisconsin Press) 1986

Lorbieki, Marybeth. *Aldo Leopold: A Fierce Green Fire* (Falcon Publishing) 2005

Mattes, Merrill J. *The Great Platte River Road.* (University of Nebraska Press) 1987

Norton, Bryan G. *Toward Unity Among Environmentalists.* (Oxford University Press) 1994

Strong, Douglas H. *Dreamers and Defenders.* (University of Nebraska Press) 1988

Walden and Other Writings of Henry David Thoreau. Edited by Brooks Atkinson. (Random House) 1950

Wilson, Edward 0. *The Future of Life.* (Alfred A. Knopf) 2002

Visits to Concord and Walden Pond.

Sources read for our brief sketch of Land Use were numerous and included:

A.F. Vass, *Range and Ranch Studies in Wyoming* University of Wyoming, Agricultural Experiment Station 1926

Bernard Devoto, *The West Against Itself* 1955

William Voigt, Jr., *Public Grazing Lands* 1976

Bernard Shanks, *This Land Is Your Land* 1984

Wesley Calef, *Private Grazing and Public Lands* 1960

Bernard Devoto, *The Easy Chair* 1955

Chapter Twenty-Eight

The Least of These

All creatures make a spiritual mark. The elk is elegant. The wolf is ethereal; the beaver sober, the eagle is inspiring. The fox is cunning. The quail is gentle and the hummingbird is mysterious. Perhaps the humming-bird carries hope on a rapid wing beat for the future of our environment.

Think of the hummingbird as the apostle of the spirituality of nature. They seem to have some quality that makes them irresistible to people. Is it their size? Are we drawn to things that dart and weave, only to hover in an iridescent halo of motion and feathers? Perhaps it is their energy and intensity, or their pugnacity? Is it because now we see them and in an instant they are gone?

Speed is their essence. Are we drawn to this as we are to the opium of speed in autos, horses or any piece of technology that promises fast results? They live by a faster clock. Is this why, increasingly, people are hanging feeders from some handy post so they can better observe these tiny bundles of motion?

Research the hummingbird and you find no one completely understands how they became the accelerated, action packed creatures they are. The biologists have begun to sort out bits of information. They have complex sets of independent adaptations. It seems to be a story of evolution. A story that would have profound implications for the care of the earth.

WATCHING HUMMINGBIRDS

This much we seem to know. They have been clocked flying at speeds up to 40 miles per hour, a rather remarkable achievement for so small a bird. When they hover, they use 52 wing beats per second. They have a short gut and can digest and eliminate food in less than half an hour. To burn food this fast, they may need 300 breaths per minute to supply enough oxygen. They have an unusually large heart for a bird their size and this may beat 500 to more than 1200 times a minute to pump out oxygen rich blood.

They do not hop nor walk like other birds, traveling almost exclusively by air. They will fight, feed and display on the wing. Even when resting on a branch, they will start their wings, let go of their position, and fly to reposition themselves. They seem to stay in pairs, evidently after they have mated, and tend to protect their spot at the feeder or the flower.

I watch hummingbirds on a deck of a cabin along a Wyoming river. Being the big screen naturalist, I have not particularized their antics in a journal. I can report they seem to become tamer as the season progresses and eventually come close enough I can hear their buzzing sound clearly. When they finish at the feeder, they fly off into the willows along the river. There is a report of an ornithologist, Oliver Pearson, who tracked the patterns of a hummingbird for two days and concluded his quarry spent "about three quarters of its time preening, perching and digesting its food."

It is the apparent evolution of the bird, and ecosystem of which it is a part, that provides a theme for this small creature. The ancestors of the bird evidently started down an evolutionary road that took them to the Wizard of Nectar. Within their long bill is an even longer tongue. It is the tongue that reaches into the flower to gather nectar. Whether the original tongue was as long as it is today is open to conjecture. But the flowers and the small birds decided they would dance together. In a way, they became "patches" in nature's crazy quilt. The botanists will tell you the flowers and the birds have a cooperative effort. The hummingbird delivers pollen from blossom to blossom. The bird gets nectar. The flower gets procreation. Most of the flowers used by the birds are red. The blossoms seem to be tube like and usually hang in loose clusters. They are positioned to discourage the bees. The longer tube of the flower keeps the short tongue of the bee from getting at any nectar. Bees around a feeder generally indicate a feeder that has not been washed clean of the sugar water it contains.

IT'S IN THE BOOK

Many of the flowers that attract the hummingbird are in the same family as other flowers that attract the bees. But the botanists believe certain species of these flowers actually evolved to accommodate the bird and discourage the bee. The bird and the environment have reached an accommodation. There is more to this complex story in a book published by the Arizona-Sonora Desert Museum. "Oddly enough, one consequence of the continent-wide cooperation between North American hummingbirds and their flowers is competition . . . among hummingbirds for flowers and among flowers for hummingbirds Natural selection often works to reduce competition among living things by causing them to divide up the resources in dispute. It goes back to the idea that nature is in a constant state of pruning. Hummingbirds reduce their competition for flowers by claiming territories, and different species may divide up the habitat by feeding and nesting in separate geographic areas . . . even different parts of the same canyon.

"For their part, hummingbird flowers may reduce their competition for hummingbirds by growing in different places, but several species share the same meadow. Unable to divide up the habitat, they may divide up the year by blooming at different times. The co-evolution of hummingbirds and their flowers is still going on today. Botanists have caught flowers midway in transition from insect to hummingbird pollination, the latest recruits to the ever-growing ranks of hummingbird plants."

My small friends on the back deck eventually came to be migratory travelers. They began their season-long journey in Mexico, went up the California coast, crossed mountains and in the late summer headed back to their wintering grounds. The Rufus specie, if they are lucky, will live about 10 years. For some of them, their round trip is a journey of some 2,200 miles. That would be a remarkable trip for anything but especially for a creature only 3 inches long. I wonder if the bird that flew into the window behind the feeders and knocked himself out, ever completed the journey. It blinked at me when I stretched out a finger to carefully inspect its condition as it lay on the deck. Then it righted itself, and buzzed off to join the congregation gathering nectar at the flower box.

In essence, flowers, bees and hummingbirds have formed an alliance. While there is competition, there is accommodation. Unlike people who divide up land to satisfy some selfish purpose, there is a balance. Without this balance, some or all of the attendant species would be pruned back. The role of the predator is minimal, either human or natural. None of the involved species has become a food source, has been trapped for trade purposes nor have they been appropriated for recreational purposes. Economics, other than bird feeders and sugar has never been a factor. Nor has the government ever issued any regulations regarding the bees, the flowers nor the hummingbirds.

Nature has provided a natural evolution. Unlike the Monarch Butterfly whose wintering grounds has been threatened by intensive logging, but is otherwise a migrant similar to the hummingbird, the ecological determinants for the bird have remained fairly stable. There has been little nor any human interference.

Further, the hummingbird has enjoyed a very benevolent social contract. Because they had little nor any market value, they have escaped man the predator. They have gained a social appreciation for being what they are.

Their value, to their admirers, is entirely esthetic. They are far more valuable as living beauty than as dead birds. So many species have never been able to attain this unique social contract. To the contrary, most species have had more value dead than as living species.

CONSIDERATIONS OF NATURAL VALUE

This strange circumstance leads to a consideration of natural value and how Americans go about placing values on our natural resources. If consumption is the only criteria, then a natural resource is a commodity to be used. But, if there are other values to be considered such as esthetics, ethics or spirituality, then the social contracts of human beings must reflect a different value system. Values would then be at the very core of the conflict between consumption and conservation, extraction and balanced or sustainable use.

While pondering one day the lesson of the interdependence of flowers, nectar and humming birds I was standing beside a wood pile. It reminded me that for all the heavy thoughts about hummingbirds, I really should address the more temporal task of splitting some logs for a fall fire. I noticed the end of one of the logs, wondering if it would split well. I looked more closely at the end of the log, the many rings it had. Values are like the rings of the tree. At the center is the heart, the core value. But the tree needed rings. These grew slowly. They had lived a long time. The rings had surrounded the center value, protecting it. The heart of a tree makes no judgments. Neither do the rings around it. Only the human can assign a value to the heart or the rings.

As a tree grows, it is protected by bark. Some trees have smooth bark. Some values are protected by smooth talk. Or, the tree can have rough bark. Values can be protected or defended by rough talk. Sometimes, the heart of the tree will die. It looks okay on the outside for a time, but the inside is rotten. Eventually the tree will die. Sometimes that is a slow process. Values are like that too. Values have many things in common with logs.

ACROSS MANY LIFE SPANS

. . . and the spaces of uncertainty, we Americans have made many choices about values. We may have been hunters, but we worked without roadmaps. We may have been settlers but we plowed up grass with little consideration of the value of that grass. Across this space of time, we labored with a burden. Without realizing it and with no Book of Conservation or Letters to the Environmentalists, we may have blasphemed. Because by the very nature of this journey, we made many choices about our earth, water and air. Many of these choices were economic, but many were moral. And if those choices were based on self preservation and extraction of the earth . . . if that was the core value, then we grew rings around that value that would protect and justify it.

This trunk, this value, grew for three centuries. From Pilgrims and Puritans, to Nordic loggers, to German farmers, to Irish canal workers, to black women who chopped cotton to Civil War survivors who found freedom astride a horse in a cattle drive . . . this value was that man, backed by a bible and a gun, was master of the earth. This was the core value. The earth would take care of man. Then we built rings of justification around the core. These rings were economic choices, some even of self survival. Some may even have been moral. This is our legacy.

EACH GENERATION LEAVES A LEGACY

If it is true each generation leaves a crazy quilt hanging on Mother Nature's clothesline, that quilt is a patchwork of conflict and confrontation with natural resources. Seldom was there a search for mutual accommodation. With rare exceptions, we simply did not consider a conservation ethic part of our national social contract. If the individual contributed to this condition, then can the individual unravel the result? More stories seem to be emerging about the efforts of individuals, and some businesses working on an accommodation with natural resources and the environment.

This story was reported by Bill Heavey in *Field and Stream* magazine in the September 2000 issue. He sketched the efforts of one individual working to preserve or restore good social contracts with natural resources. It was a story of an early day family in Halifax County, Virginia. An ancestor had received a 5,000 acre land grant from the King of England, had filed a will in 1752 that set things up for future generations. Today, the subject of the story, Hudson Reese carries on that tradition. Heavey writes:

"For as long as Reese can remember, men in his family hunted birds, which is to say quail. He caught the bug early from his older brother, who went every Saturday during the season, always with his collar buttoned up to the neck and usually wearing a tie. A love of land and the birds binds generations together here. Reese believes that hunting quail . . . the pride and humility it brings . . . is his grandson's birthright." (To this point in the story, you have natural resources, birds and land, tradition and social contracts mandating the value of hunting.)

Heavey continues Reese's words. "Quail hunting, one of the oldest traditions in this part of the country, is dying. It's being killed by winter grasses like fescue, which create a thick mat that quail can neither move through nor penetrate to get at the bugs in the soil. (Now add species substitution to the mix.) It's being killed by high monoculture crop fields, the decline of prescribed burning and the mania for "clean farming" all of which insulate the land from the periodic housecleaning that nature intended and quail need to survive. And it's being killed by simple ignorance by well intentioned people who buy a weekend place and a tractor and think of nothing better to do on a Sunday afternoon than to go out and tidy up the land, clearing old fence rows, weeds and broom sedge meadows so precious to bobwhites."

"So Reese is putting his money where his mouth is. He doesn't hunt as hard as he used to . . . he won't shoot at a covey of fewer than eight birds. A group needs that many to survive a winter's night, when the birds snuggle together, forming a circle of warmth and protection. He leaves his own field borders uncut, plants lespedeza, clover and rye on the fields he has taken out of rotation for a season. He has also represented Virginia on the National Association of Conservation District's Council for the past 25 years. His work has led to a close friendship with Steve Capel, a hunter and farm biologist for the state who is at the center of Virginia's ambitious new program to educate farmers about leaving some cover and planting quail friendly crops."

From a story of the efforts of one individual to a recent story of efforts by a smaller company. This news came from *Business* magazine, September of 2006, written by Jeff Nachtigal. He tells about a company wide effort by Adobe Systems to become "a towering example of environmentalism. This $2 billion software company had received the Platinum award from the nonprofit U.S. Green Building Council." The company had won the award by retrofitting its existing office towers, about 1 million square feet. To date, the company has "invested some $1.1 million in 45 energy efficient projects. These produce nearly $1 million in annual savings. An automated irrigation system that cost $3,610 to modify saves $10,000 annually. Motion sensors that now turn on only when needed save $105,000 and water-less urinals save $14,896 each year. Adobe has reduced its annual electricity use 35 % and gas consumption by 41%. A company official is quoted as saying, "This isn't some pie-in-the-sky kind of thing the environs are pushing. It really works"

There are more of these small stories when you search for them. More conservation magazines carry stories of private/public partnerships. There is more work by conservation organizations on the grass roots level. In almost all instances, the theme of the story involves the amount of money that is being saved. This is good news. But if all the savings being uncovered are legitimate, then it would seem there has been great waste in the past.

A SMALL TOWN IN A BIG ECO-PROBLEM AREA

Colorado is a state that has experienced a lot of environmental burps the past 20 years. Water problems have plagued the state. While the Colorado River has drawn the most attention, the Platte River has presented its share of conflict between irrigation and recreation users. Population has boomed, putting more pressure on municipal services. North of Denver, on the east slope of the Rockies, Fort Collins has been quietly pushing the idea that conservation, renewable energy, more efficient use of natural resources should be a theme in daily thinking. This effort surfaces in local businesses, municipal government and especially in the classrooms and research labs at Colorado State University. In the fall of 2005, Fort Collins was identified by the Sierra Club as one of the top four cities in the United States at promoting renewal energy and efficiency. Portland, Oregon . . . Chicago, Illinois . . . and Austin, Texas were the other three.

Talk to Hunt Lambert, Associate Vice President of Economic Development at CSU and he will tell you, "within the community infra-structure, our academic research has a lead role. We had 25 faculty people working in clean energy research and we just decided to formalize it. We have developed a new type of solar energy panel. CSU will offer this prototype to a Colorado partner and if we cannot locate one, we will go outside, but our plan is to offer these panels globally." He said they had 57 people working on water quality research. They are testing a technology to convert algae from waste carbon products and produce a bio-diesel result. And he added, "We have a number of people involved with a grid simulation research project that is involved with climate warming."

How has this academic activity translated to the grass roots? Kelly Girard started a lawn care company, Clean Air Lawn Care. He bought all-electric equipment. This cost a "little more" than going with gas powered machines. This will save $1,500 in gasoline during summer months and prevent 2,500 pounds of carbon emissions each year. Small stuff? It is all part of a growing picture that we have actually been rather wasteful in our use of natural resources. Several years ago, Fort Collins city officials were all a-twitter because it looked like an Anheuser-Busch brewery would locate there. It did. A brewery uses a lot of water in their operation. In the past, the brewery cooled heated beer in the pasteurization process using cold water. Then that water went down the sewer. Now, that water is sent through a heat exchanger, the temperature is lowered and the water is reused. Result . . . 100 million gallons of water saved annually. They are saving 10% on the water bill. Again, the word "savings" is present.

FROM SMALL BIRDS TO BIG RETAILERS

The Fort Collins story is unique in that it combines the grassroots, the infrastructure and the research resources of a university, with the university taking a scientific lead. Until recently, small conservation groups have provided the bridge, a continuum of good intentions washing against the dams of public apathy. Partnerships were not always developed. There was little unity of effort between disparate groups. Conservationists tend to be hopers and wishers. On the grass roots level, they like the company of small groups that preach good deeds and really don't rile anybody. They are even uncomfortable if their political candidates take strong positions for environmental issues. So when you are talking conservation or environmental concern and you weave from a small bird to the world's largest retailer, one that has 1.8 million employees and has 176 million shoppers at their stores every week, you need some connection for the jump. In this case, the connection is hope. If the hummingbird represents a natural evolution of natural resources, then perhaps what is beginning to happen at Wal-Mart can lead the way to an evolution of the conservation attitudes and practices of "big business" in this country.

The Wal-Mart shift in direction has been some time in the making. A high performance organization under icon Sam Walton, there were some stumbles after his death. The company stock, which had enjoyed a 1200% growth in the 1990's, began to decline. Same store sales dropped in a number of locations. Efforts to build new stores began to meet local resistance. Critics of Wal-Mart pay levels, the health care they provided employees and gender discrimination all came under increasing public discussion and even legal action. All this was brewing when Lee Scott came in as CEO in January of 2000. He later would describe the Wal-Mart reaction to mounting criticism as . . . "We would put up the sandbags and get out the machine guns."

THE INFLUENCE OF ANOTHER GENERATION

In this time period, "Rob" Walton and his brother John were assuming lower profile roles in the company. John, very much a conservationist, would die in a plane crash. Rob's son had been working as a Colorado River guide. Meantime, new CEO Scott, who was born in Baxter Springs, Kansas and was quite an angler, was beginning to question many of the Wal-Mart management practices. And many of those questions had a conservation base of sorts. The bottom line was the Wal-Mart hierarchy was now more open to some sort of eco-management. A good story by Marc Gunther in the August, 2006 issue of *Fortune Magazine* tells of the top people in Wal-Mart being exposed to conservationists or environmentalists from large organizations, of trips they made to foreign countries, of exposure to the top minds in the conservation movement.

On the operational side, Wal-Mart had opened what we will call "an Eco-friendly store" in Lawrence, Kansas in 1993. It didn't exactly get standing ovations. By 1995, it had opened two other large eco-friendly stores, one of which was in Denver. Both inside and outside the Wal-Mart organization, different factions saw the new direction differently. Wal-Mart was in the stage of Storm. You can hear the old guard muttering about how "it ain't like it was when Sam was here." You can hear suppliers, still smarting from the Wal-Mart practice of pressuring suppliers for the lowest price wondering what the Wal-Mart motive really was.

And what would our friends, Marvin and Charley have been saying while headed out on another hunting trip? Charley might say to Marvin, as he looked at his map, "Wonder, Marvin, what sort of map Wal-Mart is using these days. Sure some funny stories comin' outa Bentonville?"

"Well," says Marvin, "they've had a little management shakeup. But all business companies go through those. But that is big time business. I was reading where they have better than 27 hundred supercenters . . . this was in a utility magazine . . . power company . . . and it said they use 1.5 million kilowatt hours of electricity a year. Those big figures are just hard for me to get ahold of."

"Yeah," says Charley, "I don't worry a lot about what big business does. I've just always felt Wal-Mart was on the side of the common folks with their low prices. I buy stuff there when I want to save money."

I talked to Sophia Wilson, the store manager in the Riverton, Wyoming Super Center. I asked her what the reaction of her employees had been? "Very positive," she said. "In fact, I've heard no negatives. We are now recycling all the plastics and card board we used to throw away. We have what we call a 'sandwich bell' and we put the cardboard and all plastics in those and they haul them away" "And what about customer reaction?" I asked. "Well," she paused, "we put out bins for them to use to put the sacks . . . the plastic sacks in, when they return them, which we invite them to do. We have to empty the bins nearly every day."

According to the article in *Fortune Magazine* Wal-Mart has some goals with their new philosophy. And it does seem to be more a company philosophy than a strategic plan. CEO Scott outlined some of the goals in a speech broadcast to all facilities. They aim to increase the efficiency of their vehicle fleet 25% over the next three years, reduce the energy used in stores by 30% and reduce solid waste by 25% in three years. They have already become a major player in organic cotton, being the biggest buyer of this commodity in the world.

THE JURY IS STILL OUT

Because of the size of its retailing ability, Wal-Mart has the clout to seduce their suppliers into a more eco-friendly way of doing business. Some results have already surfaced in the suppliers of fish and coffee products. Wal-Mart is encouraging a more sustainable harvest. So the environmental thrust that Wal-Mart officials admit was more of a "defensive strategy" would seem to be turning out to be more the opposite. Quoting Gunther's article again, where he quotes Scott . . . "We stepped back from that experience . . . (the way the company handled delivery of vital supplies after the Katrina hurricane) and asked one simple question: How can Wal-Mart be that company— the one we were during Katrina—all the time?"

Not all suppliers are buying the new philosophy. Some still refuse to do business with Wal-Mart. They use the term "selling our souls." And the use of that word fits with our use of the word "philosophy" because there is an undertone of morality to what Wal-Mart has embarked upon. And in today's business climate, and in Washington, that is not always a popular word. Only time will tell. But like the story of the hummingbird, the Wal-Mart story offers hope.

Environmentalism doesn't have too many activists. There aren't too many Rachel Carsons, Gifford Pinchots or John Muirs. What's even stranger is conservationists don't have many heroes. The closest to this might be President Theodore Roosevelt. We have elevated Generals to hero status, even genuflected before some Hollywood heroes. We haven't elevated many businesses to a plateau of appreciation. We need more players on nature's team of environmental sustainability.

Yet most humans seem to have an innate need to be part of something, to belong to something. That could be family, tribe, nationality, street gang or social club. Many of our naturalists have left the singularity of the individual to band with the ultimate community of nature. E.O. Wilson uses a term "biophilia" to describe the human need to affiliate with a non-human world. Our language is loaded with terms like "dog lover," or "bird lover" and even "flower lover." People not only cultivate these "loves," they transfer great affection to the non-human species. Even in a materialistic world where society loves and accumulates "things" we find time to relate to non-human, but living things. In that idea there is hope for our earth, air and water.

For all this affection, the reality is we have lived for too many centuries in a world of extinction and expiration. Some person has estimated that every twenty minutes of every day, another animal, fish or plant species expires. Small wonder, in the face of such odds, the average conservationist looks at such a number and wonders, "What can one person do against these odds?"

WELL . . . EVERYTHING STARTS SOMEWHERE. One person could switch to a long-lasting compact florescent (CF) light bulb. Come on, you say? But if every weekly customer at a Wal-Mart store did that, it has been estimated it would save $3 billion on national electric bills. One person can reach out. They can provide experiences for young people to come in touch with nature or provide an experience for nature to touch young people. It could be as simple as watching hummingbirds. If your city is involved in "green building" or efforts to make the infrastructure more energy efficient, you can support those efforts.

And in the bigger picture, if some business is making sincere efforts to produce a better environment, you can support their efforts or at least reserve critical judgment until the jury comes in. In the following chapter I try to plant some seeds that would make care of our environment a more pro-active cause. As Lee Scott of Wal-Mart said, "We will not be measured by our aspirations. We will be measured by our actions."

REFERENCES

Gunther, Marc. *The Green Machine.* Fortune Magazine August 2006
Heavey, Bill. *The Heritage of Quail.* Field and Stream September 2002.
Lazaroff, David W. *The Secret Lives of Hummingbirds.* (Arizona-Sonora Desert Museum Press) 1995
Nachtigal, Jeff. The *Greenest Office in America.* Business Magazine September 2006
North American Fisherman. The Angling Mind. October–November 2002.
Wilson, Edward O. *The Future of Life.* Alfred A. (Vintage). 2003

Chapter Twenty-Nine

"Where There is No Vision the People Perish"

We have used ten ethics to describe the disparate thinking about conservation in our country. We have tried to separate the thinking, as much as possible, of people who have advanced the thinking of the stewardship of natural resources. We have identified twenty five people who contributed in some manner to the thinking about the environment and natural resources. We used the analogy of a "crazy quilt" to give the subject some form or continuity.

In all of this confusing narrative, there is a lurking question. What would have happened in this country if, in the beginning of the national trend toward rugged individualism and extraction, and the national agreement on that philosophy, our thinking had been as unified on the subject of the care of the earth? Instead of a national mantra of extraction, the accepted ideal had been stewardship of our national resources.

What kind of a country would we have today? What kind of a society would be the norm today? Could we actually have had a condition of sustainability, renewal and environmental caring? What shape would our land, water and air have been in today if we had adopted the spiritual view of nature that was embraced by the Native American?

Ecology is counted in decades, centuries and eras. We have traced the 18th century as a time of individual extraction. The 19th century saw the pace escalate with business and corporate extraction applied to natural resources for consumption and production. The early 20th century continued that pattern. Then in the 1970s, the continued exploitation of natural resources began to be questioned. But in none of these three periods of time was there any consensus that would resemble a national conservation ethic. The idea of conservation for a sustainable economy or condition was a minority view.

Ecological thinking would begin to focus on individual agendas in the late 1970s. Species would be singled out for protection. More wilderness would be set aside. The use of forests would make the news. Then, sometime in the 1990s, there was more news about water quantity and quality. That issue was still not more paramount than building golf courses for development purposes, but nagging issues began to crop up. The issue of the Colorado River water was reexamined. More dams were scheduled for removal. Residents near Mount Hood in Oregon where unemployment was up to 9%, would argue that environmental damage from a proposed development would outweigh the jobs it would create. Nebraska, hard pressed by drought, would pass legislation to restore and recreate riparian areas along their river systems.

THE GROWING CONCERN ABOUT WATER

The west is the magic region in the United States. Much is still primal, an epicenter of mystery and myth. It is a huge and vast collection of landscapes, alive and ageless. The west lives on in spite of man rather than because of man. Nowhere in our national scene is water more important than in the western United States, especially in arid areas. In Wyoming, in the throes of sustained drought, there is deep concern about water for irrigation of hay land and what cattle will eat and drink. But there is also growing concern for what people have to drink. The state wrestles now with the energy industry and methane polluted water that has already turned the Powder River into another "most endangered river."

In eastern Oklahoma, the poultry industry is in a face-off with water environmentalists. The concern may be more for quality water for residents to drink as the practice of spreading chicken litter on watershed land is debated. Chicken litter may be Chicken Little to some folks who relate that to the line about the sky falling. In India, the sky has already fallen. There, 200 million people are without access to clean water. A visitor to that country runs the risk of violent illness if they just brush their teeth in water from the faucet. It is estimated 90% of the country's water resources are polluted with industrial and domestic waste, pesticides and fertilizers. Figures from that country, on the verge of becoming the world's largest population, estimate 1.5 million children under age five will die each year of water borne disease.

There is hope in this coming water crisis. When people cannot get a glass of good water to brush their teeth they suddenly become more environmentally aware. There is nothing like a bad taste in your mouth to create personal awareness for the cause of the taste. This whole issue of water quality is pushing the subject of ecology down to the grass roots level where most people do most of their drinking. Lately, of course, more people are drinking their water from a plastic bottle. Another example of creating waste to avoid a problem. Mark Twain had a little perspective on this. "Whiskey is for drinking, water is for fighting." I believe water will become a major conservation issue for the American people in the 21st Century.

THIS IS THE SETTING

While our national focus has been on the land, the forests and individual specie, we have pushed concerns for water into the bleachers behind the end zone. Water will be the super bowl environmental conflict in this 21st century!

When man begins to manipulate nature, there are three fundamental questions that should be asked about the situation. They are questions to be asked before deciding upon a course of action, be it wild turkeys or dams.

How Many Do We Want?
Where Do We Want Them?
What Do We Do With Any Annual Surplus? Or,
What Do We Do About Any Annual Deficit?

Try these three questions on irrigation and isolate the questions to an area like Nebraska. How many acres of irrigated land would produce maximum efficiency of the state's natural resources? In what parts of the state would irrigation yield the highest returns? If there was a surplus, what do we do? Store it? Pass it on down? Use it for some other purpose? The same responses would apply to an annual deficit. Now, the options of dams and large reservoirs come into play. But, again, in this option, the same three questions would apply.

Try these same three questions on wolves. How many do we want? Where do we want them? What do we do with the annual surplus? Suppose these had been the questions asked when re-introduction was being considered? We want at least five packs and we want them in Yellowstone Park. If there are more than five packs and any of these are outside the park, we will allow reduction or removal. Tough questions? Yup! Result? A common sense plan!

After walking the lonely path of the conservationist, after assembling many patches of material for this book which present so much disappointment and frustration, I have reached some conclusions. The first is that it is better to reach conclusions with a deep sense of humility. Still, I have a great faith in seeds and the miracle of renewal. Thoreau had some thoughts about seeds. "Though I do not believe that a plant will spring up where no seed has been, I have great faith in a seed . . . Convince me you have a seed there and I am prepared to expect wonders." So what follows was not recovered from a burning bush in some distant place nor inscribed on a tablet of stone. These are seeds.

SEED #1. Let's assume the next conservation issue in the United States will be water. While some proactive things are happening to ensure water quality, such as the work New York City has done in their watersheds, we are retreating in the wise use of water in this country. In the arid west alone, we are irrigating marginal land for agricultural purposes. We have pointed out that 85 % of the water supply in California is used for agricultural purposes. This raises the question of the best and highest use of a valuable natural resource. That same question could be raised about parts of the prairie country, those marginal areas of 18 inches or less of rainfall. How long can the Ogallala reservoir sustain the withdrawal of water being used for irrigation? How long can the Platte River provide water for most of southern Nebraska? What will be the eventual role of the Missouri river?

Can the Snake River continue to water most of southern Idaho for the primary purpose of raising barley and potatoes? How much of a limited western water supply should be used on golf courses? From what I have observed in good river restoration projects, even restoration projects around lakes, the most successful are those that involve a reasonably large constituency. Popular support, a good social contract, will focus on the watershed. The focus is on the watershed, not the river itself. Only recently have successful group efforts crossed state lines. And therein lies one of the most immediate problems. Too often one state is pitted against another in cognitive boundaries. Still, there is hope in recent events.

UNTIL WE HAVE RIVER COMPACTS

. . . we will not make much real progress in restoring quality to our rivers. These compacts should include the entire watershed. Not much was resolved in the Platte river dispute until Colorado, Nebraska and Wyoming recognized the commonalities and rewards that result in conservation of an entire watershed. This would hold true for the Colorado River, the Snake River and other western rivers where water rights are still based on old law. There needs to be more action taken like was done in Tennessee. More state action like SB 156 in Colorado which modified old, in stream flow law and gave the state water board power to accept unconditional water rights in amounts "appropriate to the improvement of the environment."

In a best case scenario, every stream in this country would have a Watershed Compact, composed of individuals, either appointed or elected. It might have members from conservation organizations. It might need legislative approval. It would operate in the best interest of the watershed. It might secure grants, federal underwriting or foundation support. It would put water issues on the grass roots level.

SEED #2. Congress would create a *Department Of Natural Resources*. This department would be headed by a Director, appointed by the President with the nomination approved by Congress. Into this department would be folded any present department having to do with natural resources. This would include The National Forest Service, The National Park Service, U.S. Fish and Wildlife Service, the EPA, and the Bureau of Land Management. Each of these entities would have a Manager, reporting to the Director. The Director would select these managers.

This would do two things. First, it would create a very powerful department that would generate the perception that natural resources are an important part of our national thinking. Since it would involve the interests of every state in the union, it would have the attention of Congress. It should be able to generate adequate budgets.

Second, it would open the door for administrative and manpower efficiencies. Non-profit conservation organizations would know where to go to seek assistance or to provide input.

SEED #3. Every state needs a Natural Resources Confederation. This is a small group that represents conservation organizations. The Director is nominated by state conservation groups and approved by the Governor. There are two other board members. One is appointed by the State Legislature, the other selected and nominated by state conservation organizations. The Director has a six year term and can serve no more than two terms. The two board members serve four year terms and can serve only two consecutive terms. The Director is the person who supplies coordination between the members. He, or she, represents the views of all the state conservation organizations. It would be for profit, supported by member organizations. The Director would be a lobbyist. He, or she, would create an internet that alerted all members to proposed bills, legislative action or the lack thereof. They would poll their members when issues needed a consensus of opinion. This person would be supported by a contribution from the various organizations. The contribution would be based upon the size of membership of that organization.

From experience in several states, and reports from others, it would be my opinion that not many conservation organizations on the state level talk with other organizations. There is really no unified voice on the state level that addresses the needs, issues, or problems of conservation. Most conservation organizations are so focused on their individual mission they are not aware of what other organizations are doing. What's more, there is little nor any cooperation between the organizations unless it is a specific project. They are a minority preaching to a minority.

There needs to be a unified voice on the state level that speaks for, and to, conservation issues without the fear of losing their non-profit status. Whether it is ducks, pheasants, elk, quail or trout, or a huge issue like water quality and use, there is more commonality in the cause than there are differences. Most conservation organizations are toothless when it comes to a presence in the political arena. They fear for the loss of their tax exempt status. Nothing much will happen in the political world until conservation becomes a factor in that world. That will require a joint effort that gains the respect of the political reality. Natural resource issues simply must have a pro-active posture on state levels. Obviously, such a readjustment in state and national priorities would produce debate and conflict. Rural interests would be threatened. But the reality of our American culture is that the old rural community is an endangered specie. It is already leaking badly, propped up by giant buckets of federal subsidy. It is time to help people who take care of the earth, water supplies and water quality.

SEED # 4. Allow the professional managers to manage. Too often, state natural resource management plans are eventually written by the state legislatures. The reality of state control of natural resources is that political pressure is stronger on the state level than the national level. Most states still labor with a large bloc of rural legislators. The pressure of the rural interest is still very strong. Most rural people still have an innate suspicion of the professional land or forest manager. Forest management has many highly trained people. In the private sector, such people produce great return on investment. Today, the forests in the southeastern part of the United States are well managed. "Between 1953 and 2000, this region's share of world production rose from 6.3% to nearly 16%. Those forests are producing not only economic return, they are producing a substantial percentage of the wood needed for the world's supply."

Could there be differences of opinion and conflict in a scenario where professional managers were allowed to manage? Certainly. But the ultimate decisions should reside with professional management. Experience has shown, however, that managers produce the best results when there is some system of accountability. A state oversight committee could provide review and citizen input.

Any forest, should be and can be, managed to produce a system of sustained economic stability. Grazing cows on forest land will not produce that sort of balance sheet. That is subsidy disguised. Every forest ought to be a forest primarily, not a pasture land.

As this is being written, the USDA Forest Service is revising plans for the management of national forests and grasslands. They believe their proposed planning would make public involvement, particularly from individuals, easier . . . allow managing for the sustainability of natural resources . . . promote using research, not emotion, in making management decisions and . . . provide monitoring and evaluation. This quote from the Wild Turkey Federation magazine, *The Caller*. "The proposed rule however, strives to change the complexity of participating in national forest planning and reduce the seemingly endless planning process to a reasonable amount of time. It will also make it easier for regular folks to bring their ideas and wants to the table and contribute to national forest planning and management."

The point of this seed is accountability. Every conservation organization, every State Game and Fish Department, every Federal wildlife entity needs an oversight committee that evaluates the job they are doing.

SEED #5. Push the management process of natural resources as close to the grass roots as possible. Eliminate management from a distance. In the case of small watershed dams, keep building and oversight on the state level. In the case of wolf management, let the state do the management outside Federal Parks. Allow state managers to write the management plans. You cannot manage a Montana forest from Washington.

The best example of this grass roots approach I have seen is in Vermont where Mary Watzin wrestles with the Lake Champlain cleanup. She is close to the problem. Her office is on the shore of the lake in downtown Burlington. There you have university resources, two states, and one Canadian province all working on the problem. That doesn't make a solution any easier but it has reduced the time lag present in so many EPA directed cleanups. Still, it has taken this group nine years to arrive at standards that would work for each interested party. Restoration after years of neglect is not fast work. But it does work faster on the local level.

SEED #6. Encourage states to recreate a version of the depression years CCC. This effort would be of two types. One would be coordinated between Land Grant Colleges and State Fish and Game Departments. It would provide summer employment for young people and would carry college credit. On site work and class would be coordinated. These might be on small dams, watershed projects, shelter-belt restoration, riparian stream restoration, research on predator control. There are dozens of such possibilities for work-camp-learning experiences.

If the military can provide a bridge from high school to college, then why can't conservation units be established that do essentially the same thing. Courses in Natural History or Ecology could be integrated into the actual "hands-on" learning experience. College credits could be included. It would be a part education, part productive work program that would give participants a sense of pride.

Certainly, a course in Natural History or Ecology should be a requirement before a student graduated from university. There is a school of thought that people have at least Eight Intelligences. These include things like linguistics, music, interpersonal skills, and the eighth area which is nature smarts. It is not too much to hope that any college graduate would have been at least introduced to the natural world in which they spend a good part of their natural life.

SEED #7. In every state, there should be recognition for genuine effort for conservation of natural resources. This might be a "Governor's Award that provided recognition for outstanding stewardship." The state Public Television network could carry an evening program where the nominees were presented and the winner announced and presented. Conservation achievement absolutely needs the public attention supplied by mass media. North Carolina would be an example where this is now being done.

SEED #8. Support, encourage, participate in any boni fide effort to provide children and the younger generation with opportunities to experience the personal freedom of the natural world. There is a growing body of evidence that lends credence to the idea that healthy minds and bodies are more likely if there has been active exposure to nature; if there has been a nature-young person reunion.

Today's youngsters are surrounded with ersatz nature in shopping malls, TV programming or internet information. Most of their physical activity consists of programmed or organized sports. Sex is exhorted as the only sense in a diminished life of senses. We really feel we are awash in data and knowledge. The world has become small and we have all the answers. Young people often use the phrase, "been there and done that." But, as the sign above Albert Einstein's door said, "Not everything that counts can be counted, and not everything that can be counted counts." In short, our younger generation is being shorted the natural experience so important to the balanced young mind and body.

Closely related to the effort to put more green learning into the education of young people is the green effort to put more balance into the environment where most people live. David Orr is a professor of environmental studies at Oberlin College in Ohio. He is a leading exponent of "design intelligence" as part of the emerging green movement in the United States. He speaks of a "sane civilization that would have more parks and fewer shopping malls, more small farms and fewer agribusinesses, more prosperous small towns and smaller cities, more solar collectors and fewer strip mines, more bicycle trails and fewer freeways." His point is that we need "greening," not just in schools, but in our every day living. In Jacksonville, Florida, the Trust for Public Land is leading that city in a four point green printing process that includes such steps as intense public conversation, locating ways to pay for land to be bought, and then to identify actual land to be used for the greening of the city. The aim is a city wide effort to get nature back into the urban environment.

SEED #9. Greater cooperation between the private sector, media and advertising. We have all sorts of incentives for people to sign up to win a trip to Disneyland. Why not a trip to Yellowstone Park or a fishing trip on the Smith River in Montana. Or a photo safari in a state or national park? Why can't we promote a national conservation ethic? It might put a whole new perspective on taking care of the earth. Conservation needs national media exposure.

The United Campaign has a close alliance with professional football. That effort is aimed at helping a human resource. Why not a similar alliance with a sporting event that would focus on natural resources? Why not connect outdoor recreation to a national sport and make natural resource conservation a little more of a national pastime?

Make more of what we already have. We have an Arbor Day but in most parts of the country it is a well kept secret. Can you imagine how the consciousness for stewardship would be raised if every community had some sort of Arbor Day celebration. Communities could be ranked by population and awards given to the communities that did the best job of creating awareness of the value of trees?

SEED #10. And this last seed involves socio-economics. The short term way to make money is to maximize resources, either human or natural. Our economic system has provided examples of both. There isn't a Chinaman's chance . . . a phrase from the coolie legacy of western railroad lore, that a silver tongued environmentalist would convince an audience of American business people they ought to quit making money. The Puritan preference for money over natural resources is too deeply imbedded in our American culture. But, if the business community becomes convinced or can be convinced, they can do as well or better by using valid conservation practices, there is a possibility they might come to embrace another view of their bottom line.

In Short . . . We Simply Must Find More Ways To Make Conservation Profitable, Short Term And Long Term.

I think this seed has already been sown. The popular phrase now in business circles is "being green." There is actually a list available of Fortune 500 companies that are trying to improve their environmental efforts and at the same time improve their bottom line. These include the ones we have mentioned, but also would include names like Intel, UPS, GE, Dupont and Goldman Sachs. These companies are discovering something. We have clung to a long standing tradition in business circles in this country that stewardship of natural resources or conservation of natural resources is costly. Our businesses and our infrastructure are learning now there is very little truth to that bromide.

We have used a little, grass roots model in this book to explain the connected relationships of economics, social contracts and government. It simply visualizes that, singly or when social contracts, socio-economics and government put maximum gain ahead of natural resource value, the natural resources will suffer. That has been our pattern. A creative thinker once suggested we could have solved part of the industrial abuse of water by making an industrial plant take water from below their operation on the river rather than using water from above the plant. In the world of agriculture, the incentives encouraged people to use marginal land or to produce more with the use of fertilizer and pesticides. Sometimes, in retrospect, it seems the farmer was encouraged to create waste. The incentives have favored the abuse of the land, not the care of the land.

Granted when you put care of natural resources, or a wiser use of resources into a business plan, you introduce more complexity. It was easier to ignore this dimension. It was easier to fence that idea off. Then business would "circle the wagons" and fight off the hostiles. By adopting this mentality, we created, especially in the business world, cognitive and perception boundaries. There were two earmarks of this boundary of the imagination. One was the perception of "us versus them" where there was no common ground. Protection of the status quo was the natural result. This was even true of the feeling between the hunting/fishing people and certain conservation organizations.

From Seeds to Conclusions

One evening while tilting at slowly twirling aluminum cans with my cowboy friend, he took a long sip and tapping with the can as he spoke, he said, "well, old philosopher, what do you think about all this environmental stuff by now? How would an old sports announcer put it?" You need to keep the response to this fellow short. He doesn't like longwinded pronouncements. Say it like a sports announcer calling a play.

This was the response. We have now played the first half of our conservation game in this country. There has been a lot of pushing and shoving. Our ground game has not been good. We have plowed a lot of holes for ourselves. Our game through the air hasn't been much better. We have thrown some long bombs that produced a lot of smog. Made it hard to see the game if you were sitting up high. Mostly we have relied on our defense to make a statement for us. We are good at defending positions. The second half is coming up. A lot will depend on how much we want to win. Both the extractionists and the environmentalists seem to be able to stage late rallies. It just boils down to which side wants to win a big game.

He tapped his can. "I've been thinking," he said, "bout moving that horse corral back from the river. Probably is a lot of manure that washes into the water. I don't eat fish so it's of no count to me. Means moving a lot of posts, ya know. Where it is makes it easy for the horses to drink. . . . lots of work to move all those fence posts." He tapped the bar with his empty can and two fresh cans appeared. I heard echoes of bigger conservation issues.

WHAT ABOUT SPIRITUALITY?

Is it a part of any examination of natural resources? Robert Frost, in his pastoral lines about New England, makes references to "the spires." In the search for permanence and security, the sturdy Vermont settlers found time to build churches. The "spires" still dot the landscape providing inspiration for poets. While we preached state and church separation, we wove both into our relationship to natural resources.

Travel the reaches of what was once prairie grassland and it is not unusual to come upon a large, limestone church. I found such an edifice one day in the remnants of a small, western Kansas village near the Kansas/Nebraska line. It was standing guard over a single remaining convenience store and a two pump gas station. It was a testimony to the euro American who found no permanence in a land of 18 inches of average rainfall. It was a limestone tabernacle that stood like a prairie Parthenon. The bell tolled for a congregation on the underside of the grass.

Does the Anglo-Saxon mind find a certain balm in building churches of stone? We can stand back and consider them at a distance. Perhaps building things of stone and steel gives us a sense of permanence. Man seems to be able to handle coded law because it strokes his own power or comfort. Spiritual truth is harder to handle because it diminishes his arrogance, his urge to own, the power of ownership, the press to believe "it's my property and I can do with it what I want."

Now there seems to be a growing movement in this country to link nature and theology. In his book, LAST CHILD IN THE WOODS, Richard Louv traces the cross pollination of churches, religious groups and religious leaders who are speaking of something called "Biblical Ecology." It has become a faith based environmentalism. The idea is that as human beings we have dominion over the earth, whether we like that responsibility or not. Louv believes "the coming decades will be a pivotal time in Western thought and faith. For students, a greater emphasis on spiritual context could stimulate a renewed sense of awe for the mysteries of nature and science."

THE GREAT LESSON OF GRASSROOTS ECOLOGY IS RENEWAL

If you believe the supply of Red Snapper in gulf Coast waters is inexhaustible, you will have little concern for the next season. The sustainability of the resource is not part of the equation. In fact, you can destroy the smaller fish and keep only the larger ones. This has been the standard practice for Cod fishing off the coast of Maine and Red Snapper fishing in the Gulf. Basically, there was no thought given to sustaining the resource.

This lack of concern for sustainability was present in prairie country. The common idea was to extract the grass, or crop, for use this year. "Next year was another year." Not everybody practiced the ecological ethic that "you always leave it as good as, or better, than you found it."

Sustainability and renewal are the two keys to the stewardship of natural resources. This is nature's law. When mankind breaks this law of nature, things get messy. This principle of stewardship is an ecological ethic we need to hear often enough that it is renewed and sustained.

This lesson is hidden in the saga of the small visitor who uses the nectar, leaves the flower some pollen, creating a mutual stewardship that ensures renewal.

This idea of renewal is a seed. It must be sown in each generation to maintain the cycle. It must be sown in a hospitable environment that provides purpose. Renewal is not something that happens once and then is packed away. It needs constant care, nurturing and nourishment. It is much more than temporary appreciation.

Renewal is the vital link in the pattern of continuity. It is indispensable to natural resources. It becomes a basic choice. Either we elect to let the earth take care of us, or we decide to take care of the earth, the air and the water. That is the message of *Grassroots Ecology*.

REFERENCES

Albers, Jan. *Hands On The Land.* (MIT Press) 2002
Babbitt, Irving. *Democracy and Leadership.* (Houghton Mifflin Company) (1953)
Cronon, William. *Changes in the Land: Indians, Colonists, and the Ecology of New England.* (Hill and Wang) 1983

Henretta, James A. *The Evolution of American Society, 1700–1815: An Interdiscipli-nary Analysis.* (D C Heath & Co) 1973

Schorger, A. W. *The Wild Turkey: Its History and Domestication.* (University of Oklahoma Press) 1966.

Williams, Rev. Samuel. The *Natural and Civil History of Vermont.* Walpole, N. H. 1794

Richard Lour. *Lost Child in the Woods.* (Algonquin Books of Chapal Hill) 2006

Index

Abenaki Indian tribe: composition of, 64, 65–66; contrasted with Ponca/Shoshone people, 75; "dark years" options of, 69–70; European diseases and, 66; Frederick Matthew Wiseman on, 64–65, 67, 69–70; lifestyle of, 67–68; Pilgrims and, 65; prairie settlement of, 86; Samoset of, 65; Squanto of, 65

accountability and conservation, 294–95, 299, 330–31

Adams, John Quincy, 36

Adobe Systems, 319

agriculture: agribusiness subsidies, 100, 104; American compared with Russian, 178–79; culture and, 86; hardships of, 235–36

agro ecology, 180

Alaska, 258, 260–61

Albright, Horace Marden, 301, 311–12

Aldo Leopold Wilderness, Gila National Forest, 148

Algonkian Indian tribes, 16

American Conservation Program, 108

American creed of conservation, 32–33, 35–36

American culture: and ecological ethics, 28; and environmental law, 15; and natural resources, 2, 11–12, 14, 33, 146–47, 220; natural value considerations of, 317–18

American Farm Bureau, 180–81, 281

American Scholar magazine (1944), 133

amphibians, decline of, 9–11

Anaconda Copper Mining Company, 15

Anheuser-Busch, conservation effort of, 320–21

Appalousian (Palousian) horses, 87

Arapahoe Indian tribe, 89

Arbor Day, 223, 333

Archery Manufacturers and Merchants Organization (AMO), 203

arid western lands: commonality in, 245; demands of, 253; development in, 248; forces of nature in, 252–53; John Wesley Powell and, 245–46, 248, 252–53; mobility and, 250–51; natural resource extraction, 248–49; state rights and land use, 248; Taylor Grazing Act (1934), 253; water and, 244–45, 247–48

Ashley, General William H., 165

Atlantic Brant and eelgrass, 213–14

Atlantic Monthly magazine (2003) article, 104

Attwater's prairie chicken, 292

Audubon Society, 108

Babbitt, Irving on legalism in conservation, 290

barbed wire, 20

Ratt, Bruce on conservation of waterfowl, 293–94

beaver (Castor Canadensis): folklore and, 257; natural resource model and, 256; stewardship of, 256; trapping of, 255, 256–57

Beck, Russell on wolves, 276

Bennett, Hugh Hammond, 99, 120–21

Bessey, Charles Edwin: biography of, 174–75; importance of, 124–25; work of, 221–22, 223

Bigheart, Chief James, 137

Billings, Frederick, 4, 258

biological population limitations, 241

bison, 139, 140, 141–43, 259–60

Bodmer, Karl, 75

Bradford, William, 66

Brandis, Sir Dietrich, 308

Bridger, Jim, 14, 89, 93

Brokaw, Tom, 5

Brower, David, 148–49

brucellosis, 142

Bryan, William Jennings, 178

Bryson, Amy Joi, 167

Buffalo Bill, 171–72

Buffalo Bill Historical Center, Cody, Wyoming, 172

Bullock, Selph, 224

burning of natural resources: economic decisions in, 154–56, 157–58; ecosystem benefits of, 154–55; effect of in Vermont, 153, 154; ethics and, 156; of forest land, 153, 156; grapes and, 154; impact of, 154; lesson learned from, 158; patterns of, 153; of prairie land, 153, 156–57; public indifference to, 158–59; social contracts and, 156, 157; stewardship of natural resources and, 156; in Tall Grass Prairie Preserve, 138–39; U.S. Forest Service on, 157–58; wildfire, 157–58

business: after cost of, 335; conservation and profits of, 333, 341; conservation efforts of, 320–24

Business magazine, 319

Cadillac Desert, Marc Reisner, 105

California: California Water Project, 100–101; Stanislaus River dam, 238; water use in, 103

Callenbach, Ernest on the conservation ethic, 218–19

The Caller magazine, 331

capitalism and land use, 5, 7, 60–62, 66, 75–76, 79–81

Carson, Rachel, 31, 182–83, 208–9, 310

Carter, A. J., 76

Catlin, George, 16, 40, 75, 144, 163–64

cattle, reduction of inventory of, 106

Cayuse Indian tribe, 87

Center for Great Plains Studies, 118

A Century of Dishonor, Helen Hunt Jackson, 20

Chabot, Anthony, 258

Champlain, Samuel De, 66

Changes in the Land, William Cronan, 60

Chapman Barnard Ranch, 137–38

Chapman, John, 170, 170–71

Chase, Alston on Yellowstone Park, 53

Cherokee Advocate newspaper, 21

Cherokee Indian tribe, 86, 90

Cherokee Phoenix newspaper, 21

Chesapeake Bay cleanup, 102

Cheyenne Indian tribe, 86, 88

Chickasaw Indian tribe, 86

Chisholm, Jesse, 166

Choctaw Indian tribe, 86

Chronic Wasting Disease (CWD), 236, 270–71

Civilian Conservation Corps, 226

Clements, Frederic, 124, 125, 126

Clements, Frederic study of prairie land, 111, 125–27

Cleveland, Grover, 222

climate change, 11, 192, 207

The Closing Circle, Barry Commoner, 310

Cody, William F., 172

colonialism and land use, 60–61

Colorado: cattle inventory, reduction of, 106; Colorado State University, 320; eco-tourism and, 240; fish management in, 193; Fort Collins, environmental efforts of, 320, 321; Greeley, Colorado, 247; South Platte River, damming of, 240; water law and, 247; water use legislation, 328

Colorado River, 102

Colt, Samuel, 16

Colter, John, 17

Comanche Indian tribe, 88

Commerce of the Prairies, Josiah Gregg, 21

common good ethics, 30

Commoner, Barry, 194–95, 310

communications and ecology, 58–59

Conrad, C. E., 259

conservation: Aaron Montgomery Ward and, 15; accountability and, 294–95, 299, 330–31; American culture and, 2, 11–12; boundaries and, 59–60; capitalism and, 5, 7; Chesapeake Bay cleanup success, 102; conclusions concerning, 334–35; Conservation Reserve Program, 116; continuity and perpetuation, 32–33, 35–36; current land use and, 103–4; economic consumption and, 31–32, 218–19, 259; of elk (Wapiti), 266, 267; ethic, as paramount, 218–19; ethic of spiritual conservation, 30–31; ethic viewpoint of, 67; ethics, muddle of, 32; ethics of, search for, 11–12, 27; Forest Preserve Act (1891), 222; grassroots support for, 289; "greatest generation" and, 4, 7; greed and, 31–32; Henry David Thoreau and, 84; hunters/fisherman supported, 6–7; Inland Waterway Commission, 103; input/output mentality in, 291–92;

John Muir and, 81–83; land ownership and, 118–19; legalism in, 290, 299; local chapters, necessity of, 290; moral order in, 31; Native Americans and, 94; organizations, transition of, 290–92; origin of ethics of, 31; religion, commonality with, 289–90; spirituality and, 133–34, 231; sportsman, evolution to, 28–30; stages of, 6; sustainable and, 215, 226–27, 235; Timber-Culture Act (1873) quoted, 222; Wal-Mart and, 321–23, 324; water conservation, early efforts in, 102–3; water conservation, efforts to implement, 114–16; of water in prairie country, 99–101; water measurement systems and, 101–2; of waterfowl, 206, 291, 292, 293–95

Conservation Reserve Program, 116

conservation, suggestions for improvement of: create a Department of Natural Resources, 329; create a state Natural Resources Confederation, 329–30; create river compacts, 328–29; eliminate management from a distance, 331; encourage cooperation between private sector, media and advertising, 333; find ways to make conservation profitable, 333–34; improve accountability of management, 330–31; improve environmental education, 332–33; recognize efforts for natural resource conservation, 332; re-create a version of the CCC, 331–32

conservationist, evolution of, 28–30

consumption mandate, 31

Corps of Engineers, 79, 80, 81

Cowboy symbol, 165–66

Creek Indian tribe, 86

Crocker, Bull, 19

Cronan, William on colonialism and land use, 60, 61

Crow Indian tribe, 89

dams: American Falls Dam, 239;
America's fixation with, 237–38;
failure of, 239–40; fundamental
questions involving, 327; irrigation
and, 240–41; Jackson Lake Dam,
239; on the Missouri River, 238;
Snake River dams, 238–40; on the
Stanislaus River, 238; Teton Dam,
238, 239–40; unnecessary, 238, 239;
value of water and, 298
Dawes Act, 18
DDT (dichloro-diphenyl-
trichloroethame), 182–83, 208. *See
also* insecticide spraying, effects of
Delta Waterfowl organization, transition
of, 290–92
Delta Waterfowl Report, 206, 291
Democracy in America, Alexis de
Tocqueville, 190
Denali National Park, 52–53, 269
Desert Land Act (1877), 16
Deutschlands I-fanzengeographie, Oscar
Drude, 125
Devney, John L. on waterfowl
management, 291, 294
DeVoto, Bernard, 15, 245
Dieffenbach, Bill on flood control, 80
Ding Darling, 285–86
Dismal River Reserve, 223, 226
Dodge, Major General Granville, 14
Downes, William, 281
Drude, Oscar, 125
Ducks Unlimited: evolution of, 52,
215–16; "greatest generation" and, 5;
Greenwing program, 36; new focus
of, 193; social contract and, 191;
sports/harvest ethic and, 28, 32;
Waterhen Marsh and, 192; work of, 7
Dutton, Ron, 157
"Duty of Governments in the
Preservation of Forests", Franklin B.
Hough, 221

E. W. Carter Oil Company, 140
Earth Day, 195

Earth Island Institute, 148
Echo Park dam, 105
ecological ethics: allegory of the birds,
25; American culture and, 33; Bambi
ethic of, 28; burning of natural
resources and, 156; common good
ethics, 30; conservation ethic and,
67, 215, 218–19; economic gain
viewpoint of, 26, 154–56, 180;
Leopold land ethic, 188; list of, 34;
material satisfaction viewpoint of,
26–27; Missouri River ecosystem
and, 35; morality and, 27, 28, 188;
Old Testament viewpoint of, 26;
property rights viewpoint of, 27, 50,
128, 190–91; regional ethic, 34, 180;
restoration and, 43; science and, 27,
33; search for, 33–34; social contract
and, 26–28; space and, 218–19;
spiritual, 30–31, 124; sports/harvest
viewpoint of, 28; tourism and, 35,
41; utilitarian viewpoint of, 27;
weapons and, 33; wildlife
management and, 218. *See also*
theoretical ecological ethics
ecological model: components of, 48;
Denali National Park and, 52–53;
eco-tourism and, 51–53; evolution in,
52; evolving nature of, 3; extraction
and, 259; five stages of, 44, 46–47;
government and, 48, 51–52, 53;
management versus alignment in, 52;
of Native Americans, 54, 56–58;
natural resources defined in, 48;
organizational model for, 44, 45;
original idea of, 44, 46; original
tallgrass prairie and, 47–48; prairie
model, 49, 131, 138; social contracts
and, 50, 51, 53; socio-economics in,
48, 50–51; validity of, 54; in wildlife
management, 216–18; wildlife
populations and, 52–53
ecology: agro ecology, 180; burning
and, 153, 154–56; business
relationship to, 51; communications,

role in, 58–59; compressed issues of, 7–8; concept of (emergent), 8–9; connectedness in, 58; definition of, 3; determinants of, 3; economics and, 154–56, 169–70; education and, 36–37; empathy and, 188; first ecological issue, 69–70; first text on, 8; grassroots ecology and renewal of, 336; grassroots philosophies of, 241–42; humanity and the principles of, 9; hunting/fishing, impact on, 205; morality and, 30–31; parable of the hunters and, 2, 9, 155, 186–87, 193, 322; of prairie land, 234; restoration and, 43; scientific ecology, 128–33; Shelterbelt Program, 16, 99, 226–27; sociology and, 128–29; space and, 110–12; sustainable, 31; symbols and, 166; weapons and, 33, 66

Ecology and Field Biology, Smith, 8
ecology and measurement: American dream and, 96–98; ecological patterns of, 110; pattern of squares in, 97–99; population shifts and, 103–4; water and, 99–102, 106–7
ecology and sociology, 231
ecology and water: agribusiness subsidies and, 100, 104; California Water Project, 100–101; current land use involving, 103–4; dam building, 79–81, 101–2, 105, 238–41; development in the western states and, 248; eco-tourism and, 240–41; Interior Department mismanagement, 105; irrigation, 101, 104, 115–16, 240–41, 327, 328; John Wesley Powell and, 246, 248; land measurement and, 101; legal issues in, 101–2, 246–47, 249; mentality involved in, 101; mismanagement of, 105; Nebraska, shift in social contract of land use, 114; Ogalala aquifer, 100; in prairie country, 99–101, 105; ratio of water to

pounds of cow, 100; salt problem, 102; Shelterbelt Program, 99; silt management, 102; water conservation, early efforts in, 102–3; water law and, 101, 246–47; water restoration cost, 102; water rights, 247, 249; wetlands drainage, 116. *See also* arid western lands; water
Ecology, Ernest Callenbach, 218
Ecology of a Cracker Childhood, Janisse Ray, 231
The Ecology of Commerce, Paul Hawken, 51
economics and ecology, 154–56, 333
economics and forest management, 212
economics and wildlife management, 212–13
ecosystem manipulation, 261–62
ecosystem model components, 48
eco-tourism: in Colorado, 240; effect on the environment, 35, 41; ecological model and, 51–53; elk (Wapiti) and, 267; forest management and, 228–29; in the Great Plains, 104; in Vermont, 51, 237; water and, 240–41; in Wyoming, 51
eelgrass, 213–14
Egleston, Nathaniel H., 221
elk (Wapiti): adaptability of, 267–68; conflicts in natural resource thinking and, 266, 270–72; criticism of present management systems of, 271–72; economic opportunities and, 272–73, 280–81; eco-tourism and, 267; hunting of, 266, 271–72; management of, 267–73; restoration of, 266; revenue from hunting of, 272–73, 280–81; Rocky Mountain Elk Foundation and, 267; social contracts and, 267; wolves and, 271, 273; Wyoming and, 268–69
Emerson, Kirk, 55
Emerson, Ralph Waldo, 83, 306
empathy and ecological ethics, 188

Endangered Species Act, 35, 278, 279–80, 283, 309
endangered species process, 213
environmental law and American culture, 13
environmental quality, 3
Environmental Quality Incentives Program (EQIP), 184
environmental seminars and conferences, 3
extraction: American patterns of, 32; limit to, 220; of minerals/gold, 15, 18, 31, 248–49, 257–58; of oil, 14; of timber, 258–59

faith based environmentalism, 335–36
Farm Bill (1985), 116
Farm Bill (1995), 117
Federal Duck Stamp program, 286
Fellers, Gary, 11
Fernow, Bernard E., 221
Field and Stream magazine, 108, 282, 318
Fire in America: A Cultural History of Wildland and Rural Fire, Stephen J. Payne, 157
fish conservation efforts, 193, 296–97
fishing: conflict in, 200; ecological impact of, 205; natural resources and, 5–6; social contract in, 201
the Five Civilized Tribes, 86, 86–87
Fontaine, Joe on wolves, 284
Forest and Stream magazine, 108, 214
forest management: Albert C. Worrell on, 161; Appropriations Act (1876) and, 221; burning forests, 153, 156, 157–58; college education in, 221; decentralized management, 160–61; determinants in ecosystem and, 160; early Forest Service, 224–25; economics and, 212; eco-tourism and, 228–29; forest fires, cost of, 212; Gifford Pinchot's view of, 159–60; Halsey National Forest and,

223, 225, 226, 227, 229–31; lack of consensus on, 160; management philosophy of Forest Service, 157–58, 160, 230–31; models for, 212; Nebraska and, 221 24; public apathy for forest resources, 225–26; public indifference to, 158–59; question of purpose of, 159; science and, 230; Timber-Culture Act (1873), 222. See *also* stewardship of natural resources
Fort Randell dam (South Dakota), 81
Fortune Magazine on Wal-Mart, 322
Foster, Henry, 137
Francis, Charles on argo ecology, 180
Fraser, James Earl, 172
Fried, Jeremy on forest management, 212
Friends of the Earth, 148
Frost, Robert, 335
Fund for Animals, 55

game management, quandaries of, 213–15
Garrison Dam (North Dakota), 80
Gerhardt, Herb, 224, 228
Gilbert, Paul, 115
Girard, Kelly, 320
Glidden, J. F., 20
global warming, 192
Goterl, Fred, 10
government and ecology, 48, 51–52, 53
Graft, Darrell on wolves, 281
grandmother's quilt: Aaron Montgomery Ward patch, 15; Alice Fletcher patch of, 18; Asa Whitney patch of, 22; barbed wire patch of, 20; Cherokee Phoenix patch of, 21; commonality in, 22; corporate exploiters patch, 15; Cottonwood Creek patch of, 20; David Love patch, 17; Delaware Indian tribe patch, 16; Desert Land Act (1877) patch of, 16; frontier myths and symbols patch, 14;

George Catlin patch of, 16; "glory day" patch of, 19; gold patch of, 18; Helen Hunt Jackson patch of, 20; Jim Bridger coal fired electric plant patch of, 19; John Colter patch of, 17; John L. Mason patch of, 21; John Muir patch of, 22; John Wesley Powell patch of, 21; Josiah Gregg patch of, 21; legal patch of, 15; Mary Elizabeth Lease patch of, 15; mystery patch of, 18; nature as a version of, 23–24; packers/stickers patch of, 15; petroleum patch of, 14; Press Clay Southworth patch of, 17; railroad patch of, 20; Samuel Colt patch of, 16; Shelter Program patch of, 16; stone fence builders patch of, 18; story of, 13; vigilante law patch of, 19; westward expansion patch of, 14, 19; Yellowstone Park patch of, 14
grasses, preservation of, 118
grassland evolution, 124–26
grassland soils, complexity of, 120
grassroots ecology lesson, 336
"greatest generation" and conservation, 4, 7
greed and economics, 31–32
Greenwood, Ken, 339, 340
Grenell, George Bird, 107–8
Guernsey Reservoir, 101
Guess, George (Sequoyah), 21
Gustafson, Ron on reduction of cattle inventory, 106

habitat conditions, determinants of, 126
Hall, William L., 222, 223
Halsey National Forest, 223, 225, 226, 227, 229–31
Hamilton, Bob, 138, 140, 144
Hammit, Frank N., 225
Hands on the Land, Jan Albers, 70
Harrison, Benjamin (U.S. President), 222

Hawken, Paul on natural resources, 51
Heavey, Bill on social contracts, 318–19
Hebard, Grace Raymond on Chief Washakie, 93
Hochbaum, H. Albert, 289, 291
Hogenson, Jeff, 269
Horak, Michael on trade-offs in conservation, 293
horse culture, 87
Hough, Emerson, 16, 214, 221, 263–64
Howard, Joseph, 15
Howe, Elias, 16
hummingbirds: attributes and habits of, 315; bees and, 315; co-evolution with flowers, 315–16; the environment and, 315–16; human interference with, 316, 317; implications of research on, 314; longevity of, 316; migration of, 316; predation of, 316; spiritual mark of, 314
hunting: conservation and, 204–5; dissonance in, 199–200; ecological impact of, 205; economics and, 201; future of, 206–7; important issues in, 207; licenses for, 6, 204; mania for, 200–201; for market, 191; public perception of, 204; reasons for, 203–4; shift in purpose of, 202–3; social contract in, 201; space limitations for, 202; technological changes in, 202; waterfowl management and, 206
Hutcheson Memorial Forest, 151–52
Hutchins, Thomas, 97
Hutchinson, Wallace I., 226

Ickes, Harold, 105, 301–2
Idaho: Snake River irrigation, 328; water use and, 239, 240, 247
The Inconvenient Truth, Al Gore, 192
Indian treaties, 76
Inland Waterway Commission, 103
insecticide spraying, effects of, 182–84

Interior Department mismanagement of water, 105
irrigation, 101, 104, 115–16, 240–41, 327, 328
Izaak Walton League, 5, 52

Jackson, Helen Hunt, 20
Jackson Hole National Monument, 301
Jamenez, Mike on wolves, 284
Jefferson, Thomas: biography of, 72–73; environmental creed of, 36; on importance of ideals, 57; as a naturalist, 73; pattern of squares and, 97, 106; portrait of, 72; prairie land and, 233–34; his view of history, 289
Jim Bridger coal fired electric plant, 19
John L. Mason, 21
Johnson, Lyndon Baines, 309
Johnston, Velma B. ("Wildhorse Annie"), 167
Josiah Gregg, 21
Journal of a Mountain Man, Russell Osborne, 89
Joyce Kilmer Memorial Park, 152

Kansas: Algonkian Indian tribes and, 16; raising corn in, 15; stone fence builders of, 18
Kaw Indian tribe, 86
Kay, Charles on natural regulation, 271
Keigley, Richard on overgrazing in Yellowstone National Park, 270
Kellogg, Royal S., 223
Kelly, Matt on wolves, 282
Kemble, Edward C., 76
Kincaid Act (1904) effect, 115
Kings Canyon national Park, California, 301
Kroegel, Paul, 205

Lambert, Hunt on clean energy research, 320
land ethic of Aldo Leopold, 28
land giveaways, 177

Land Institute's Prairie Writers Circle, 179
land retirement programs, effects of, 117–18
land use: American viewpoint of, 79–80; capitalism and, 60–62, 66, 75–76; colonialism and, 60–61; dam building and, 79–81, 238–41; federal programs and, 116; land leasing, 248; Mississippi River Commission, 79; Missouri Basin Program, 79–80; Native Americans and, 79; in Nebraska, 113–16; Standing Bear legal decision and, 78; westward expansion and, 60–62
Landis, John on wolves, 276
Last Child in the Woods, Richard Louv, 335
law, abuse of, 15
Lease, Mary Elizabeth, 15, 177
Leff, David K. on natural resources and reality, 311
legalism, 129, 290, 299
Leopold, Aldo: importance of, 230, 310; life and work of, 147–48; Wisconsin sand country and, 309
Leopold land ethic, 28, 30, 188
Lewis and Clark expedition, 198, 234
Lind, Michael on Great Plains farmers, 104
Little Santeetlah Forest, 152–53
Living on Principle, Wallace Stegner, 244
Locke, John, 155
Lococo, Andrea on wolves, 282–83
Love, David, 17

malaria eradication, 183
Man and Nature, George Perkins Marsh, 4, 32, 71, 126, 236
Manderville, John, 227
Mann, Luther (Shoshone Indian agent), 90
Marsh, George Perkins: biography of, 70–72; ecological foresight of, 4, 31,

63, 237, 242, 306; environmental
creed of 36; portrait of, 71; on waste
of natural resources, 236–37
Marshall, Robert, 161–62
Mather, Stephen Tyng, 262–63
Mathews, John Joseph, 136
McCormick, Cyrus, 16
McCormick, Stephen J. on trade-offs in
conservation, 293
McGee, W. J., 103
McKay, Douglas, 105
McKinstry, Kathie, 168
Michigan, 202
Migratory Bird Hunting Act, 286
Miller, George L., 77
Miller, L. C., 224
Mississippi River Commission, 79
Missouri Basin Program, 79–80
Missouri River dams, 238
Missouri River ecosystem, 35
Mobil Oil Company, 292
Montana: fish management in, 193;
Kalispell, Montana, 259–60;
Madison River, protection of,
297–98
Montgomery Ward, 15
Moody, Dave on wolves, 283–84
moose and wolves, 284
morality, 27, 28, 30–31, 188, 190
Morine, David on trade-offs in
conservation, 293
Mormons, 238–39
Morton, J. Sterling, 223
Mound Builder culture, 86
Mountain States Legal Foundation,
281
Muir, John: birthplace and year of, 81;
death of, 82; his ethic of
preservation, 31; history of, 22,
81–83, 222: portrait of, 82
Mussman, Dave on his financial losses
in farming, 236
myths: Farm Bureau myth, 180–81; of
promises and performance, 180–85;
of the small freeholder, 177–78

Nachtigal, Jeff on Adobe Systems
environmentalism, 319
National Academy of Science, 35
National Wildlife Federation, 52, 286
National Wild Turkey Federation
(NWTF): "greatest generation" and,
5; Jakes program of, 36; on
management of national
forests/grasslands, 331; restoration of
the wild turkey, 295–96;
sports/harvest ethic and, 28, 32
Native Americans: agriculture and,
86–87; conservation and, 94; corn,
their use of, 78–79; cultural shift of,
87; as expendable, 58; George
Washington and, 56–58; horse
culture of, 87; land use of, 59–60,
74, 79; military actions against,
87–88; rights to water, 101; social
contracts and, 91; social contracts of,
92–93; social structure of, 91;
spirituality of, 30; status of children,
92; status of women, 92; use of
natural resources, 54, 56–58, 91;
view of natural resources, 299–300;
warfare of, 86
natural resources: absolute use of, 5;
American culture and, 2, 11–12, 14,
15, 33, 220; application of science
to, 128–33; boundaries and, 59–60;
business aftercosts and, 51; depletion
of, 179–80; determinants to use of,
94; economic consumption and,
31–32, 61, 259, 317; federal support
for, 6; fundamental questions
involving manipulation of, 327;
hunters/fisherman supported
conservation for, 6–7; hunting and
fishing and, 5–6, 205; legalism and,
129; Native American use of, 54,
56–58, 91; naturalist thinking, 311;
public apathy for protection of,
214–15, 225–26; reality and, 311;
stewardship of, 143–44, 218;
suggestions for management of,

218–19, 328–34; sustaining of, 215, 226–27, 235; westward expansion and, 60–62
natural value considerations, 317–18
naturalist thinking: of Barry Commoner, 310; differences in, 306–7; evolution of, 306–8; of Frederick Law Olmsted, 306–7; of George Perkins Marsh, 306; of Gifford Pinchot, 307–8; government and, 307–10; of Henry David Thoreau, 304–6, 311; of John Wesley Powell, 307; lack of cohesion in, 308–9; science and, 310; of Stephen Mather, 307; of Theodore Parker, 305; of Theodore Roosevelt, 307
The Nature Conservancy: evolution of, 52, 143, 292–93; greed and, 293; Mobil Oil Company and, 292; selling of bison, 142; shift in philosophy of, 143; Tallgrass Prairie Preserve and, 136, 138, 143; work of, 139
Nebraska: agricultural economic loss (2002), 236; alteration of space in, 113–14; Arbor Day and, 223; botanical study of, 128–29; Center for Great Plains Studies, 118; compared with Vermont, 237; Conservation Reserve Program, 116; deer hunting in, 227; Dismal River Reserve, 223, 226; ecological conservation future in, 118–19; Environmental Trust, success of, 119; Farm Bill (1985), 116; Farm Bill (1995), 117; floodplains converted to croplands, 114; forest management and, 221–24; Game, Fish and Park Department of, 117; Halsey National Forest, 223, 225, 226, 227, 229–31; Holt County experiment, 223; improved acreage in (1900), 127; irrigation in, 101, 115–16, 327; Kincaid Act (1904) effect, 115; land ownership in, 118; land retirement programs and, 117–18; Paul Gilbert and, 115; preservation of grasses, 118; reduction of cattle inventory, 106; Russian Thistle in, 127; Salt Creek Project, 114; Sandhill Crane Refuge, 212; shift in social contract of land use, 114; small farmers of, 117; University of Nebraska, 127, 132; virgin prairie destruction, 127–28; wetlands drainage, 116; wildlife management in, 115
NebraskaLand magazine on environmental issues, 119
Nevada, Storey County, 166, 167
New American Foundation, Washington, DC, 100
New Jersey hunting license law, 6
New Yorker magazine, 108–9
Noble Foundation, 44–45
North American Grouse Partnership, 216
North American Prairie, J. E. Weaver, 120
North Dakota, 80
Norwood, Jay. *See* Ding Darling

Ogalala aquifer, 100
oil sites, remediation of, 140
Oklahoma: concern for water, 326; Magnum, 227; number of dams and, 235; Oklahoma State at Stillwater, 139; Pawhuska, Oklahoma, 135; Ponca Indian tribe and, 76–78; Shelterbelt Program, 16; Tallgrass Prairie Preserve, 135; University of Tulsa, 69
Old Jules, Marie Sandoz, 113
Olmsted, Frederick Law, 27, 108–9, 152, 306–7
Olympic National Park, Washington, 301
Omaha Indian tribe, 74, 76, 86
Omaha World Herald, 77
Ordway, John, 75
Oregon, 169, 240

The Oregon Trail, Francis Parkman, 251–52

Orr, David on "design intelligence", 332–33

Osage Indian tribe, 86, 135–36, 137

Our Natural History, Daniel B. Botkin, 80, 211

Out of Doors magazine on Chronic Wasting Disease, 270

Owens, Wayne (Congressman), 281

Palmer, Dr. Michael, his work on plant community, 139–40

Palmer, Eric on ecological ethics, 28

Palouse Indian tribe, 87

Parker, Theodore, 305

Pawnee Indian tribe, 86

Payne, Harvey, 138, 142

Payne, Stephen J. on forest fires, 157

Pelican Island, Florida, 206

Petera, Pete on wildlife management, 268, 272

Phillips, Frank, 138

Physical Geography As Modified by Human Action, George Perkins Marsh, 71

phytogeography, 125

Pinchot, Gifford: Biltmore experiment and, 152; close association with Theodore Roosevelt, 225; dedication to the Forest Service, 230; importance to Forest Service, 225; influence on Theodore Roosevelt, 307–8; life and work of, 196–97; naturalist thinking and, 308; Society of American Foresters and, 225; viewpoint on forest use, 27, 102, 222

Pittman-Robinson Federal Aid Act, 6

Platte River Recovery Implementation Program, 119

Playing God in Yellowstone Park, Alston Chase, 53

Ponca City, Oklahoma, 78

Ponca Indian tribe: A. J. Carter and, 76; authority of women in, 75; composition of, 64; contrasted with Abenaki/Shoshone people, 75; Edward C. Kemble and, 76; Helen Hunt Jackson and, 20; history of, 20, 76; lifestyle of, 74; Oklahoma and, 76–78; prairie settlement of, 86; resettlement of, 76–77; smallpox and, 74–75; Standing Bear of, 76–78; suffering of, 76

population limitations (biological), 241

Pound, Roscoe, 124, 125, 128

Powell, John Wesley: biography of, 286–87; importance of, 21, 245–46; irrigation and, 128, 248; naturalist thinking of, 307; understanding of arid country, 252–53

Powell's Geographical Survey (1879), 105, 246

prairie chicken populations, 114, 115, 144, 145–46, 292

prairie land: agribusiness subsidies and, 100, 104; American dream and, 96–97; application of science to, 128–31; Arthur George Tansley's view of, 130–31; botanical study of, 128–29; burning of, 47, 49–50, 138–39, 154–55, 156–57; Charles Edwin Bessey's view of, 124–25, 127; determinants of its health, 111, 125–27; Dismal River Reserve, 223; ecology of, 234; as eternal, 133; forests on, 221–25; Fredric Clement's view of, 111, 124–27; insecticide spraying, 183–84; irrigation and, 104, 240–41; life of early settlers of, 129–30; Marc Reisner on prairie water, 105; model for, 49; original condition of, 47; the pattern of squares and, 15; Paul Sears on prairie water, 105; replacement of, 133; Roscoe Pound's view of, 124, 125, 128; shelterbelt programs, 234; short grass, 48; social contract in, 180; Soil Conservation Act (1935), 120; soil productivity,

104; the soul of the prairie, 133–34; sustainability of farming on, 235–37; tallgrass, 47–50; Tallgrass Prairie Preserve, 44–45; Thomas Jefferson and, 233–34; virgin prairie destruction, 127–28; water and, 99–101, 105, 234, 235–37

prairie lore, 74

prairie model considerations: determinants of prairie health, 111–12, 126, 131; grasses, preservation of, 118; grasslands, complexity of, 113, 120; land ownership and, 118–19; land retirement programs and, 117–18; lessons from, 112–14; Nebraska and, 113–16; original condition replication, 112; preservation of grasses, 118; space model in, 111; space models extended, 112; space variable in, 110–12; variables in prairie country, 112–13; water conservation, 114–16; wetlands drainage, 116; wildlife, 113–15

Principles of Forest Policy, Albert C. Worrell, 161

Principles of Sociology, Herbert Spencer, 128

property rights and ecology, 36, 38–40, 42

Putten. Mark Van on wolves, 281–81

quail populations, 114

quails/quail hunting, 318–19

Randolph, Ken, 102

Reese, Hudson, 318–19

Reisner, Marc on prairie water, 105

Reiswig, Barry on elk (Wapiti) management, 271

religion, 33, 289–90, 335–36

Report on the Arid Regions of the West, John Wesley Powell, 21

Research Methods in Ecology, Frederic Clements, 125

The Right Place, E. O. Wilson, 242

Rinehart, Mike on the Endangered Species Act (1973), 284–85

river compacts, 328–29

river restoration (Esopus River, New York), 297

Roberts, Reverend John, 93

Robertson, A. Willis, 6

Robinet, Deb on wolves, 284

Rockefeller, John D., Jr., 300–301

The Rocky Mountain Elk Foundation, 5, 28

Rocky Mountain Wolf Recovery Team, 277–78

Rolfe, John, 54

Roosevelt, Franklin Delano, 99, 103, 173–74, 301

Roosevelt, Theodore: Dismal River Reserve and, 223; Federal Bird Reservation establishment, 206; George Bird Grinnell and, 107; Gifford Pinchot influence on, 307–8; importance of, 121–22; on land use for profit, 309; naturalist thinking of, 307; Nebraska forest reserves, 125

Ross, William Potter, 21

Rousseau, Jean Jacques on social contracts, 187–88, 190, 191

Russell, Osborne on the Shoshone Indian tribe, 89

Russia, agriculture in, 178–79

Russian Thistle, 127

Rutgers University, 152

Samoset, 65

A Sand County Almanac, Aldo Leopold, 148, 309

Sand Hill cranes, 114–15

Santo, Bob on Waterhen Marsh, 192

Saturday Evening Post (June 5, 1915), 214

Sauer, Carl on yield and loot, 9

science, its application to natural resources, 27–28, 33, 128–33, 211, 212, 230, 310

Scientific Monthly article by Robert Marshall, 162

Scott, Charles C., 223, 224

Scott, Lee, 321, 322, 324

The Sea Around Us, Rachel Carson, 208

Sears, Paul on prairie water, 105

Seattle, Chief on natural resources, 299–300

Seminole Indian tribe, 86

Sequoyah (George Guess), 21

Shelborne Farm, 4

Shelterbelt Program (USDA soil conservation), 16, 99, 226–27

Sherman, General William Tecumseh, 88

Shoshone Indian tribe: Chief Washakie, 88, 89, 90, 92–94; contrasted with Abenaki/Ponca people, 75; groups of, 88; intermarriage and, 89; Jim Bridger and, 89; Luther Mann and, 90; natural resources loss, 89–90; nomadic life style of, 64; prophet-chief Ohamagwaya, 88; tribal evolution of, 89–90; US government and, 89–90; Wind River Reservation of, 88, 90

Sierra Club, 52, 148

Silcox, Ferdinand A. on evolving philosophy of the Forest Service, 230

Silent Spring, Rachel Carson, 31, 182–83, 208

Sioux Indian tribe, 74, 76, 78

Smith, Bruce on elk (Wapiti) management, 272–73

Snake River dams, 238–40

Snake River Land Company, 301

social contracts: Alexis de Tocqueville on, 190, 194; character and, 50; consensus and, 187; defined, 46; ecological ethics and, 26–28, 37–38, 188; ecological model and, 48, 50, 51, 53; economic consumption and, 61, 259; elk (Wapiti) and, 267; equality of opportunity and, 190; as evolving, 188; extended to animals, 188–89; extensions of, 191–92; fish management and, 193; hunting and, 201, 206; Jean Jacques Rousseau on, 186–87, 190; law and, 78; Leopold land ethic and, 188; Mayflower Compact, 66; morality and, 188; Native Americans and, 91, 92–93; new social contract, 193–94; property rights and, 42, 50, 190–91; shift for land use in Nebraska, 114; stages of, 50; unbridled development and, 51; wolves and, 277

society as an organism, 128

sociology and ecology, 128–29

The Sod House (anonymous author), 75

Soil Conservation Act (1935), 120

Soil Conservation and Domestic Allotment Act, 174

Soil Conservation Service, 130, 234–35

Soil Erosion Service, 120

soil productivity, 104

Soper, Fred, 182–83

South Dakota: Charles Mix County, 198; Farm Bureau myth in, 180–81; Fort Randell dam, 81; Nemo Burn area, 224; South Dakota Wildlife Federation, 270; Wildlife Federation, 181; Wildlife Federation Magazine article quoted, 184

South Dakota Wildlife Federation, 35, 80

Southworth, Press Clay, 17

spiritual conservation ethic, 30–31, 133–34

spirituality: conservation and, 133–34, 231; faith based environmentalism, 335–36; and the hummingbird, 314–17; of Native Americans, 30; timber management and, 231; and vision, 325–26

sportsman, evolution to conservationist, 28–30

spotted owls, economic impact of, 169–70

Squanto, 65

Standing Bear of Ponca Indian tribe, 76–78

state rights and land use, 248

stewardship of natural resources: conservation ethic and, 215; economics of, 143–44; empathy and, 188; Henry David Thoreau and, 304–6; moral ethic of, 188; paramount ethic in, 218–19; renewable cycles as a consideration in, 156. See *also* forest management; wildlife management

Stone Age people, 85–86

Storebridge, James, 19

The Story of Man in Yellowstone, Merrill D. Beal, 171

"strong father" morality ethics, 30

Stubbendieck, James, 118

Sublette, Kerry on oil remediation, 140

Sutter, John, 257–58

Sutton Avian Center, 145

symbols of America: the bison, 259–60; Buffalo Bill, 171–72; the cowboy, 165–66; and ecology, 166; Johnny Appleseed, 170–71; sculptures of James Earl Fraser, 172; showmanship and, 171; Smokey the Bear, 168; spotted owls and, 169–70; timber harvesting, 259; wild horses and, 166–69; the wolf, 275

Talking to the Moon, John Joseph Mathews, 136

Tall Grass Prairie Preserve: acceptance of, 143; benefits of, 143; bison and, 139, 140, 141–43; burning of, 138–39; changes in, 140–41; Chapman Barnard Ranch, 137–38; description of, 135–36; directors of, 138; investment in, 143; lessons learned from, 146–47; Nature Conservancy and, 136, 138, 143; owners and management of, 136; prairie flowers return, 138;

remediation of oil sites, 140; scientific work in, 139–40

Tansley, Arthur George, 130–31

Teton Dam, 238, 239–40

Texas: Bryson, 99; cowboys, 166; reduction of cattle inventory, 106

Texas Longhorn cattle, 260

theoretical ecological ethics: continuity and perpetuation creed, 35–36; economic shifts and, 41–42; education and, 36–37; extractionist ethic, 36; grassroots ethic, 38–40; harvesting of game/fish, 40; property rights and, 36, 38–40, 42, 128; social contract ethic, 37–38; utilitarian viewpoint of, 36–37

Thomas, Jack Ward, 231

Thoreau, Henry David: birthplace and date of, 83; death of, 84; ecological thinking and, 304–6; environmental creed of, 36; portrait of, 83; Ralph Waldo Emerson and, 83; Walden Pond, 84

timber management and spirituality, 231

Timber-Culture Act (1873), 4, 222

Tocqueville, Alexis de on a quality of conditions in America, 190, 194

Trout Unlimited, 5, 193, 296

Truman, Harry, 301

Turner, Frederick Jackson, 98

Twain, Mark (allegory of the birds), 25

Udall, Stewart, 8, 209–10

United Brotherhood of Carpenters and Joiners, 152

University of Nebraska, 127, 132

US Bureau of Reclamation, 102

U.S. Forest Service: early Forest Service, 224–25; establishing a policy of fire suppression, 157–58; Gifford Pinchot and, 27, 102, 225, 230; management philosophy of, 160, 230–31; plans for management of forests/grasslands, 331

US News and World Report (May 2002), 10–11

values, considerations of natural value, 317–18
Varley, John on natural regulation, 271
Vermont: Abenaki Indian tribe, 4, 64–66, 67–68, 70; compared with the prairie states, 237; ecological challenge in, 237; ecological changes in, 61; eco-tourism and, 51; forest burnings, impact of, 153, 154; George Perkins Marsh and, 4, 31, 32, 36, 63, 237; Shelborne Farm, 4; as a template for use of natural resources, 4
virgin forest remains, 151–53
The Virginian, Owen Wister, 165
The Voice of Dawn, and Autohistory of the Abenaki Nation, Frederick Matthew Wiseman, 64–65

Walden, Henry David Thoreau, 84
Wal-Mart and conservation, 321–23, 324
Walters, Colonel Ellsworth E., 135
Walton, Rob, 321–22
Wampanoag Indian tribe, 65
Wapiti. *See* elk
Ward, Aaron Montgomery, 15
Washington, George: Native Americans and, 56; vision of, 57
water: arid western lands and, 244–45, 247–48, 328; bottled water, 327; in California, 100–101, 103; conservation of, 99–101, 102–3, 114–16; eco-tourism and, 240–41; fundamental questions involving, 327–28; growing concern about, 326–27; in Idaho, 239, 240, 247, 328; irrigation, 240–41, 327; measurement of, 99–102, 106–7; Native Americans right to, 101; river compacts, 328–29; value and, 298–99; water law, 247. See *also* ecology and water

waterfowl management, 291, 292, 293–95
Waterhen Marsh story, 192
Watzin, Mary, 331
weapons: and ecology, 33, 66; stone points, 85
Weaver, John E.: on complexity of grasslands soils, 120; on destabilizing effect of man on nature, 132; on scientific management of wildlife, 132–33
Webb, Lila Vanderbilt, 4
Webb, Walter Prescott, 118
Webb, William Seward, 4
Weber, H. J., 125
westward expansion and land use, 60–62
wetlands: drainage of, 191–92, 298; importance of, 298
Whitney, Asa, 22
wild horses: Bureau of Land Management and, 167, 168–69; economics, 168–69; Johnston, Velma B. ("Wildhorse Annie") and, 167; state of Nevada and, 166–67; Wild Free-Roaming Horse and Borrow Act, 167
wild turkey restoration, 295–96
Wilderness Act (1964), 309
Wilderness Society, 162–63
Wildhorse Annie (Velma B. Johnson), 167
Wildlife Habitat Incentives Program (WHIP), 184
wildlife management: biological stage of, 215; Chronic Wasting Disease (CWD) and, 236, 270–71; conservation ethic in, 67, 215; criticism of present management systems, 270–71; ecological stage of, 215; economics of, 212–13; ethic in (paramount), 218–19; example of model for (eelgrass), 213–14; game management, initial stage of, 214;

game management, quandaries of, 213–15; grassroots apathy for, 214–15; organizational model for, 216–18; poaching and, 214; scientific inquiry and, 211; scientific management of, 132–33; sociological level of, 216; stages of, 213; sustainability of natural resources, 215; technology and, 211; valuable models for, 211–13; value of, 218; Yellowstone National Park and, 214–15. See *also* elk (Wapiti); wolves
Wilkerson, Bill on Texas ranching, 106
Willamette National Forest, 169–70
Williams, Beth on Chronic Wasting Disease, 270
Willie Handcart Company, 20
Wilson, E. O.: "biophilia" and, 323; on the human sense of place, 242–43
Wilson, Sophia, 322
Wiseman, Frederick Matthew on Abenaki Indian nation, 64–65, 67, 69–70
wolves: as an absolute in the ecosystem, 275; bounty hunting of, 276–77; compensation for wolf killings, 280–81; control of, 278–81; economics and local interests pressure on, 285; elk (Wapiti) and, 271, 273, 280–81, 284; Endangered Species Act (1973) and, 278, 279–80, 283; eradication of, 191, 276-77; fundamental questions involving, 327; Gray Wolf, 277–78, 281; habits of, 278; increase in number of, 278–79, 283; killings of, 276–77, 280–81, 284; legal battle over, 281–82, 283; moose and, 284; natural dispersion of, 278–79; our social contract with, 277; press accounts of, 282–83; restoration of,

277, 280–81, 283, 285; Rocky Mountain Wolf, 277, 278; Rocky Mountain Wolf Recovery Team, 277–78; in Yellowstone National Park, 277–79
The Wonderful Wizard of Oz, Frank Baum, 177–78
Worrell, Albert C. on demand for forest resources, 161
Worster, Donald on depletion of natural advantages, 179
Wyoming: blizzard (1856), 20; Buffalo Bill Historical Center, Cody, 172; the Cheyenne Club, 166; compressed ecological issues and, 7–8; Dubois, Wyoming, 281; eco-tourism and, 51; elk (Wapiti) and, 55, 267–69, 272–73; extraction of minerals/goals, 248–49; fish management in, 193; Fort Washaski, 93; growing concern for water, 326; Guernsey Reservoir, 101; Jim Bridger coal fired electric plant, 19; Midvale project, 240, 249–50; Pinedale, Wyoming, 249; The Riverton Project, 250; Spence-Moriarty acquisition, 269; Sweetwater County, 165; Sweetwater River, 165; Wildlife Federation questionable reports, 268; Wind River Indian Reservation, 88, 90

Yamparika Indian tribe, 88
Yellowstone National Park: Denali National Park compared with, 53, 269; elk (Wapiti) management in, 269, 270; esthetics and, 157; facts about, 14; game management and, 214–50; wolves and, 277–79
Yellowstone Park Timberland Reserve, 222
Yosemite National Park, 11

About the Author

Ken Greenwood

Ken Greenwood says, "I've been a very lucky guy to have worked in two worlds, human resources and natural resources. I have found the connections between the two worlds amazing."

Ken is considered by many to be the master sales trainer in media. He was honored by the Radio Advertising Bureau as "The Dean of Radio Sales Trainers." He pioneered formal sales training in radio, television and cablevision. His work with interactive video tape has resulted in two of the top selling training courses in the industry. All this followed a successful career as part owner and President of Swanson Broadcasting.

His lifelong study of the psychology of selling led to further study of social psychology. He was, for several years, Head of the Communications Department at the University of Tulsa in Tulsa, Oklahoma.

A lifelong love of hunting and fishing took him into the study of ecology and the natural world. Here again, he has received many honors including the NatureWorks Wildlife Stewardship Award, the Conservation Service Award from Ducks Unlimited and the Communicator of the Year Award from the Oklahoma Department of Wildlife Conservation. He served 2½ years as the Director of development of the Oklahoma Chapter of the Nature Conservancy and ten years as the volunteer Executive Director of the NatureWorks Wildlife Art Show and Sale in Tulsa during which time it became recognized as one of the top five wildlife art shows in the U.S.

This is his fifth published book, following the sold out edition of *High Performance Selling, Seven Strategies for High Performance Selling, High Performance Leadership and Casting For Value.*